T0187875

Excipient Toxicity
and Safety

DRUGS AND THE PHARMACEUTICAL SCIENCES

Executive Editor

James Swarbrick

AAI, Inc.
Wilmington, North Carolina

Advisory Board

DRUGS AND THE PHARMACEUTICAL SCIENCES

A Series of Textbooks and Monographs

1. Pharmacokinetics, *Milo Gibaldi and Donald Perrier*
2. Good Manufacturing Practices for Pharmaceuticals: A Plan for Total Quality Control, *Sidney H. Willig, Murray M. Tuckerman, and William S. Hitchings IV*
3. Microencapsulation, *edited by J. R. Nixon*
4. Drug Metabolism: Chemical and Biochemical Aspects, *Bernard Testa and Peter Jenner*
5. New Drugs: Discovery and Development, *edited by Alan A. Rubin*
6. Sustained and Controlled Release Drug Delivery Systems, *edited by Joseph R. Robinson*
7. Modern Pharmaceutics, *edited by Gilbert S. Banker and Christopher T. Rhodes*
8. Prescription Drugs in Short Supply: Case Histories, *Michael A. Schwartz*
9. Activated Charcoal: Antidotal and Other Medical Uses, *David O. Cooney*
10. Concepts in Drug Metabolism (in two parts), *edited by Peter Jenner and Bernard Testa*
11. Pharmaceutical Analysis: Modern Methods (in two parts), *edited by James W. Munson*
12. Techniques of Solubilization of Drugs, *edited by Samuel H. Yalkowsky*
13. Orphan Drugs, *edited by Fred E. Karch*
14. Novel Drug Delivery Systems: Fundamentals, Developmental Concepts, Biomedical Assessments, *Yie W. Chien*
15. Pharmacokinetics: Second Edition, Revised and Expanded, *Milo Gibaldi and Donald Perrier*
16. Good Manufacturing Practices for Pharmaceuticals: A Plan for Total Quality Control, Second Edition, Revised and Expanded, *Sidney H. Willig, Murray M. Tuckerman, and William S. Hitchings IV*
17. Formulation of Veterinary Dosage Forms, *edited by Jack Blodinger*
18. Dermatological Formulations: Percutaneous Absorption, *Brian W. Barry*
19. The Clinical Research Process in the Pharmaceutical Industry, *edited by Gary M. Matoren*
20. Microencapsulation and Related Drug Processes, *Patrick B. Deasy*
21. Drugs and Nutrients: The Interactive Effects, *edited by Daphne A. Roe and T. Colin Campbell*

22. Biotechnology of Industrial Antibiotics, *Erick J. Vandamme*
23. Pharmaceutical Process Validation, *edited by Bernard T. Loftus and Robert A. Nash*
24. Anticancer and Interferon Agents: Synthesis and Properties, *edited by Raphael M. Ottenbrite and George B. Butler*
25. Pharmaceutical Statistics: Practical and Clinical Applications, *Sanford Bolton*
26. Drug Dynamics for Analytical, Clinical, and Biological Chemists, *Benjamin J. Gudzinowicz, Burrows T. Younkin, Jr., and Michael J. Gudzinowicz*
27. Modern Analysis of Antibiotics, *edited by Adjoran Aszalos*
28. Solubility and Related Properties, *Kenneth C. James*
29. Controlled Drug Delivery: Fundamentals and Applications, Second Edition, Revised and Expanded, *edited by Joseph R. Robinson and Vincent H. Lee*
30. New Drug Approval Process: Clinical and Regulatory Management, *edited by Richard A. Guarino*
31. Transdermal Controlled Systemic Medications, *edited by Yie W. Chien*
32. Drug Delivery Devices: Fundamentals and Applications, *edited by Praveen Tyle*
33. Pharmacokinetics: Regulatory • Industrial • Academic Perspectives, *edited by Peter G. Welling and Francis L. S. Tse*
34. Clinical Drug Trials and Tribulations, *edited by Allen E. Cato*
35. Transdermal Drug Delivery: Developmental Issues and Research Initiatives, *edited by Jonathan Hadgraft and Richard H. Guy*
36. Aqueous Polymeric Coatings for Pharmaceutical Dosage Forms, *edited by James W. McGinity*
37. Pharmaceutical Pelletization Technology, *edited by Isaac Ghebre-Sellassie*
38. Good Laboratory Practice Regulations, *edited by Allen F. Hirsch*
39. Nasal Systemic Drug Delivery, *Yie W. Chien, Kenneth S. E. Su, and Shyi-Feu Chang*
40. Modern Pharmaceutics: Second Edition, Revised and Expanded, *edited by Gilbert S. Banker and Christopher T. Rhodes*
41. Specialized Drug Delivery Systems: Manufacturing and Production Technology, *edited by Praveen Tyle*
42. Topical Drug Delivery Formulations, *edited by David W. Osborne and Anton H. Amann*
43. Drug Stability: Principles and Practices, *Jens T. Carstensen*
44. Pharmaceutical Statistics: Practical and Clinical Applications, Second Edition, Revised and Expanded, *Sanford Bolton*
45. Biodegradable Polymers as Drug Delivery Systems, *edited by Mark Chasin and Robert Langer*
46. Preclinical Drug Disposition: A Laboratory Handbook, *Francis L. S. Tse and James J. Jaffe*

47. HPLC in the Pharmaceutical Industry, *edited by Godwin W. Fong and Stanley K. Lam*

48. Pharmaceutical Bioequivalence, *edited by Peter G. Welling, Francis L. S. Tse, and Shrikant V. Dinghe*

49. Pharmaceutical Dissolution Testing, *Umesh V. Banakar*

50. Novel Drug Delivery Systems: Second Edition, Revised and Expanded, *Yie W. Chien*

51. Managing the Clinical Drug Development Process, *David M. Cocchetto and Ronald V. Nardi*

52. Good Manufacturing Practices for Pharmaceuticals: A Plan for Total Quality Control, Third Edition, *edited by Sidney H. Willig and James R. Stoker*

53. Prodrugs: Topical and Ocular Drug Delivery, *edited by Kenneth B. Sloan*

54. Pharmaceutical Inhalation Aerosol Technology, *edited by Anthony J. Hickey*

55. Radiopharmaceuticals: Chemistry and Pharmacology, *edited by Adrian D. Nunn*

56. New Drug Approval Process: Second Edition, Revised and Expanded, *edited by Richard A. Guarino*

57. Pharmaceutical Process Validation: Second Edition, Revised and Expanded, *edited by Ira R. Berry and Robert A. Nash*

58. Ophthalmic Drug Delivery Systems, *edited by Ashim K. Mitra*

59. Pharmaceutical Skin Penetration Enhancement, *edited by Kenneth A. Walters and Jonathan Hadgraft*

60. Colonic Drug Absorption and Metabolism, *edited by Peter R. Bieck*

61. Pharmaceutical Particulate Carriers: Therapeutic Applications, *edited by Alain Rolland*

62. Drug Permeation Enhancement: Theory and Applications, *edited by Dean S. Hsieh*

63. Glycopeptide Antibiotics, *edited by Ramakrishnan Nagarajan*

64. Achieving Sterility in Medical and Pharmaceutical Products, *Nigel A. Halls*

65. Multiparticulate Oral Drug Delivery, *edited by Isaac Ghebre-Sellassie*

66. Colloidal Drug Delivery Systems, *edited by Jörg Kreuter*

67. Pharmacokinetics: Regulatory • Industrial • Academic Perspectives, Second Edition, *edited by Peter G. Welling and Francis L. S. Tse*

68. Drug Stability: Principles and Practices, Second Edition, Revised and Expanded, *Jens T. Carstensen*

69. Good Laboratory Practice Regulations: Second Edition, Revised and Expanded, *edited by Sandy Weinberg*

70. Physical Characterization of Pharmaceutical Solids, *edited by Harry G. Brittain*

71. Pharmaceutical Powder Compaction Technology, *edited by Göran Alderborn and Christer Nyström*

72. Modern Pharmaceutics: Third Edition, Revised and Expanded, *edited by Gilbert S. Banker and Christopher T. Rhodes*

73. Microencapsulation: Methods and Industrial Applications, *edited by Simon Benita*

74. Oral Mucosal Drug Delivery, *edited by Michael J. Rathbone*

75. Clinical Research in Pharmaceutical Development, *edited by Barry Bleidt and Michael Montagne*

76. The Drug Development Process: Increasing Efficiency and Cost Effectiveness, *edited by Peter G. Welling, Louis Lasagna, and Umesh V. Banakar*

77. Microparticulate Systems for the Delivery of Proteins and Vaccines, *edited by Smadar Cohen and Howard Bernstein*

78. Good Manufacturing Practices for Pharmaceuticals: A Plan for Total Quality Control, Fourth Edition, Revised and Expanded, *Sidney H. Willig and James R. Stoker*

79. Aqueous Polymeric Coatings for Pharmaceutical Dosage Forms: Second Edition, Revised and Expanded, *edited by James W. McGinity*

80. Pharmaceutical Statistics: Practical and Clinical Applications, Third Edition, *Sanford Bolton*

81. Handbook of Pharmaceutical Granulation Technology, *edited by Dilip M. Parikh*

82. Biotechnology of Antibiotics: Second Edition, Revised and Expanded, *edited by William R. Strohl*

83. Mechanisms of Transdermal Drug Delivery, *edited by Russell O. Potts and Richard H. Guy*

84. Pharmaceutical Enzymes, *edited by Albert Lauwers and Simon Scharpé*

85. Development of Biopharmaceutical Parenteral Dosage Forms, *edited by John A. Bontempo*

86. Pharmaceutical Project Management, *edited by Tony Kennedy*

87. Drug Products for Clinical Trials: An International Guide to Formulation • Production • Quality Control, *edited by Donald C. Monkhouse and Christopher T. Rhodes*

88. Development and Formulation of Veterinary Dosage Forms: Second Edition, Revised and Expanded, *edited by Gregory E. Hardee and J. Desmond Baggot*

89. Receptor-Based Drug Design, *edited by Paul Leff*

90. Automation and Validation of Information in Pharmaceutical Processing, *edited by Joseph F. deSpautz*

91. Dermal Absorption and Toxicity Assessment, *edited by Michael S. Roberts and Kenneth A. Walters*

92. Pharmaceutical Experimental Design, *Gareth A. Lewis, Didier Mathieu, and Roger Phan-Tan-Luu*

93. Preparing for FDA Pre-Approval Inspections, *edited by Martin D. Hynes III*

94. Pharmaceutical Excipients: Characterization by IR, Raman, and NMR Spectroscopy, *David E. Bugay and W. Paul Findlay*
95. Polymorphism in Pharmaceutical Solids, *edited by Harry G. Brittain*
96. Freeze-Drying/Lyophilization of Pharmaceutical and Biological Products, *edited by Louis Rey and Joan C. May*
97. Percutaneous Absorption: Drugs–Cosmetics–Mechanisms–Methodology, Third Edition, Revised and Expanded, *edited by Robert L. Bronaugh and Howard I. Maibach*
98. Bioadhesive Drug Delivery Systems: Fundamentals, Novel Approaches, and Development, *edited by Edith Mathiowitz, Donald E. Chickering III, and Claus-Michael Lehr*
99. Protein Formulation and Delivery, *edited by Eugene J. McNally*
100. New Drug Approval Process: Third Edition, The Global Challenge, *edited by Richard A. Guarino*
101. Peptide and Protein Drug Analysis, *edited by Ronald E. Reid*
102. Transport Processes in Pharmaceutical Systems, *edited by Gordon L. Amidon, Ping I. Lee, and Elizabeth M. Topp*
103. Excipient Toxicity and Safety, *edited by Myra L. Weiner and Lois A. Kotkoskie*
104. The Clinical Audit in Pharmaceutical Development, *edited by Michael R. Hamrell*

ADDITIONAL VOLUMES IN PREPARATION

Pharmaceutical Emulsions and Suspensions, *edited by Francoise Nielloud and Gilberte Marti-Mestres*

Oral Drug Absorption, *edited by Jennifer B. Dressman*

Drug Stability: Principles and Practices, Third Edition, Revised and Expanded, *edited by C. T. Rhodes and Jens T. Carstensen*

Excipient Toxicity and Safety

edited by

Myra L. Weiner

FMC Corporation
Princeton, New Jersey

Lois A. Kotkoskie

National Starch and Chemical Company
Bridgewater, New Jersey

CRC Press
Taylor & Francis Group
Boca Raton London New York

CRC Press is an imprint of the
Taylor & Francis Group, an **informa** business

CRC Press
Taylor & Francis Group
6000 Broken Sound Parkway NW, Suite 300
Boca Raton, FL 33487-2742

First issued in paperback 2019

© 2005 by Taylor & Francis Group, LLC
CRC Press is an imprint of Taylor & Francis Group, an Informa business

No claim to original U.S. Government works

ISBN-13: 978-0-8247-8210-8 (hbk)
ISBN-13: 978-0-367-39931-3 (pbk)

Visit the Taylor & Francis Web site at
http://www.taylorandfrancis.com

and the CRC Press Web site at
http://www.crcpress.com

Foreword

Since the 1960s the producers of excipients and the drug dosage form formulation industry and regulators have slowly gained an appreciation of excipients and their specialized needs as distinct entities. There is a growing appreciation of the role that pharmaceutical excipients play in the production, shelf stability, dispensability, patient dosage acceptability, bioavailability, and delivery of the active pharmaceutical ingredient to the target organ.

As the late Dr. Shangraw, Professor of Pharmaceutical Sciences, School of Pharmacy, University of Maryland, pointed out (*Pharmaceutical Technology*, June 1997), when I was starting out in the field, years ago, virtually all excipients were of natural origin—as either foods, food additives, or simple inorganics. They were supplied primarily by food producers or chemical suppliers. During the 1950s and 1960s development of excipients began to accelerate in the industry. The need for reproducible disintegration, dissolution, and bioavailability began to be recognized. Since available excipients were natural or well-known compounds, without obvious physiological activity, excipients were universally considered inactive and inert. Work during the 1960s alerted us to tablets, which transversed the gastrointestinal tract intact. Thus, the "reproducible" rate and extent of disintegration became an issue. When research indicated little or no absorption through the gut wall for certain excipients, dissolution became an issue. Bioavailability entered our active vocabulary. Polymers, sustained release agents and absorption modifiers exhibited interesting properties, but very little was known about their safety. Therefore, the need for adequate toxicological testing became apparent.

Over the years, few new potential excipients found use beyond utilization as food additives and cosmetic ingredients. As changes were made in regulations controlling pharmaceutical production, the excipients were still not considered separate entities but only components of a final drug dosage form. The lack of regulatory status was very evident when Robert Pinco, Esq., presented a paper to a United States Pharmacopeia's Joint Pharmacopeia Open Conference on International Harmonization of Excipient Standards in 1991. He discussed this lack of regulations governing the requirements for safety evaluation of possible new excipients in Europe, Japan, and the United States. Many attendees at that conference agreed that there was a need for a regulatory road map. The time had come to recognize the unique properties of excipients and the need for appropriate and scientifically valid regulations for toxicological testing, specifications, and GMPs.

During that conference the International Pharmaceutical Excipients Council (IPEC) was conceived as a voluntary industry association of excipient producers and users. A key objective was to develop the basis for a regulatory road map, including new excipient safety evaluation guidelines for those willing to work on new excipients desired by the pharmaceutical industry. By 1992 the IPEC New Excipients Safety Evaluation Committee came into existence with members from excipient manufacturers and excipient users, academia, and the Food and Drug Administration. Many members of that committee have authored chapters in this volume, particularly the chapters in Part II. It is most significant that the committee decided to present a guideline for use by competent professionals rather than a checklist or a set of protocols. The regional IPEC-Americas, IPEC-Europe, and JPEC (Japanese organization) agreed on the principles and have, to a greater extent, harmonized the guideline. The next several years are expected to reveal progress by the regulators in the development of individual and harmonized guidelines for toxicological evaluation of excipients, for excipient drug master files, and for independent evaluation of drug master files, as well as a regulatory acceptability decision to foster the development of new materials.

This book meets a need at all levels of the pharmaceutical industry from undergraduate student through senior management, including regulators and regulatory scientists. It reviews the basics of pharmaceutical excipients, the specifications, and the regulatory status. Safety evaluation and risk assessment are reviewed. Finally, the key areas of risk communication and global harmonization are discussed. The book provides a pragmatic overview of excipients and excipient safety in pharmaceuticals.

The pharmaceutical industry is globalizing, and, therefore, the development of new concepts and new approaches to drug therapy is accelerating. A drug must be safe and effective. An excipient must be not only safe, but also suitable. My thanks to the editors and contributors for a very timely and valuable book.

Mr. Louis Blecher
Chairman
The International Pharmaceutical Excipients Council

Preface

The objective of this book is to familiarize the reader with the safe and legal use of pharmaceutical excipients. We hope to provide the reader with a comprehensive understanding of the current scientific basis for safety evelution of excipients. Excipients have not received much attention as separate entities. Until recently, excipients had been evaluated for toxicity as part of new drug formulations, with the active ingredient and all the various excipients of the formulation tested together. Until the proposal for safety evaluation procedures for new excipients by the Safety Committee of the International Pharmaceutical Excipients Council, no procedures existed for testing excipients alone. Historically, many excipients have been food additives and "Generally Recognized As Safe" (GRAS) by the United States Food and Drug Administration. The safety studies for such excipients have been reviewed as part of their GRAS designation. Now, the new IPEC guidelines will allow excipient manufacturers to approach their products in a scientific framework developed for this class of products.

Part I of the book defines excipients and discusses their historical use in drug formulations as inactive ingredients with specific functional properties, the requirements and importance of purity specifications, and the current regulatory requirements for excipients in the United States and Europe. Part II covers all aspects of safety evaluation of excipients as a unique class of products. The guidelines for safety evaluation of pharmaceutical excipients by various routes is an outgrowth of the Safety Committee of the International Pharmaceutical Excipients Council. The principles on which these guidelines are based and the technical details for conducting studies via each route of exposure are expounded in this section.

Part III of the book explains how data generated in toxicity studies are used to identify hazards for use in drug formulations. Exposure assessment, a new area of excipient evaluation, is necessary to link hazard identification to risk. Principles of exposure assessment from other types of chemicals and products are used as a lens through which to view exposures to excipients, taking into consideration some of the typical exposures to drug products. Risk assessment and risk communication are the final steps in the overall understanding of the safe use of excipients. Finally, Part IV describes harmonization of existing issues for pharmaceutical excipients.

We hope that this book will be a valuable resource to pharmaceutical scientists in industry and academia, regulators, toxicologists, and risk assessors.

Myra L. Weiner
Lois A. Kotkoskie

Contents

Foreword   *iii*

Preface   *v*

PART I: INTRODUCTION TO EXCIPIENTS

1. What are Excipients?   1
    Thomas A. Wheatley

2. Purity of Excipients   21
    Dankward Jäkel and Martin Keck

3. History of Excipient Safety and Toxicity   59
    Charles L. Winek

4. Regulation of Pharmaceutical Excipients   73
    Christopher C. DeMerlis

PART II: SAFETY EVALUATION OF PHARMACEUTICAL EXCIPIENTS

5. Development of Safety Evaluation Guidelines   101
    Joseph F. Borzelleca and Myra L. Weiner

6. Routes of Exposure: Oral   123
    Lois A. Kotkoskie

7. Routes of Exposure: Topical and Transdermal   141
    Matthew J. Cukierski and Alice E. Loper

8. Routes of Exposure: Inhalation and Intranasal   185
    Charmille B. Tamulinas and Chet L. Leach

9. Routes of Exposure: Parenteral   207
    David B. Mitchell

10. Routes of Exposure: Other   231
    Carol S. Auletta

PART III: RISK ASSESSMENT OF EXCIPIENTS

11. Toxicokinetics and Hazard Identification   267
    Frank M. Sullivan and Susan M. Barlow

12. Exposure Assessment   283
    David J. George and Annette M. Shipp

13. Risk Assessment and Risk Communication   305
    Anthony D. Dayan

PART IV: FUTURE ISSUES

14. Harmonization of Excipient Standards   321
    Zak T. Chowhan

Index   355

Contributors

Carol S. Auletta Director of Toxicology, Huntingdon Life Sciences, Inc., East Millstone, New Jersey

Susan M. Barlow Independent Consultant in Toxicology, Harrington House, Brighton, East Sussex, United Kingdom

Joseph F. Borzelleca Professor of Toxicology and Pharmacology, Medical College of Virginia, Richmond, Virginia

Zak T. Chowhan Pharmaceutical Development Consultant, Cockeysville, Maryland

Matthew J. Cukierski* ALZA Corporation, Mountain View, California

Anthony D. Dayan Department of Toxicology, St. Bartholomew's and the Royal London School of Medicine and Dentistry, London, United Kingdom

Christopher C. DeMerlis Manager, Regulatory Affairs, Pharmaceutical Division, FMC Corporation, Philadelphia, Pennsylvania

---

* *Current affiliation:* Director, Drug Safety Evaluation, Coulter Pharmaceutical, Inc., South San Francisco, California

David J. George Senior Director, Regulatory Toxicology, Whitehall–Robins Healthcare, Madison, New Jersey

Dankward Jäkel Novartis Pharma AG, Basel, Switzerland

Martin Keck Novartis Pharma AG, Basel, Switzerland

Lois A. Kotkoskie Toxicology Associate, FMC Corporation, Princeton, New Jersey

Chet L. Leach Division Scientist, Department of Toxicology and Pathology, 3M Pharmaceuticals, St. Paul, Minnesota

Alice E. Loper Director, Biopharmaceutical Chemistry, Pharmaceutical Research and Development, Merck Research Laboratories, West Point, Pennsylvania

David B. Mitchell Section Head, Regulatory and Clinical Development Division, Health Care Research Center, The Procter & Gamble Company, Mason, Ohio

Annette M. Shipp Vice President, ICF Kaiser, The K. S. Crump Group Inc., Ruston, Louisiana

Frank M. Sullivan Independent Consultant in Toxicology, Harrington House, Brighton, East Sussex, United Kingdom

Charmille B. Tamulinas Manager, Department of Toxicology and Pathology, 3M Pharmaceuticals, St. Paul, Minnesota

Myra L. Weiner Manager, Toxicology Programs, FMC Corporation, Princeton, New Jersey

Thomas A. Wheatley* Research Fellow, Pharmaceutical Division, Chemical Research and Development Center, FMC Corporation, Princeton, New Jersey

Charles L. Winek Professor of Toxicology, School of Pharmacy, Duquesne University, Pittsburgh, Pennsylvania

---

* Retired

1
What Are Excipients?

Thomas A. Wheatley*
FMC Corporation, Princeton, New Jersey

I. INTRODUCTION

A. What Are Excipients?

There are several definitions for an excipient. In some cases, the definition is simple; in others, the definition is more encompassing and complex. Webster (1) defines an *excipient* as "inert substance (as gum arabic or starch) that forms a vehicle (as for a drug)."

The *National Formulary* (2), a book of standards that provides monographs for pharmaceutical ingredients used in drug dosage forms, defines an *excipient* as any component, other than the active substance(s), intentionally added to the formulation of a dosage form. It is not defined as an "inert" commodity or an "inert" component of the dosage form. *The United States Pharmacopeia (USP)* and *National Formulary (NF)* are recognized in the Federal Food, Drug and Cosmetic Act. According to section 501 of the act, assays and specifications in the monographs of the *USP* and *NF* constitute legal standards. Most commonly recognized are *USP* and *NF* standards for determining the identity, strength, quality, and purity of the articles ("excipients").

The *Handbook of Pharmaceutical Excipients*, originally published in 1986, was the first English-language publication to comprehensively and systematically describe the physical and chemical properties of pharmaceutical excipients. The second edition (1994) of the *Handbook of Pharmaceutical Excipients (3) defines excipients* as the additives used to convert pharmacologically active compounds into pharmaceutical dosage forms suitable for administration to patients.

In the spring of 1991, the International Pharmaceutical Excipients Council (IPEC) was formed. IPEC membership includes both companies that manufacture

*Retired

1

pharmaceutical excipients and companies that use excipients in the manufacture of drug dosage forms. *Pharmaceutical excipients*, as defined by IPEC (4), are any substance other than the active drug or prodrug that has been appropriately evaluated for safety and is included in a drug delivery system for any of the following purposes:

1. Aid processing of the system during manufacture
2. Protect, support, or enhance stability and bioavailability
3. Assist in product identification
4. Enhance any other attribute of the overall safety and effectiveness of the drug product during storage or use

As can be seen from the foregoing paragraphs, there are many definitions for an excipient. However, the intent in all cases is to define an excipient as a material(s) that has been evaluated for safety, aids in the manufacture of the dosage form, and protects, supports, or enhances stability and bioavailability of the drug or active ingredient.

This chapter will provide the reader with an appreciation and understanding of excipients: what they are; how they are employed; and what is their role in turning active ingredients into efficient and effective medicines.

II. EXCIPIENTS FOR USE IN ORAL MEDICINES

By far the most frequently employed dosage form used today throughout most areas of the world is the compressed tablet. *Tablets* may be defined as solid pharmaceutical dosage forms containing drug substances, with or without suitable diluents (excipients), and prepared either by compression or molding methods (5). The use of a tablet as a dosage form can be traced to well over 1,000 years ago when a procedure for molding solid forms containing medicinal ingredients was recorded (6). Tablets have been in widespread use since the latter part of the 19th century, and their popularity continues. Tablets remain popular because of the numerous advantages over other oral medicines, some of which are (a) accuracy of dosage, (b) compactness and portability (c) ease of administration, (d) durability of physical characteristics for extended periods of storage, and (e) stability of the chemical and physiological activity of the drugs.

For purposes of this chapter, excipients used mainly in the manufacture of compressed tablets will be discussed. Many of these excipients are also used in other oral dosage forms, including capsules and other types of tablets, which include chewables, effervescent, bilayer, multiple compressed, topical tablets, and tablets for solution. They are also used in tablets for specific modes of action (i.e., buccal or sublingual release and modified or controlled release).

Excipients perform very important functions in tablet formulations (7) specifically as

Fillers or diluents
Binders
Disintegrants or super disintegrants
Lubricants
Antiadherents
Glidants
Wetting and surface-active agents
Colors and pigments
Flavors, sweeteners, and taste maskers

Excipients may be classified according to the role they play in the finished tablet. Those excipients that help impart satisfactory processing and compression characteristics to the formulation include fillers–diluents, binders, glidants, and lubricants. The second group of excipients helps impart additional desirable physical characteristics to the finished tablet. Included in this group are disintegrants, colors and pigments, and wetting and surface-active agents. For chewable tablets, flavors, sweeteners, and taste-modifiers are employed. For controlled- or modified-release tablets, polymers or waxes or other solubility-retarding or modifying excipients are used. Chowan (7) recently provided a list (Table 1) of excipients commonly used in the manufacture of compressed tablets. Although not all-inclusive, the excipients are listed according to their intended use: direct-compression excipients, wet-granulation excipients, and those excipients that help to impart additional desirable physical characteristics to the finished tablet. The choice of excipients in a tablet formulation depends on the active ingredient, the type of tablet, the desired tablet characteristics, and the process used to manufacture the tablet. Compacted or compressed tablets are produced from powder mixtures or granulations made by one of the following general techniques:

Direct Compression.   Direct compression consists of compressing tablets directly from powdered material without modifying the physical nature of the material itself. The process consists of mixing and blending the active ingredient with the appropriate excipient(s) before compression.

Wet Granulation.   Wet granulation consists of weighing and mixing the active ingredient and excipient(s), granulation with a binder (low- or high-shear mixing), screening the damp mass (granulation), drying of the granulation, dry screening, lubrication, and compression. The wet granulation method is labor-intensive and time-consuming relative to tablets prepared by the direct compression technique.

Dry Granulation (by Roller Compaction or Slugging).   The third process for making the ''running'' powder blend for tableting is the dry granulation process. This process requires five steps: (a) mixing, (b), roller compaction or slug-

Table 1 Tablet Excipients

Direct compression excipients	Wet granulation excipients
Cellulose	*Binders*
Avicel PH Microcrystalline Cellulose NF, Ph.Eur., JP, BP	Avicel PH Microcrystalline Cellulose NF, Ph.Eur., JP, BP
Microfine cellulose	Cellulose derivatives
Lactose	*Povidone USP*
Super-Tab Spray-Dried Lactose Monohydrate NF, Ph.Eur., JP, BP	*Gelatin NF*
Anhydrous lactose	Natural gums
Other sugars	Starch paste
Compressible Sugar NF	*Pregelatinized Starch NF*
Dextrose Excipient NF	*Sucrose NF*
Dextrates NF	Other binders
Starch and starch derivatives	Others
Native starches	*Disintegrants*
Pregelatinized Starch NF	Ac-Di-Sol Croscarmellose Sodium NF, Ph.Eur, JPE
Sodium Starch Glycolate NF	*Sodium Starch Glycolate, NF, Explotab, Primojel*
Inorganic salts	*Crospovidone NF*
Dibasic Calcium Phosphate USP	*Lubricants*
Tribasic Calcium Phosphate NF	Magnesium stearate
Calcium Sulfate NF	Calcium stearate
Polyols	Stearic acid
Mannitol USP	Sodium stearyl fumarate
Sorbitol NF	Hydrogenated vegetable oils
Xylitol NF	Mineral oil
	Polyethylene glycols
	Antiadherents
	Glidants

Source: Ref. 7.

ging, (c) milling, (d) screening, and (e) final blending. The same excipients that are used in direct compression can also be used in dry granulation.

A. Direct Compression Excipients

The direct compression process generally involves mixing a drug with excipients before compression. Direct compression excipients must have good flow and compression characteristics. In addition, direct compression excipients must exhibit low lubricant sensitivity to compression; have good stability; promote tablet disintegration and drug dissolution; and exhibit noninterference with bioavailability of the active ingredient.

I. Cellulose

The process of direct compression was revolutionized by the introduction of Avicel PH microcrystalline cellulose (MCC) in the early 1960s, although spray-dried lactose had been introduced at an earlier data. In combination, MCC and spray-dried lactose are used together in varying ratios in most direct compression formulas. Microcrystalline cellulose is described in the *National Formulary* (*NF*) as a purified, partially depolymerized cellulose prepared by treating α-cellulose, which is obtained as a pulp from fibrous plant material, with mineral acids. Avicel PH MCC is the excipient most often used in tableting as a filler, disintegrant, flow aid, and dry binder in directly compressible tablets.

MCC has extremely good binding properties as a dry binder. It improves flow and has good lubrication and disintegration properties. Tablets prepared with MCC generally exhibit excellent hardness and low friability. Aivcel PH MCC is available in various particle size, density, and moisture grades (Table 2) to meet the various tablet requirements. Microfine cellulose (Elcema) is a mechanically produced cellulose powder. The granular grade (G-250) may be used in direct compression because of its improved flow and compression properties. Microfine cellulose possesses a poor dilution potential relative to MCC. *Dilution potential* is defined as the ability of a given quantity of an excipient to bind a specified amount of an active ingredient to form an acceptable tablet (7). The greater the quantity of active ingredient the excipient is able to bind or carry, the better is its dilution potential.

2. Lactose

Lactose is the most commonly used filler in tablet formulations. it is a natural disaccharide produced from cow's milk. Some forms of lactose meet the require-

Table 2 Avicel PH Microcrystalline Cellulose—Typical Average Particle Size, Bulk Density, and Loss on Drying at the Time of Shipment

Grade	Typical average particle size (μm)	Bulk density (g/mL)	Loss on drying (%)
PH-101	50	0.28	4
PH-102	90	0.30	4
PH-103	50	0.28	2
PH-105	20	0.25	3
PH-112	90	0.30	1
PH-113	50	0.30	1
PH-200	180	0.32	4
PH-301	50	0.38	4
PH-302	90	0.39	4

ments for a direct-compression excipient. Hydrous lactose does not flow, and its use is limited to tablet formulations prepared by wet granulation. Spray-dried lactose monohydrate (Super-Tab) and anhydrous lactose (Sheffield) have good flowability and compressibility. Spray-dried lactose monohydrate is specifically engineered for direct compression and is ideally suited for drugs that do not compress well.

3. Other Sugars

Large crystals of sucrose flow very well through a hopper, but their compaction properties are poor. *Compressible Sugar NF* consists mainly of sucrose that is processed to have properties suitable for direct compression. It also may contain small quantities of starch, dextrin, or invert sugar. Compressible sugar is a sweet, white crystalline powder and is complete water solubility. Because of the high water solubility, tablets containing compressible sugar as an excipient do not disintegrate, but rather, the sugar dissolves, releasing the drug. It is widely used for chewable vitamin tablets because of its natural sweetness.

Dextrose Excipient NF is available in the anhydrous and monohydrate forms. The compression properties are poor and the tablet compacts are soft.

Dextrates are prepared from a controlled enzymatic hydrolysis of starch. Because their sweetness and negative heat of solution, dextrates are recommended for chewable tablets.

4. Starch and Starch Derivatives

Starch and starch derivatives are among the most commonly used excipients in tablet formulations. They can function as disintegrants, binders or fillers. Native starches used as excipients are obtained from corn, wheat, rice, tapioca, and potatoes, but cornstarch is most commonly used. Native starches are used as disintegrants, but with introduction of the super disintegrants in the late 1970s, they are no longer the disintegrant of choice. Because of poor flow, loss of binding and compressibility in the presence of a lubricant, they are less suitable for direct compression tablet formulations. Starch, when used as a paste, makes a good binder, particularly when the drug is insoluble and in high concentrations. The native starches are used as a binder that comes in the form of a 5%–10% paste cooked in a double boiler, and the concentration of starch in the formulation may vary between 2 and 5%.

Pregelatinized starch is obtained by a chemical or mechanical process that ruptures the starch granules in the presence of water. Partially pregelatinized starch acts as a binder as well as a distintegrant. If starch is fully pregelatinized, it loses its disintegrant properties and acts only as a binder.

Sodium starch glycolate is the sodium salt of a carboxymethyl ether of starch. It is used as a super disintegrant, which will be discussed later in this chapter.

5. Inorganic Salts

The most commonly used direct compression inorganic salts are dibasic calcium phosphate, tricalcium phosphate, and calcium sulfate.

Dibasic Calcium Phosphate Dihydrate USP is the most commonly used directly compressible filler-bind. *Dicalcium Phosphate Dihydrate* (Di-Tab), in its unmilled form, has good flow properties and compressibility. Because it has no inherent lubricating or disintegrating properties, other excipients must be added to prepare a satisfactory tablet formulation.

Tribasic calcium phosphate is available as a directly compressible filler–binder for tablets. Tribasic calcium phosphate has shortcomings in that it has a high tendency to adhere to punches and dies, and it has a deleterious effect on dissolution, especially after aging of the tablets.

Calcium Sulfate NF (terra alba) is available as a specially processed grade of excipient for direct compression. It is an inexpensive filler.

6. Polyols

Polyols for pharmaceutical use include sorbitol, mannitol, and xylitol.

Sorbitol NF is closely related to glucose, which can be obtained from starch or sucrose. Direct compression grades of sorbitol, available from several manufacturers, can be used for the preparation of chewable tablets, lozenges, and disintegrating tablets. However, the hygroscopicity of sorbitol limits its use in tableting.

Mannitol is a popular excipient for chewable tablets, owing to its pleasant taste and mouthfeel, resulting from its negative heat of solution. Mannitol powder has poor flow and compression properties. It is available in granular form, for direct compression, which has good flow and compression properties, and it is not hygroscopic.

Xylitol is used as a noncariogenic sweetening agent in tablets, syrups, and coatings.

B. Wet Granulation Excipients

The most widely used and most general method of tablet production is the wet-granulation method. The excipients that agglomerate drug, filler, and other excipients together and cause them to form granules are the binders. There is a twofold purpose for agglomeration of the drug and excipients: (a) to improve the powder flow to minimize weight variation and content uniformity problems; and (b) to improve compressibility, resulting in tablets with low friability and good tensile strength. The choice of binding agent depends on the binding force required to form granules. Most binders are hydrophilic and soluble in water. Natural gums and polymers function by forming a thin film on the surface of the particles, which then agglomerate during the granulation step. Table 3 provides a partial listing of binders commonly used in wet granulation.

Microcrystalline cellulose (MCC) functions as a wet granulation binder. MCC permits faster addition of the granulation solution through a rapid wicking action in the wet phase. It also produces less screen blockage during wet screening, speeds drying, minimizes or prevents case hardening, and eliminates or reduces color mottling. Microcrystalline cellulose is the only wet granulation binder that also works well in directly compressible formulations.

Polymers (cellulose derivatives), including *Carboxymethyl Cellulose Sodium USP*, *Hydroxypropyl Cellulose NF*, *Hydroxypropyl Methylcellulose USP*, *Methylcellulose USP*, and *Hydroxyethyl Cellulose NF*, all are examples of excipi-

Table 3   Binders Used in Wet Granulation

Name	Concentration used (% of formulation)	Solvents
Microcrystalline cellulose	10–50	Water
Polymers (cellulose derivatives)	1–5	Water
Carboxymethyl cellulose sodium		
Hydroxypropyl cellulose		
Hydroxypropyl methylcellulose		
Methyl cellulose		
Hydroxyethyl cellulose		
Ethyl cellulose	2–7	Alcohol
Povidone (PVP)	2–5	Alcohol, water
Gelatin	1–3	Water
Natural gums	1–5	Water
Acacia		
Tragacanth		
Guar		
Pectin		
Starch	2–5	Water (paste)
Pregelatinized starch	10–25	Water
Sucrose	2–20	Water
Others		Water
Corn syrup		
Polyethylene glycols		
Sodium alginate		
Magnesium aluminum silicate		

ents used as wet binders. The concentration of the binder is 1–5% of the formulation, and the solvent is water. *Ethylcellulose NF* is also employed as a binder. It is soluble in solvents such as alcohol. In some formulations, and depending on the drug, ethylcellulose may provide controlled release of drug.

Povidone USP (PVP) is one of the most commonly used binders. PVP or polyvinyl pyrrolidone is a synthetic polymer with several grades available, differing only in the molecular weight of the polymer. The most common grade used is povidone K-29/32. It is normally used in concentrations of 2–5% of the formulation.

Gelatin NF as a binder has largely been replaced by synthetic polymers. However, when used, its level of use is 1–3% of the formula. Natural gums, such as acacia, tragacanth, guar, and pectin, are still employed at 1–5% concentrations. Natural gums have been largely replaced by the synthetic polymers owing to variability in quality of the gums.

Cornstarch is widely used as a binder as starch paste. It is prepared by suspending 5–10% starch in cold water, followed by heating in a double boiler until fully gelatinized. The concentration of starch may vary between 2 and 5% in the formulation. A significant improvement to starch paste is *Pregelatinized Starch NF* (i.e., partially pregelatinized [Starch 1500] or fully gelatinized starch). The concentration of pregelatinized starch in the formula will vary depending on the type used, but is usually in the 10–25% range. Pregelatinized starch provides good binding properties and acts as a disintegrant.

Sucrose NF is the form of a 50–70% solution is used as a binder. The actual concentration of sucrose in the formula may vary between 2 and 20%. Generally, granules formed using sucrose as a binder are hard, and excessive tablet machine pressure is required to make a tablet.

C. Other Tablet Excipients

Excipients may be classified according to the role they play in manufacture of the finished tablet. As discussed previously, fillers–diluents and binder help to impart satisfactory processing and compression characteristics to the formulation. Lubricants, glidants, and antiadherents also help to impart satisfactory processing and compression characteristics. A second group of excipients helps impart additional desirable physical characteristics to the finished tablet. Included are disintegrants, colors and pigments, wetting and surface-active agents and, for chewable tablets, flavors, sweeteners, and taste modifiers.

1. Disintegrants

A *disintegrant* is an excipient added to a table formation to cause the tablet to break apart or disintegrate after administration. The drug must be released from the tablet matrix as quickly as possible to permit its rapid dissolution.

Starch is the oldest and was the first most commonly used disintegrant in compressed tablets (Table 4). Because of requirements for faster dissolution and problems with compression and tablet softening, starch is being largely replaced with the newly developed "super disintegrants."

The name *super disintegrant* comes from the low use levels (2–8%) at which they are effective. Croscarmellose sodium, sodium starch glycolate, and crospovidone are examples of a cross-linked cellulose, cross-lined starch, and a cross-linked polymer. Cross-linking serves to greatly reduce water solubility, while allowing the excipient to swell and absorb many times its weight of water, causing the tablet to break apart or disintegrate.

2. Lubricants

Lubricants have various functions in tablet manufacture. To prevent adhesion of the tablet material to the surface of the dies and punches (Fig. 1), reduce interparticle friction, and facilitate ejection of the tablet from the dye. Commonly used lubricants include magnesium stearate, calcium stearate, stearic acid, hydrogenated vegetable oils, polyethylene glycols, and sodium stearyl fumarate.

Magnesium stearate is the most commonly used and effective lubricant for tablets. Its use level is 0.2–2.0% (max). Calcium stearate is employed at the same use level, but is not as popular as magnesium stearate. Stearic acid, hydrogenated vegetable oils, and mineral oil are frequently used if there is a chemical and physical incompatibility of the active ingredient with magnesium stearate. In some formulations, however, stearic acid is used in combination with magnesium stearate.

Table 4 Most Commonly Used Disintegrants and Recommended Levels in a Formulation

Disintegrant	Example	Level (%)
Starch NF	Corn, wheat, potato, rice (Corn most commonly used)	5–10 in dry granulation
	Pregelatinized starch (Starch 1500)	Binder and a disintegrant, 5–20 in wet granulation
Croscarmellose Sodium NF	Ac-Di-Sol	2–4 in wet or dry granulation
Sodium Starch Glycolate NF	Primogel, Explotab	2–8 in dry granulation
Crospovidone NF	Crospovidone	2–5 in wet or dry granulation

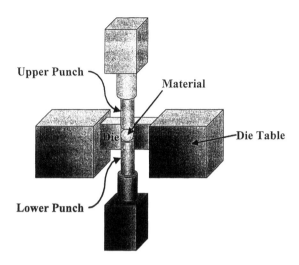

Figure 1   Simple tableting system.

3.   Antiadherents

Antiadherents prevent sticking of the tablet blend to the die wall and punch face. They are used in combination with magnesium stearate when sticking becomes a problem. Commonly used antiadherents and use levels include cornstarch (5–10%) and talc (1–5%).

4.   Glidants

A glidant is an excipient used in tablet formulations to improve flow of the powder mixture. Glidants are mixed, in low concentrations, into the final tablet blend in dry form just before compression. Colloidal silicon dioxide (Cab-O-Sil, Syloid, and Aerosil) is the most commonly used gliant. It is used in low concentrations (0.1–0.2%). Talc (asbestos-free) is also used (0.2–0.3%) and may serve the dual purpose of lubricant and glidant. In certain formulations, the alkali stearates and starch are employed.

5.   Coloring Agents

Colors in compressed tablets serve several functions: (a) making the dosage form more esthetic in appearance; (b) helping the manufacturer to control the product during its preparation; and (c) serving as a means of identification to the patient. Any of the approved, certified water-soluble FD&C dyes, mixtures of the same, or their corresponding lakes may be used to color tablets. A color lake is a combination of adsorption of a water-soluble dye to a hydrous oxide of a heavy metal (usually aluminum), resulting in an insoluble form of the dye. Each country has its own list of approved colorants that must be taken into consideration when designing a formulation for international markets.

6. Wetting Agents

Sodium lauryl sulfate, in combination with disintegrants, such as starch, is an effective disintegrant. It has been suggested that effectiveness of surfactants in improving tablet disintegration is due to an increase in the rate of wetting.

7. Flavors and Sweeteners

Flavoring agents and sweeteners are seldom found in standard compressed tablets, but frequently in chewable tablets. Flavors are available as oils, liquid mixtures, and spray-dried products from several flavor houses. In addition to the sweetness added by the excipients mannitol, sorbitol, or sucrose, artificial sweeteners may be added to the chewable tablet formulation. Aspartame (Searle), a new synthetic sweetner, has found applications in pharmaceutical tablet and liquid formulations.

D. Excipients for Other Oral Dosage Forms

Capsules are solid oral dosage forms in which the drug is enclosed in either a hard or soft, soluble shell of gelatin. Excipients commonly used in manufacture of hard gelatin capsule dosage forms include microcrystalline cellulose, lactose, and starch. Depending on the formulation and equipment used to encapsulate the powder blend, lubricants, such as magnesium stearate or stearic acid, and glidants, such as colloidal silicon dioxide or talc, are also employed.

Liquid-filled soft gelatin capsules are a popular dosage form for delivery of vitamins (e.g., vitamin E). Excipients commonly used in the liquid fill are natural vegetable oils which, in some cases, are in a water-dispersible form. Preservatives, parabens, are used on occasion, depending on the formulation requirements.

Effervescent tablets are used to deliver oral medications, such as antacids and analgesics. In addition to the active ingredient, excipients are sodium bicarbonate and organic acids, such as tartaric or citric acid. In the presence of water, these excipients react, liberating carbon dioxide, which acts as a disintegrant and produces effervescence.

III. EXCIPIENTS FOR VARIOUS ROUTES

The objective of the following discussion of excipients for various routes is to provide the reader with an appreciation of excipients, how they function, and how they are employed. The list of excipients is quite extensive. In the interest of brevity, only the more commonly employed excipients in each category of use are discussed. The reader is referred to the standard pharmaceutical texts, such as *Remington's Pharmaceutical Sciences*, 18th ed. (8) for a more thorough discussion of excipients and their categories of use.

A. Topical and Transdermal Delivery System

Drugs are applied to the skin or inserted into various body orifices in liquid, semisolid, or solid form. Semisolid preparations generally refer to therapeutic ointments, creams, salves, or pastes. These preparations are generally viscous in consistency when intended for application to the skin. Newer modes of drug delivery include transdermal delivery systems. These systems have been developed to optimize drug delivery or to overcome the shortcomings of some of the earlier delivery preparations.

The USP recognizes four general classes of ointment bases. Block (9) categorized the ointment bases into five classes for purposes of showing differences in the performance properties of the bases (Table 5).

Petrolatum and mineral oil are perhaps the best examples of hydrocarbon (oleaginous) bases. *Petrolatum USP* is tasteless, odorless, smooth, and greasy in texture and appearance. It is often used externally for its emollient properties. Petrolatum, when used as an ointment base, has exhibited a high degree of compatibility with a variety of active ingredients. Mineral oil is obtained from petroleum, as petrolatum, by collection of a particular viscosity-controlled fraction. The lower viscosity grades of mineral oil are preferred for semisolid products, because they are less tacky and greasy. The main disadvantage of the hydrocarbon or oleaginous ointment bases is that they are greasy. The greasy or oily material, when used topically, may stain clothing, and it is difficult to remove the stain.

Absorption bases are hydrophilic, anhydrous materials that have the ability to absorb additional water. The word absorption refers only to the ability of the base to absorb water. There are two types of absorption bases: the anhydrous form and the emulsion form. *Hydrophilic Petrolatum USP* and anhydrous lanolin are examples of anhydrous bases that absorb water to form water-in-oil emulsions. Anhydrous lanolin is an example of a hydrous base that is a water-in-oil emulsion having the ability to absorb additional amounts of water.

Water-washable bases or emulsion bases are commonly referred to as creams. Vanishing cream bases fall into this category. These preparations are the most commonly used type of ointment base. The vast majority of commercial dermatological products are formulated as an emulsion or cream base. Emulsion bases are washable and can be removed easily from the skin or clothing. The list of excipients used to prepare water-washable bases is numerous and includes stearic acid, stearyl alcohol, cetyl alcohol, glycerol monostearate, lanolin, glyc-

Table 5 Classification and Properties of Ointment Bases

Hydrocarbon bases (oleaginous)	Emulsion bases (water/oil type)
Example: White petrolatum	Examples: Lanolin, cold cream
Properties:	Properties:
1. Emollient	1. Emollient
2. Occlusive	2. Occlusive
3. Non–water-washable	3. Contain water
4. Hydrophobic	4. Some absorb additional water
5. Greasy	5. Greasy
Absorption bases (anhydrous)	Emulsion Bases (Oil/Water Type)
Examples: Hydrophilic petrolatum; anhydrous lanolin	Example: Hydrophilic ointment
Properties:	Properties:
1. Emollient	1. Water-washable
2. Occlusive	2. Nongreasy
3. Absorb water	3. Can be diluted with water
4. Anhydrous	4. Nonocclusive
5. Greasy	Water-soluble bases
	Example: Polyethylene glycol ointment
	Properties:
	1. Usually anhydrous
	2. Water-soluble and washable
	3. Nongreasy
	4. Nonocclusive
	5. Lipid-free

Source: Ref. 9.

erin, and others. Frequently, preservatives (methyl and propyl paraben) are added to maintain potency, and integrity of the product and to control microbial growth. Emulsifiers, anionic, cationic, and nonionic, are important components of water-washable bases. Sodium lauryl sulfate is an example of anionic emulsifier. Cationic emulsifiers are used infrequently owing to irritation to skin and eyes and to considerable incompatibility problems. Many nonionic surfactants are condensation products of ethylene oxide groups with a long-chain hydrophobic compound. Examples of nonionic surfactants are the Span and Tween products.

Water-soluble bases are prepared from mixtures of high and low molecular weight polyethylene glycols, which have the general formula:

$$HOCH_2 (CH_2OCH_2)_n CH_2OH$$

Polyethylene glycols of interest include the 1500, 1600, 4000, and 6000 products, ranging from soft, waxy solids (1500 is similar to petrolatum in consistency) to hard waxes. Polyethylene glycol 6000 is an example of a hard wax-like material.

Suitable combinations of high and low molecular weight polyethylene glycols yield products that have ointment-like consistency. They soften or melt when applied to the skin. No water is required for their manufacture.

In addition to preservations, antioxidants are frequently added to semisolid ointment bases whenever oxidative deterioration is expected. Often two antioxidants are used, because the combination is often synergistic. Common antioxidants include butylated hydroxyanisole (BHA), butylated hydroxytoluene (BHT), and propyl gallate.

B. Parenteral Systems

Parenteral products are intended for use by injection under or through one or more layers of the skin or mucous membranes. Most frequently they are solutions or suspensions. Because this route of administration circumvents the highly efficient protective barriers of the human body, exceptional purity of the parenteral dosage form must be achieved. Products for the eye and ear, although not introduced into internal body cavities, are placed in contact with tissues that are very sensitive to contamination. Thus, similar standards of sterility and purity are required for ophthalmic and otic dosage forms.

The excipient of greatest importance for parenteral products is water. Water of suitable quality for parenteral administration must be prepared either by distillation or reverse osmosis. *Water for Injection USP* is a high-purity water intended to be used as a vehicle for injectable preparations. It is manufactured by exacting standards and meets stringent monograph requirements. *Sterile Water for Injection USP* (SWFI) is an excipient intended to be used as a packaged and sterilized product.

Certain aqueous vehicles are used as isotonic vehicles to which an active ingredient may be added at the time of administration. These vehicles include sodium chloride injection, Ringer's injection, and others. Several water-miscible solvents are used primarily to improve solubility of certain active ingredients and to reduce hydrolysis. The most important solvents in this group are ethyl alcohol, propylene glycol, and the liquid series polyethylene glycols. The most important group of nonaqueous vehicles are the fixed oils, including corn oil, cottonseed oil, peanut oil, and sesame oil. Fixed oils are used particularly as vehicles for certain hormone preparations (i.e., testosterone injection).

Buffers are employed to stabilize a solution against the chemical degradation that may occur if the pH changes significantly. Acetate, citrate, and phosphate are the most common buffers used in parenteral products. Antioxidants are frequently required to preserve products because of the ease with which many drugs are oxidized. Sodium bisulfide is the most frequently used antioxidant.

Antimicrobial agents in bacteriostatic or fungistatic concentrations must be added to parenteral preparations contained in multiple-dose containers. Their purpose is to prevent the multiplication of microorganisms inadvertently introduced into the container while withdrawing a portion of the contents with a hypodermic needle and syringe. Benzyl alcohol is the most commonly used antimicrobial. Parabens are the next most common preservatives.

For a thorough review of ''Excipients and Their Use in Injectable Products,'' the reader is referred to a recent review article by Nema et al. (10).

C. Emulsions and Suspensions

Emulsions may be defined in any number of ways, but essentially an emulsion is a two-phase system prepared by combining two immiscible liquids, one of which is dispersed uniformly throughout the other phase. Generally, one of the liquids is water and the other is some type of lipid or oil. Most emulsions incorporate an aqueous phase into a nonaqueous phase (or vice versa).

The list of excipients used to prepare emulsions is quite extensive. Choice of excipients for the oil phase includes various grades of mineral oil, a number of edible vegetable oils, and other such. Many emulsifying agents (or emulsifiers) are available, including natural emulsifying agents, finely divided solids, and synthetic emulsifying agents. Again, the list is too cumbersome for this presentation. Standard pharmaceutical texts (i.e., *Remington's Pharmaceutical Sciences*, 18th ed.; (8), can be consulted for more detailed information on excipients used for preparation of emulsions.

A *suspension* is a dispersion or dispersed system in which the internal, or suspended, phase is dispersed uniformly throughout the external phase, called the suspending medium or liquid. It is a two-phase system consisting of a finely divided solid (active ingredient) dispersed in a liquid, suspending medium. There are three general classes of pharmaceutical suspensions: (a) orally administered suspensions, (b) externally applied suspensions (topical lotions), and (c) injectable (parenteral) suspensions.

Suspending agents are used to impart greater viscosity and retard sedimentation. Suspending agents include cellulose derivatives, clays, natural gums, and synthetic gums. The list of agents is too extensive to be covered in this presentation. Excellent reviews of pharmaceutical suspensions (11,12) contain more detailed information on suspending agents (excipients).

D. Intranasal and Inhalation Delivery Systems

Nasal solutions or suspensions are usually aqueous systems designed to be delivered to the nasal passages in drops or sprays. Many of the excipients used to prepare pharmaceutical solutions or suspensions are used in the preparation of nasal products. In addition to water, cellulosics, surfactants, and buffering agents are commonly employed. Aqueous nasal solutions usually are isotonic and slightly buffered, for example, sodium chloride and dextrose, to maintain a pH of 5.5–6.5. Antimicrobial preservatives similar to those used in ophthalmic preparations are employed on occasion.

Aerosol dosage forms for oral and topical use were developed in the mid-1950s. The aerosol product itself consists of two components: (a) concentrate (containing the active ingredient(s); and (b) propellant(s) (13). The propellant provides the internal pressure that forces the product out of the container when the valve is opened and delivers the product in its desired form. Excipients for aerosols are divided into two categories: (a) those for the drug concentrate; and (b) those for the propellant.

1. Drug Concentrate

The drug(s) may be solubilized or micronized and suspended in the concentrate. Antioxidants (i.e., ascorbic acid) and dispersing agents (i.e., sorbitan trioleate, oleic acid, and such), are employed, especially if the drug is micronized. Solvent blends include water, ethanol, and glycols.

2. Propellant

Compressed gases, such as nitrogen, nitrous oxide, and carbon dioxide, have been used as aerosol propellants. Unlike the liquified gases, compressed gases possess little, if any, expansion power and will produce a fairly wet spray. Liquified gas compounds are widely used as propellants because they are extremely effective in dispersing the active ingredients into a fine mist or foam. The fluorinated hydrocarbons (fluorocarbons) are nonflammable relative to the flammable hydrocarbons. Because of environmental issues, fluorinated hydrocarbons have limited use in specifically exempted metered-dose inhalers and contraceptive vaginal foams (i.e., metered-dose steroid drugs for intranasal or oral inhalation, and such). Alternatives to the fluorocarbons are now under study and development. Hydrocarbons, *n*-butane, propane, and *iso*-butane, have largely replaced the fluorocarbons for topical pharmaceutical aerosols. Although of low order toxicity, flammability tends to limit their use.

E. Mucosal, Vaginal, and Rectal Preparations

Suppositories are solid dosage forms of various sizes (weights) and shapes, usually medicated, for insertion into the rectum, vagina, or the urethra. After insertion, they soften, melt, or dissolve in the cavity fluids. Typically, a suppository consists of a dispersion of the active ingredient(s) in an inert matrix, generally composed of a rigid or semirigid base. The *USP* lists the following as usual suppository bases: cocoa butter, glycerinated gelatin, hydrogenated vegetable oils, mixtures of polyethylene glycols of various molecular weights, and fatty acids esters of polyethylene glycol.

Cocoa butter, or theobroma oil, is a fatty material composed of a mixture of C_{16}–C_{18} saturated and unsaturated fatty acid triglycerides from the roasted seed of Theobroma cacao Linné. Cocoa butter is used extensively in manufacture of suppositories. It is well tolerated, but presents several problems when formulated in suppositories, including its unique melting point, slow rate of crystallization, and changes in the marketplace (i.e., pricing and availability can be erratic).

Glycerinated gelatin is usually used as a vehicle for vaginal suppositories. These suppositories typically contain preservatives, such as the parabens.

Water-soluble or dispersible suppository bases are of comparatively recent origin. Most are composed of polyethylene glycols or glycol–surfactant combinations. Because they are not dependent on a melting point approximating body temperature, they have a distinct advantage over cocoa butter or cocoa butter-like bases. Suppositories of varying melting points and solubility can be prepared by blending polyethylene glycol polymers (Carbowax) of various molecular weight. Water-miscible or water-dispersible suppositories are also prepared by using selected nonionic surfactant excipients. For example, Polyoxyl 40 stearate is a white, water-soluble solid, melting above body temperature. Water-dispersible suppository bases may also include other surfactants that are either soluble (Tween, Myrj) or water-dispersible (Arlacel), used either alone or in combination with other wax or fatty materials.

IV. SUMMARY

To reiterate, an excipient is a material that aids in the manufacture of the dosage form and protects, supports, or enhances stability and bioavailability of the drug. Excipients play many roles in turning active ingredients into efficient and effective dosage forms. There are numerous examples discussed in the chapter. Some excipients are used in distinctly different dosage forms. A good example is hydroxypropyl methylcellulose (HPMC). HPMC is used as a binder in the preparation of tablet granulations. It is also used as suspending–thickening agent in numerous pharmaceutical suspensions. HPMC is also used as a polymeric film coating for granules, pellets, and tablets (not discussed in this chapter). Another example is sodium lauryl sulfate. It is employed as a wetting agent in tablets to improve tablet disintegration, and as an emulsifier for pharmaceutical emulsions, creams, ointments, and such.

REFERENCES

1. Webster's New Collegiate Dictionary, A Merriam-Webster. Springfield, MA: G&C Merriam Co, 1980, pp. 395.

2. The National Formulary NF 18. Rockville, Maryland: United States Pharmacopeial Convention, 1995, pp. 2201, pp. liii, pp. 1982.
3. Handbook of Pharmaceutical Excipients, 2nd ed. Washington, DC: American Pharmaceutical Association, 1994, Preface XI.
4. L Blecher. Excipients—the important components. Pharm Proc, January 1995, p. 6.
5. E Rudnic, JB Schwartz. Oral solid dosage forms. In: AR Gennaro, ed. Remington's Pharmaceutical Sciences. 18th ed. Easton, PA: Mack Publishing, 1990, pp. 1633–1635.
6. WC Gunsel, CJ Swartz, JL Kanig. Tablets. In: The Theory and Practice of Industrial Pharmacy. Philadelphia: Lee & Febiger, 1970, pp. 305.
7. ZT Chowan. Tablet ingredients. In: Problem Solver and Reference Manual. Philadelphia: FMC Corp, 1998, pp. 1–18.
8. AR Gennaro. Remington's Pharmaceutical Sciences. 18th ed. Easton, PA: Mack Publishing, 1990.
9. LH Block. Medicated applications. In: AR Gennaro, ed., Remington's Pharmaceutical Sciences. 18th ed. Easton, PA: Mack Publishing, 1990, pp. 1602–1607.
10. S Nema, RJ Washkuhn, RJ Brendel. Excipients and their use in injectable products. PDA J Pharm Sci Technol 1997; 51(no. 4), pp. 166–171.
11. RA Nash. Pharmaceutical suspensions. In: HA Lieberman, MM Rieger, GS Banker, eds. Pharmaceutical Dosage Forms Disperse Systems. vol. 1. New York: Marcel Dekker, 1988, pp. 151–185.
12. CM Ofner, RL Schnaare, JB Schwartz. Oral aqueous suspensions. In: HA Lieberman, MM Rieger, GS Banker, eds. Pharmaceutical Dosage Forms Disperse Systems. vol. 1. New York: Marcel Dekker, 1988, pp. 231–252.
13. JJ Sciarra, AJ Cutie. Aerosols. In: AR Gennaro, ed. Remington's Pharmaceutical Sciences. 18th ed. Easton, PA: Mack Publishing, 1990, pp 1694–1712.

2
Purity of Excipients

Dankward Jäkel and Martin Keck
Novartis Pharma AG, Basel, Switzerland

I. INTRODUCTION

Excipients have long been considered to be inert materials, with no significant adverse effects on the safety and efficacy of pharmaceutical preparations. That excipients can have a significant influence on safety and efficacy of pharmaceutical preparations has been demonstrated in numerous examples, ranging from bioequivalence differences, to stability problems, to transfer of critical impurities.

Among the properties relevant in this context, the purity of excipients plays an important role. The use of excipients with insufficient purity can have dramatic consequences. An example is the 1996 incident in Haiti when at least 49 children died (1) because a technical grade of glycerin, contaminated with high amounts of diethylene glycol, was used in a pharmaceutical preparation. Similar accidents, causing the death of many people, occurred in 1990 in Nigeria and in 1993 in Bangladesh with 1,2-propylene glycol (again owing to a mix up with diethylene glycol; 1–3) and in 1981 in Spain, with the use of a technical grade olive oil (denatured with rape seed oil containing fatty acid anilides as denaturing additive) for food purposes.

When we consider the high number of excipients in use today for the development and production of drugs and the great variety of starting materials and production processes from and by which excipients are produced, it is critical to understand and assess the purity of excipients. This chapter will focus on the general aspects relevant for a reliable assessment of the purity of excipients, and impurities that are of major concern from the toxicological or stability point of view.

II. GENERAL ASPECTS OF PURITY

A. Origin and Production of Excipients

With the aid of excipients, substances with pharmacological activity are transferred into dosage forms applicable to humans. The number of different excipients in use today for the development and production of drugs amounts to approximately 1000. They are produced from a great variety of different source materials, for example:

Palm kernel, coconut, beef tallow (e.g., fatty acids, stearates, fat alcohols, glycerine, polysorbates)

Maize, potato, wheat, sugar beet, sugar cane (e.g., starches, dextrins, sucrose, mannitol, sorbitol, dextrose, fructose, sodium starch glycolate)

Wood (e.g., microcrystalline cellulose, powdered cellulose, various types of methyl-, ethyl-, hydroxyethyl-, propyl-, hydroxypropyl methylcelluloses and sodium carboxymethylcelluloses differing in degree of polymerization and molecular substitution, cellulose acetate, cellulose acetate phthalate, croscarmellose)

Crude oil (e.g., petrolatum, paraffin oil, and the huge range of excipients based on petrol crack products such as polyglycols and polyacrylates)

Sheep wool (lanolin)

Milk (lactose)

Excretes of insects (beeswax, shellac)

Minerals (talc, kaolin, calcium and sodium phosphates, sodium chloride)

Bones and hides from cattle and pig (gelatin).

The foregoing compilation could be easily extended by various additional source materials. The processing of these source materials into excipients requires the application of a great variety of, in part, very complex technologies (see Figs. 1 and 2 for fat derivatives; 4).

In contrast with active ingredients, excipients, resulting from these processes, frequently consist of a mixture of homologues (e.g., fatty acids, fat alcohols, cellulose derivatives, or synthetic polymers). The degree of refinement of excipients varies widely. If we take the grade of refinement (i.e., the number of process steps required to transfer the source material into the final excipient) as a criterion, we have, for example, at the lower end of the scale, talc (a mineral simply mined, dried, and milled) and at the upper end, sorbitol powder, which undergoes more than 20 process steps from the source material (maize) to the final product.

B. Technical and Pharmaceutical Grades

The pharmaceutical industry procures excipients from production facilities of various industries not subject to the same Good Manufacturing Practice (GMP)

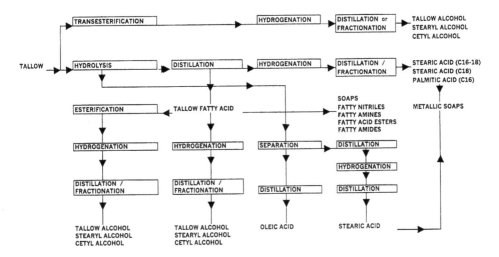

Figure 1   Production routes to tallow derivatives.

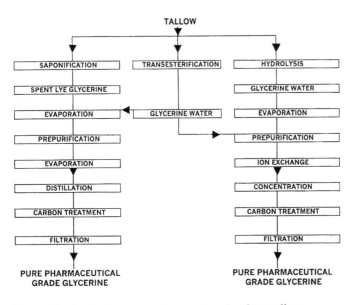

Figure 2   Production routes to pure glycerine from tallow.

regulations as the pharmaceutical industry. Frequently, the pharmaceutical industry uses only a very low percentage of the whole production output of materials predominantly designed and used for technical applications. To optimize processing properties, technical grades often contain additives, which can create problems in pharmaceutical applications owing to toxicity or interaction with the active ingredient. They are frequently not refined to the purity required for pharmaceutical applications.

Titanium dioxide is a typical example of an excipient used predominantly in nonpharmaceutical applications (Table 1). The amount used in pharmaceuticals is estimated to be considerably less than 1% of total production output. The production and analysis of titanium dioxide are summarized in Tables 2 and 3, respectively.

Excipient production processes and plants designed to meet the specific

Table 1   Uses of Titanium Dioxide

Worldwide production p.a. (1988)	2.96 million tons
Paints and lacquers	55–60%
Plastics	15–20%
Paper	Approx. 15%
Other applications[a]	Approx. 10%

[a] Includes printing inks, rubber, textiles, leather, synthetic fibers, ceramics and electroceramics, white cement, glass, catalysts, mixed metal oxide pigments, cosmetics, pharmaceuticals.
Source: Ref. 5.

Table 2   Production of Titanium Dioxide

Processes	Two
Sulfate	Ore + sulfuric acid/precipitation
Chloride	Ore + chlorine/vapor phase
Crystal modifications	Two (Anatas, Rutil)
Surface coating[a]	
Inorganic	e.g., Ti, Zr, Si, Na, K, Al, B, Sn, Zn, Mn, Ce, Sb, V-compounds
Organic	Silicones, amines, organophosphates, alcohols, alkylphthalates
Producers	Approx. 50 worldwide

[a] Purpose of the various coating processes is to achieve optimal application properties for the individual applications.
Source: Ref. 5.

Table 3 Titanium Dioxide Analysis

Samples	39 from 17 countries
Results indicating technical grades	
Arsenic	6–12 ppm (5 samples)
Antimony	22–200 ppm (10 samples)
Iron	60–480 ppm (3 samples)
Lead	25–75 ppm (2 samples)
Loss on ignition	1.9%
Acid solubles	3.2%
Assay	93.8%

needs of the pharmaceutical industry are the exception today. However, growing awareness is found throughout the supplying industry for the specific quality requirements for materials used in the manufacture of pharmaceuticals. In the 1995 publication of the *GMP Guide for Excipients*, edited by the International Pharmaceutical Excipients Council (IPEC), the concentration of the production of excipients for pharmaceutical application in only one specific plant of a company, the use of different cross-linking agents (phosphates vs. epichlorohydrin) for potato starch used in the manufacture of sodium starch glycolate, or the use of additional processing steps (vacuum stripping or heat treatment) to reduce the residual content of ethylene oxide in ethoxylates are examples of this growing awareness at the suppliers' side. However, in turn, it also happens that excipient manufacturers may decide, in view of the increased workload and risks associated with the pharmaceutical use, to stop their pharmaceutical activities and focus on technical applications.

III. QUALITY STANDARDIZATION

A. Legal Aspects

According to European legislation the requirements laid down in the *European Pharmacopoeia* (*EP*) are binding. In effect the same applies in the United States for the *United States Pharmacopeia/National Formulary* (*USP/NF*) and in Japan for the *Japanese Pharmacopoeia/Japanese Pharmaceutical Excipients* (*JP/ JPE*). By far the largest part of these monographs pertains to purity requirements. Thus, at a first glance, the problem of the quality standardization of excipients, especially relative to purity, seems to be solved by the existing, binding pharmacopeial monographs. However, numerous discrepancies exist among pharmacopeial monographs for identical substances. These discrepancies inevitably lead

Table 4 Ethanol—Comparison of Pharmacopeial Monographs

Assays	USA	Japan	Switzerland	Germany	United Kingdom	France
1. Identity	−		−	−	−	−
1.1. Color reaction with nitroferricyanide/piperazin	+	−	+	−	+	+
1.2. Jodoform reaction	+	+	+	−	+	+
1.3. Ethyl acetate reaction	−	+	−	−	−	−
1.4. Distillation range	−	−	77.8–79.0°C	78.0–79.0°C	−	78.0–79.5°C
1.5. Color reaction with nitroferricyanide/piperdin	−		−	+	−	−
1.6. Melting point derivative with 3,5-dinitrobenzoyl chloride	−		−	90–94°C	−	−
2. Relative density	0.812–0.816 (at 15.56°C)	0.814–0.816[a] (at 15°C)	0.8064–0.8089[a] (at 20.0°C)	0.804–0.809 (at 20.0°C)	−	0.8050–0.8140 (at 20.0°C)
3. Acidity	Max. 0.90 mL 0.02 N NaOH/50 mL	Max 0.10 mL[a] 0.1 N NaOH/20 mL	various limits using different color indicators	Max 0.4 mL[a] 0.01 N NaOH/50 mL	Max 0.2 mL[a] 0.1 N NaOH/20 mL	Same as Swiss Pharmacop.
4. Residue on evaporation	Max. 0.0025%	Max. 0.0025%	Max. 0.001%	Max. 0.0015%	Max. 0.005%	Max. 0.002%
5. Aldehydes, foreign substances	n.d.	n.d.[a]	n.d.[a]	n.d.[a]	n.d.[a]	−
6. Amyl alcohol, non-volatile carbonizable substances	n.d.	−				
7. Fusel oils	n.d.	n.d.[a]	−	< Color standard[a]	−	−
8. Acetone, Isopropanol	Not more intense as color standard (~10 ppm)	See 15.	−		−	−
9. Methanol	n.d.	n.d.[a]		Max. 500 ppm[a]		
10. Assay (by density)	92.3–93.8% w/w 94.9–96.0% v/v	95.1–95.6% v/v	93.8–94.7% w/w 96.0–96.6% v/v	93.8–95.6% w/w 96.0–97.2% v/v	93.8–94.7% w/w 96.0–96.6% v/v	92.0–94.7% w/w 94.7–96.6% v/v

11. Aqueous solution, Clarity	—	Clear	Clear^a	—	Clear	Clear
12. Alkalinity	—	n.d.	n.d.	Max. 0.1 mL 0.01 N HCl/15 mL^a Bromocresyl green	—	n.d.
13. Chloride	—	n.d.	—	—	—	Max. 1.25 ppm
14. Heavy metals	—	Max. 1.2 ppm	Max. 0.3 ppm^a	Max. 2 ppm	—	Max. 0.25 ppm
15. Ketones, Isopropanol, t-Butanol	—	n.d.	—	—	—	—
16. Clarity, tel quel	—	—	Clear	Clear	—	Clear
17. Color, tel quel	—	—	Colorless	Colorless	—	Colorless
18. Fluorescence	—	—	< Standard	—	—	—
19. Foreign odour	—	—	n.d.	—	—	—
20. Aldehydes and Ketones	—	—	Max. 100 ppm (oximreaction)	—	—	= Max. 100 ppm acetaldehyde
21. Methanol, Homologous alcohols and esters (GC)	—	—	Max. 0.01% v/v methanol Max. 0.02 area% Other impurities	—	Max. 0.04% in sum^a	Max. 0.02% methanol Max. 0.03 area% Other impurities
22. UV Absorption	—	—	5 cm 240 nm Max. 0.35 250–260 nm Max. 0.20 270 nm Max. 0.12	1 cm 220 nm max. 0.30 230 nm max. 0.18 240 nm max. 0.08 270 nm max. 0.02	—	1 cm 240 nm max. 0.10 250–260 nm max. 0.06 270 nm max. 0.03
23. Furfural	—	—	—	n.d.	—	—
24. Zinc	—	—	—	n.d.	—	—
25. Iron	—	—	—	n.d.	—	—
26. Density (absolute)	—	—	—	—	803.8–806.3 (kg m^{-3})	—
27. Benzene	—	—	—	—	Max. 2 ppm	Max. 5 ppm
28. Reducing substances	—	—	—	—	< Color standard	—

^a Method different, results not comparable nor convertable.
n.d., not detectable.

to the question: What is a reliable quality standardization from the scientific and technical point of view? Simultaneously, they highlight the problems associated with legally binding standards in different world regions and emphasize the necessity and importance of the International Harmonization of Excipient Quality Standards initiated by the *EP*, *JP*, and *USP* Commission (for more details, see Chapter 14). Two widely used excipients, ethanol and talc, will serve as examples to highlight these problems.

Table 5 Pharmacopeial Requirements for Talc

Requirement	Ph. Eur.II	USP XXII/ NF XVII	JP XII
1. Identity test A	+	−	−
2. Identity test B	+	+	+
3. Identity test C	+	−	−
4. Identity test D	−	+	−
5. Identity test E	−	−	+
6. Iron, soluble in 1 M H_2SO_4	Max. 250 ppm		
7. Magnesium, soluble in 1M H_2SO_4	Max. 0.4%	−	−
8. Calcium, soluble in 1M H_2SO_4	Max. 0.6%	−	−
9. Calcium, insoluble in 1M H_2SO_4	Max. 500 ppm	−	−
10. Carbonate	Not detectable	−	−
11. Chlorides, soluble in 2 M HNO_3	Max. 140 ppm	−	−
12. Readily carbonizable substances	Not detectable	−	−
13. Loss on drying, 180°C	Max. 1%	−	−
14. Microbes per gram	−	Max. 500	−
15. Loss on ignition, 1000°C	−	Max. 6.5%	−
16. Loss on ignition 450–550°C	−	−	Max. 5.0%
17. Acid-soluble substances (3N HCl, 50°C)	−	Max. 2%	Max. 2%
18. Aqueous extractables (100°C)	−	Max. 0.1%	Max. 0.1%
19. Aqueous extract, pH	−	Neutral	Neutral
20. Water-soluble iron	−	Not detectable	Not detectable
21. Arsenic, acid-soluble (0.5 N HCl, 100°C, 30 min)	−	Max. 3 ppm	−
22. Arsenic acid-soluble (H_2SO_4 10%, 100°C ca. 1 min)	−	−	Max. 4 ppm
23. Heavy metals, acid-soluble	−	Max. 40 ppm	−
24. Lead acid-soluble	−	Max. 10 ppm	−

1. Ethanol

Because of a state monopoly for alcohol in many countries, harmonization of pharmacopeial monographs for ethanol has not been realized yet at the European Union (EU) level. A comparison, including the *British Pharmacopoeia* (*BP*), the *German Pharmacopoeia* (*DAB*), the *French Pharmacopoeia* (*FP*), the *Swiss Pharmacopoeia* (*Ph. Helv.*), the *JP*, and the *USP* is compiled in Table 4. Results for a number of nominally identical testing criteria cannot be converted into each other owing to different test methods (e.g., whereas the *BP* test for aldehydes uses fuchsin, the *DAB* uses 3-nitrobenzaldehyde and aniline, the *FP* dinitrophenylhydrazine, the *Ph. Helv.* hydroxylamine, the *JP* and the *USP* different versions of a permanganate test). This means that in total 47 different tests must be performed to confirm compliance with the six foregoing pharmacopeias.

2. Talc

Talc, a natural magnesium silicate with a special physical structure, is widely used as a filler and lubricant in the manufacture of solid oral dosage forms. A comparison of the *EP*, the *JP*, and the *USP/NF* monographs is compiled in Table 5. A review of this information leads to the surprising result that the *EP* and the *USP* monograph have nearly nothing in common because they apply completely different testing criteria to specifiy the quality or purity of the same substance.

B. Scientific and Technical Aspects

1. General Considerations

The title of the article "The formidable task to set meaningful standards for excipients. A case study: magnesium stearate," published in 1988 in the *USP Pharmacopeial Forum* (6) captures the complexity of a reliable quality standardization in a nutshell. From a scientific point of view, the suitability of a specification for excipients must be assessed on the basis of the following criteria for meaningful quality standards:

1. Reliable identification
2. Detection and limitation of critical impurities from the toxicological point of view
3. Detection and limitation of impurities that can have an adverse effect on stability and efficacy of drugs owing to interaction with the active ingredient or other excipients
4. Differentiation of grades for pharmaceutical and technical application
5. Confirmation of the quantitative composition (in case of mixed substances)

6. Characterization and specification of physical properties with technological relevance, to ensure a frictionless, economic manufacture of
dosage forms and to guarantee their constant physical–galenical properties, in vitro dissolution rate, and bioavailability

From the foregoing criteria items 2–5 are directly associated with the purity
of excipients. The key elements for a reliable assessment and a meaningful standardization of the purity of an excipient are

Origin (source material) of an excipient
Production process applied to transform the source material into an excip-
ient
The needs of the individual application of an excipient

As outlined in Sec. IV.B.1. of this chapter in more detail, knowledge of
the origin and the production process of an excipient is indispensable to assess
the quality of fat derivatives or gelatin relative to transmissible spongiform encephalopathy. Even though modern analytical techniques such as X-ray fluorescence and optical emission spectroscopy with inductive coupled plasma (ICP
OES) or gas chromatography combined with mass spectrometry (GC–MS) would
usually permit identification of inorganic catalysts or solvents used in a production process, information on such details obtained from the supplier, if necessary
under secrecy agreement, is a prerequisite for a meaningful design of specifications.

Numerous materials used as pharmaceutical excipients are also used as
direct food additives. There are two fundamental differences between these applications:

1. Whereas food additives are administered exclusively by the oral route,
pharmaceutical excipients are administered by various other routes of
application.
2. Contrary to food additives, pharmaceutical dosage forms contain active
ingredients for the treatment of a disease.

Because of these differences, food additive and pharmaceutical excipients standards frequently differ considerably. Whereas it is obvious that some applications
(e.g., by the intravenous route) require higher purity standards (e.g., for pyrogens
in excipients manufactured by fermentation processes, such as dextrose, citric
acid, mannitol, trehalose), higher-purity standards for excipients used in oral or
even topical drugs seem not to be justified at a first glance. However, one reason
for concern is a potential interaction between impurities in the excipient with the
active ingredient, leading to degradation and loss of efficacy. In most formulations the active ingredients are outnumbered by excipients for the quantitative

Table 6 Ratio of Excipients to Active
Ingredient in Pharmaceutical Preparations:
Example Corticosteroid Ointment

Composition for 100 g of ointment:	Weight
Active ingredient	0.02g
Excipients	
White petrolatum	40.00g
Liquid paraffin	20.00g
Water purified	19.78g
Sorbitan sesquioleate	10.00g
5 additional excipients	10.22g
Ratio by number	9 : 1
Ratio by quantities	5000 : 1

and numeric ratio. This is especially true for highly potent active ingredients, where only milligram or even microgram amounts are applied, as for instance, in the example of a corticoid ointment shown in Table 6. From this table one can see that the quantities of excipients in a formulation far exceed the quantity of the active ingredient. In such cases, interaction between the active ingredient and impurities in the excipient can lead to a complete or partial deterioration of the active ingredient (see Sec. IV.B.3).

2. Practical Example

Practical application of the foregoing key elements for a reliable standardization of the purity of an excipient will be discussed in more detail, using ethanol/alcohol as a complex model. The following reflects the work of the IPEC Europe Harmonization Committee to develop an international harmonized, meaningful pharmacopeial monograph for ethanol/alcohol. The following companies participated actively in this work: Astra Haessle, British Petrol, Ciba Geigy, Hoffmann La Roche, Sandoz, and the Swiss Ethanol Board.

When addressing the reliable quality standardization of ethanol/alcohol, the following facts should be considered:

1. Ethanol is globally negotiated as a commodity. By far the largest percentage of the world production is used for technical applications and only a very minor part is used as a pharmaceutical excipient.
2. Ethanol is produced by a large number of different processes. Basically one must distinguish between fermentation processes and synthesis on basis of ethylene.

3. Fermentation processes start from many different sources, frequently from directly fermentable materials such as sugar cane or sugar beet as well as their molasses. Other source materials must first be converted into a fermentable form by means of enzymes, such as various starches, or even waste materials such as lignocellulosic compounds contained in the waste water of cellulose produced by the sulfite process.

4. For ethanol sold on the global market, approximately 70% is obtained by fermentation with consecutive distillation, and approximately 30% is produced by synthesis with consecutive distillation.

5. Waterfree ethanol (> 99% purity) is produced by azeotropic distillation. Benzene, toluene and cyclohexane are among the solvents used for this purpose.

6. Ethanol is subject to a state monopoly and tax system in most countries. As a consequence, one is confronted at present with a large number of national standards, requirements, and different grades of material.

7. Because of taxes, denaturation plays an important role, and a considerable number of denaturation agents (methanol, isopropanol, ethyl acetate, methyl ethyl ketone, denatonium benzoate [at 5–10 ppm only], sucrose octaacetate, t-butanol, n-butanol, crotonaldehyde, acetone) are used, varying from country to country.

8. Crude products contain numerous impurities, which vary in species, number, and concentration, depending on the source material or production process. Table 7 illustrates both the wide range of impurities found in crude ethanol and the variations among samples.

A monograph for a reliable purity standardization of ethanol must cover, in particular, the detection and limitation of toxic by-products (e.g., benzene or methanol); the detection and limitation of impurities that may interact with active ingredients (e.g., acetaldehyde or diacetal); and the differentiation between pure and denatured grades. A highly effective GC method, together with testing of the substance by ultraviolet (UV) and infrared (IR) spectroscopy are the core elements for such a monograph. In an evaluation performed under the presumption to use in the monograph one GC column only, among phases of cyanopropyl methylpolysiloxane, cyanopropyl phenyl methylpolysiloxane, and polyethylene glycol, a phase with 6% cyanopropyl phenyl methylpolysiloxane (e.g., DB 624, column length: 30 m, diameter: 0.53 mm, film thickness: 3.0 μm) provided the best selectivity.

This GC system was applied to 28 samples of pure ethanol from different sources all over the world, stemming from different source materials and production processes. The results compiled in Table 8 demonstrate that the final process purification steps cut down the impurities contained in crude ethanol drastically and also indicate that ethanol from the synthetic route meets pharmaceutical stan-

Table 7 Impurities in Spirits

No.	Acetaldehyde ppm	Methanol ppm	Acetone ppm	Isopropyl alcohol ppm	tert-Butyl alcohol ppm	n-Propyl alcohol ppm	sec-Butyl alcohol ppm	iso-Butyl alcohol ppm	tert-Amyl alcohol ppm	n-Butyl alcohol ppm	Acetaldehydetdiacetal ppm	iso-Amyl alcohol ppm	n-Amyl alcohol ppm	3-Hexyl alcohol ppm	Furfurol ppm	1-Hexyl alcohol ppm	Unknown number	Unknown ppm	Ethanol Area%	Spirit name
1	140	9660	—	30	—	200	190	640	20	20	390	1000	10	10	20	80	15	590	98.7	Gentian
2	40	10	—	20	—	250	2130	800	—	10	—	2240	10	—	150	—	1	10	99.4	Arrack
3	70	8760	—	10	—	1050	20	870	—	400	310	2390	20	10	250	70	23	5070	98.0	Plum
4	130	4540	—	30	20	920	410	810	50	120	280	680	10	230	170	150	10	1070	99.0	Grappa
5	60	1090	—	—	—	1310	1250	1500	—	270	50	3080	10	20	220	260	8	2140	98.9	Calvados
6	220	470	—	50	—	700	40	1980	10	20	230	6020	30	—	160	60	10	920	98.9	Cognac
7	40	90	—	—	—	800	270	1440	—	10	40	530	110	—	10	—	3	110	99.7	Whisky
8	20	—	—	—	—	—	—	—	—	—	—	—	—	—	—	—	—	—	100.0	Vodka
9	80	9120	—	30	—	590	1720	1400	—	240	120	4660	10	—	190	180	—	200	98.1	Apple
10	60	70	—	—	—	910	150	3140	—	10	170	2610	340	—	10	—	—	30	99.4	Rum
11	70	170	30	20	—	1290	10	1490	—	10	80	650	110	—	20	—	—	500	99.6	Whisky
12	60	672	—	40	10	910	130	450	—	170	30	1120	340	—	—	20	—	7820	98.8	Plum
13	170	820	—	20	—	530	20	1620	10	10	170	2820	—	—	70	30	—	600	99.4	Wine Brandy

Table 8   Impurities in Ethanol

No.	Acetaldehyde ppm	Methanol Area%	Acetone Area%	Isopropyl alcohol Area%	tert-Butyl alcohol Area%	n-Propyl alcohol Area%	sec-Butyl alcohol Area%	iso-Butyl alcohol Area%	tert-Amyl alcohol Area%	n-Butyl alcohol Area%	Acetaldehydetdiacetal Area%	iso-Amyl alcohol Area%	n-Amyl alcohol Area%	3-Hexyl alcohol Area%	Furfurol Area%	1-Hexyl alcohol Area%	Unknown Number	Unknown ppm	Ethanol Area%	Country	Source
1	<10	—	—	—	—	0.020	—	—	—	—	—	—	—	—	—	—	2	12	99.98	Peru	
2	<10	—	—	—	—	—	—	—	—	—	—	—	—	—	—	—	11	164	99.98	Turkey	Beet molasses
3	<10	0.006	—	—	—	—	—	—	—	—	—	—	—	—	—	—	2	11	100.0	Greece	Raisins
4	<10	—	—	—	—	0.050	—	—	—	—	—	—	—	—	—	—	5	31	100.0	Egypt	Sugarcane
5	<10	—	—	—	—	—	—	—	—	—	—	—	—	—	—	—	5	28	99.95	Chile	Molasses
6	<10	0.012	—	—	—	—	—	—	—	—	—	—	—	—	—	—	5	86	99.99	Austria	
7	<10	—	—	0.011	—	—	0.011	—	—	0.007	—	—	—	—	—	—	6	54	99.97	Korea	Synthetic
8	<10	—	—	—	—	—	—	—	—	—	0.001	—	—	—	—	—	2	23	100.0	Indonesia	
9	<10	—	—	—	—	—	—	—	—	—	—	—	—	—	—	—	2	5	100.0	Spain	Beet molasses
10	<10	—	—	—	—	—	—	—	—	—	—	—	—	—	—	—	7	105	99.99	South Africa	Sugar cane
11	<10	—	—	—	—	—	—	—	—	—	—	—	—	—	—	—	3	21	100.0	Thailand	*Oryza sativa*
12	<10	—	—	—	—	—	—	—	—	—	—	—	—	—	—	—	3	8	100.0	Japan	Sugar cane, cereals
13	<10	—	—	—	—	—	—	—	—	—	—	—	—	—	—	—	2	7	100.0	England	Synthetic
14	<10	—	—	0.004	—	—	—	—	—	—	—	—	—	—	—	—	2	16	100.0	United States	Synthetic
15	<10	0.002	—	0.007	—	—	—	—	—	—	—	—	—	—	—	—	2	21	99.99	Switzerland	Sulfite waste liquor, from cellulose production
16	<10	—	—	—	—	—	—	—	—	—	—	—	—	—	—	—	2	13	100.0	France	Molasses
17	<10	0.002	—	—	—	—	—	—	—	—	—	—	—	—	—	—	2	14	100.0	United States	Cereals
18	<10	0.004	—	0.003	—	—	—	—	—	—	—	—	—	—	—	—	2	5	99.99	Poland	Potatoes (refined in France)
19	<10	—	—	—	—	—	—	—	—	—	—	—	—	—	—	—	2	9	100.0	Finland	Cereals
20	<10	—	—	—	—	—	—	—	—	—	—	—	—	—	—	—	1	30	100.0	Hungaria	Beet molasses
21	<10	0.001	—	0.002	—	0.023	0.001	—	—	—	—	—	—	—	—	—	1	20	99.97	Taiwan	Sugar cane molasses
22	<10	—	—	—	—	—	—	—	—	—	—	—	—	—	—	—	1	10	100.0	France	Sugar beet molasses
23	<10	0.001	—	0.001	—	0.026	—	0.001	—	—	—	—	—	—	—	—	1	10	99.97	Philippines	Sugar cane molasses
24	<10	0.003	—	0.022	—	—	—	—	—	—	—	—	—	—	—	—	—	10	99.97	Colombia	Sugar cane molasses
25	<10	0.038	—	0.002	—	—	—	—	—	—	—	—	—	—	—	—	1	10	99.96	Germany	Potatoes
26	<10	3.61	—	0.002	—	—	—	—	—	—	—	0.001	—	—	—	—	—	10	96.39	United States	Synthetic
27	<10	0.004	—	0.001	—	—	—	—	—	—	—	0.001	—	—	—	—	1	10	99.99	Brasil	Sugar cane molasses
28	<10	0.003	—	0.001	—	0.003	0.001	0.001	—	—	0.003	—	—	—	—	—	—	—	99.98	Uruguay	Sorghum

dards. The total amount of impurities varied between 5 and approximately 600 ppm; the most important impurities were isopropanol, methanol, and *n*-propanol. No methanol was found in ethanol produced by the synthetic route. Sample 26 turned out to be a material denatured with methanol.

Given the considerations for a reliable purity standardization specified in the foregoing and the results obtained in comprehensive laboratory investigations, a harmonized pharmacopeial ethanol monograph was proposed by the IPEC Europe Harmonization Committee (Table 9). The combination of testing criteria shown in Table 9 permit a quick identification (IR), a reliable and quick purity assessment of the impurities and denaturing agents of practical relevance (GC + UV), and determination of the ethanol concentration (relative density). It was adopted on a European level by the European Pharmacopoeia Commission, with some minor modifications, in June 1997 and will become official at latest by January 1, 1999 (7).

Table 9   Tests for a Harmonized Pharmacopeial Ethanol Monograph Proposed by the IPEC Europe Harmonization Committee

Identification
    A. Oxidation to acetaldehyde + color reaction (*USP/BP* method)
    B. Jodoform reaction (*USP/JP* method)
    C. Relative density (d 20/20):
        Ethanol absolute: max. 0.794
        Ethanol 96%: 0.804 to 0.813
    D. IR spectrum
If identity test D is carried out tests A and B can be omitted
Tests
    1. Appearance (clarity and color) clear and colorless (LS method)
    2. Acidity or alkalinity (JP method)
    3. Related substances (GC)
        Methanol: <0.02% (v/v)
        Total amount of impurities other than methanol: max. 0.03% (area%)
    4. Acetaldehyde (GC): max 10 ppm (v/v)
    5. Acetaldehyde diethyl acetal (GC): max 30 ppm (v/v)
    6. Acetaldehyde—equivalents (sum 4 + 5) (GC): max 10 ppm
    7. Benzene (*BP* UV-method): max 2 ppm (v/v)
    8. UV-absorbing compounds 235–340 nm, 5 cm cell (modified *USP* method):
        Absorption 240 nm: max. 0.40; 250 nm: max. 0.30; 260 nm: max. 0.30; 270–340 nm: max. 0.10
    9. Residue on evaporation: max 0.0025% (m/v)

IV. SPECIFIC IMPURITIES

A. Inorganic Impurities

1. Heavy Metals

If we consider the frequency of the heavy metal test (8) in pharmacopeial monographs, it is the most important test for inorganic impurities in excipients and active ingredients, together with the test for sulfated ash. However, what this test exactly covers and what it does not cover have not been exactly defined in the pharmacopeia. Therefore, it is a source of considerable uncertainty and, for this reason, requires a careful evaluation and assessment of the meaningfulness of this test for today's problems. The history of the pharmacopeial heavy metals test is a key for this assessment.

   a. History and Objectives of the Compendial Heavy Metals Test.   Over the hundreds of years of their history, pharmacopeias have evolved from compendia on drug manufacture into compendia for drug testing. One of the first analytical tests introduced in the early phase of this development (at the turn of the century) was the test for heavy metals.

   USP VIII (1905), published the first general test for heavy metals, "Time-Limit Test for Heavy Metals." The aim of this test was defined as follows: "This test is to be used to detect the presence of undesirable metallic impurities in official chemical substances or their solutions; these should not respond affirmatively within the stated time." The test had two steps:

1. Sulfide precipitation in a strongly acidic range
2. Sulfide precipitation in an ammonia–alkaline medium

The metals listed as undesirable were antimony, arsenic, cadmium, iron, lead, copper, and zinc. A general test for the separate, specific determination of arsenic had already appeared in *USP VI* (1893).

   In *USP XII* (1942), there was a change to an acetic acid medium. Simultaneously, a comparison solution for lead was also introduced, and it was the "darkness" of this weakly acidic solution that served as a permissible limit. The Swiss and German pharmacopoeias (*Ph. Helvetica, Deutsches Arzneibuch*) underwent a very similar development, only at a later time.

   In addition to general heavy metals testing, pharmacopeias also require in part specific testing for individual heavy metals, such as

   nickel: in polyols and hardened fats
   iron: in diverse substances
   lead: in sugars

An analysis of the history of the pharmacopeial heavy metals testing shows that the following objectives were pursued. In the 19th century and in the early part of the present century, several heavy metal compounds, considered to have medicinal value, were commonly used in pharmaceutical products (e.g., arsenic in Liquor kalii arsenicosi and in Salvarsan, mercury in Unguentum hydrargiri album, flavum, rubrum, and cinereum; antimony in Kalii stibii tartaricum and Stibium sulfuratum aurantiacum; bismuth in Bismutum subnitricum and subgallicum; and lead in Emplastrum lithargyri and Unguentum diachylon). The test, therefore, was originally very broad in scope (detection of all colored and dark sulfides precipitated in acidic and alkaline solutions) to prevent the use of mislabeled products or products containing inadvertent admixtures of heavy metal compounds.

The later restriction to dark sulfides precipitated from weakly acid solution, with lead as a comparison standard, and the additional specific tests for individual elements, such as arsenic and iron, imply a fundamental change of perspective. Clearly, the purpose now was to detect contamination caused by toxicologically significant heavy metals coming from manufacturing equipment and processes. The conditions of detection chosen show that the focus was now on lead and copper, two elements formerly widely used in factory equipment (e.g., in water pipes, in copper and brass kettles, and in the lead chamber process used in the manufacture of sulfuric acid, an essential basic chemical substance required in numerous synthetic processes).

This historical review leads to the conclusion that the heavy metals test in its present form was clearly neither designed to be a universal test, nor meant to be understood as one, and it clearly does not allow for such interpretation.

b. Scope and Limitation of the Present Pharmacopeial Heavy Metals Test. The test in its present form (i.e., sulfide precipitation in a weakly acidic medium and comparison against a lead comparison solution at a concentration of usually 10 ppm) is theoretically suitable for the determination of bismuth, copper, gold, lead, mercury, ruthenium, silver, and tin (II). In practice, however, this method has several serious limitations.

Elements such as cadmium, antimony, and arsenic are not covered by this test because of the different color of their sulfides (the test is suitable for black or dark brown sulfides only), respectively, they are only partially covered in presence of very high concentrations without providing a reliable information about the true amount of impurity present (source for wrong conclusions).

Second, a substance must frequently be ignited before it is tested for heavy metals. This usually leads to a considerable loss of analyte. This loss is matrix-dependent. The average recovery rates found for hydroxypropylmethylcellulose (HPMC) are shown in Table 10 (9). It can be seen that the recovery rate is usually

Table 10 Determination of Heavy Metals in Hydroxypropylmethylcellulose

| Element | % Recovery after ignition[a] | | Precipitation | Color |
	USP	EP		
Sn	66	0	Yes	Dark
As	63	70	Yes	Yellow
Hg	0	0	Yes	Dark or red
Sb	57	61	Yes	Orange
Cd	60	57	Yes	Yellow
Pb	56	46	Yes	Dark
Bi	62	56	Yes	Dark
Cu	69	54	Yes	Dark
Cr			No	
Ni			No	
Fe			No	

only approximately 50–60%. Some elements (such as mercury) are lost completely during the ignition process. Although the recovery rate for tin according to *USP* is 66%, it is completely lost in the *EP* method. Reasons for this are the different ignition temperatures (*USP*: 550°C, *EP*: 750°C). Consequently, pharmacopeias recently changed this method requesting now a wet digestion procedure to avoid losses.

Important metals used in modern production equipment or as catalysts, such as iron, chromium, and nickel, are missed completely because they do not precipitate under the test conditions. There is virtually no way to differentiate between highly toxic and less toxic metals. The test is nonselective and barely semiquantitative.

c. Practical Experience. Over the years, thousands of samples taken from several hundred substances have been investigated by numerous companies, using the compendial heavy metals test method. According to the data and information available to us from various sources, there are hardly any cases in which the use of the official test has led to actual detection of heavy metals.

On the other hand, problems occurred in a number of cases with materials that had passed the pharmacopeial heavy metals test but in fact, were contaminated, sometimes massively, with heavy metals. Examples include the following:

440 ppm tin in polylactic acid
30 ppm platinum in a lactam (antibiotic precursor)

30 ppm nickel in a polymeric amine

108 ppm cadmium and 215 ppm nickel in magnesium stearate

High concentrations of cadmium and nickel in several other magnesium stearate samples

2500 ppm zinc in a magnesium stearate sample

52–530 ppm tin in cetylpalmitate samples.

In conclusion, based on the very comprehensive database, catalysts are currently the major source of contamination with heavy metals. The second source of contamination are excipients of mineral origin, not carefully selected for purity, such as iron oxides (with up to 70 ppm arsenic, 24 ppm lead, 2300 ppm barium, 205 ppm copper, and 141 ppm chromium) or talc (with up to 300 ppm nickel and 150 ppm copper). Based on practical experience gained with heavy metals in excipients, substances that belong to one of the following categories require specific consideration when testing for critical elements:

Minerals (e.g. talc, kaolin)

Inorganic compounds directly derived from minerals (e.g., calcium phosphates, precipitated silicon dioxide)

Organic compounds produced with the aid of metals (e.g., hydrogenated fats)

Natural compounds (e.g., gelatin, acacia, soybean lecithin)

Substances with a risk of cross-contamination from the production process (e.g., stearates)

Substances containing additives or additive residues (e.g., polymers with tin as stabilizer, shellac with arsenic compounds sometimes used as processing aid, or titanium dioxide with antimony as a stabilizer)

Liquid organics stored in metal containers and contaminated by uptake from those containers (e.g., iron in liquid glycols).

If heavy metal contamination is of practical relevance, element-specific methods, such as atomic absorption spectroscopy (AAS), optical emission spectrometry in combination with inductive coupled plasma (ICP OES), or X-ray fluorescence (X RF), in combination with adequate sample preparation techniques must be applied to ensure correct results and reliable information.

2. Sulfites

Sulfur dioxide in aqueous solution is used as a processing agent in the production of maize starch to promote the swelling of the grains. It is occasionally used in the production of starches as an antimicrobial agent to ensure adequate microbiological purity of the final starch product. Sulfites cause allergic reactions and are a source for stability problems owing to interactions with active ingredient or other excipients in a formulation (e.g., indigotin). Although the *USP* limits sulfur dioxide in maize starch to maximum 80 ppm, it has been proposed in the framework of the international harmonization to adopt the 50 ppm limit required by the European Food Additive legislation (10). In general, this limit can be met; however, occasionally results up to 130 ppm have been observed.

3. Radioactive Nuclides

Contamination of pharmaceutical excipients by radioactive nuclides did not give cause for discussions until the blast of the nuclear reactor at Chernobyl in April 1986. The radioactive cloud emerging from the blast contained as main components the following nuclides with a long half-life: cesium 134, cesium 137 (half-life 30 years), and ruthenium 103, as well as iodine 131 (half-life 8.09 days), known for its accumulation in the thyroid gland. Strontium 90 (half-life 28 years), which accumulates in the bones, was important only in the Chernobyl region. This cloud caused a radioactive fallout in wide parts of Europe, Asia Minor, and North Africa. Results of several thousand bequerel per kilogram (Bq/kg) in the food sector (e.g., in milk powder from European sources and nuts from Asia Minor), also led to an in-depth investigation of pharmaceutical excipients.

Potential risk materials in this context are excipients manufactured from vegetable, mineral, or animal origin with a low degree of raffination and a high ash content (e.g., starches, cellulose, talc, gelatin, phosphates, milk products, and titanium dioxide). Excipients manufactured from source materials of vegetable or animal origin with a high degree of processing or raffination and low ash content (e.g., fatty acids, fat alcohols, mannitol, sorbitol, or dextrose) are unlikely to present a problem for contamination by radioactive nuclides.

In a collaborative investigation performed in Basel in the course of 1986–1987, including approximately 1000 supplies of more than 100 potential risk materials, selected according to the foregoing criteria, the following results were obtained:

1.   In approximately 98% of all cases the results of radioactivity measurements were below 10 Bq/kg (limit of detectability of the applied method).
2.   Compared with this low basic level for most excipients, slightly elevated results were observed for talc (18–34 Bq/kg) and skimmed milk powder (15–63 Bq/kg). Results of a similar magnitude were also found in gelatin and agar.
3.   Results of 240–290 Bq/kg, close to the limit of 370 Bq/kg adopted from the EU requirements from May 30, 1986 (e.g., for milk products and baby food) were found for several samples of kaolin. Because kaolin samples from 1983 revealed similar high results, reason for the elevated radioactivity results is obviously of natural origin and not a

result of the Chernobyl blast. Presumably, this is also the reason for the slightly elevated results observed for talc.

4. Whereas samples of materials processed from cultures that were exposed to radioactive fallout (e.g., thyme from Eastern Europe, sugar beet molasses and wheat flour from Poland) showed results higher than 1000 Bq/kg, results for material from later harvests dropped quickly back to the low base levels. Thus, it can be concluded that in this case radioactive cesium (half-life of cesium 137: 30 years), which conceivably could be absorbed from the surface layer of the ground, was not taken up by the plants. However, in other cases, selective absorption and accumulation with increased radioactivity levels have been observed (e.g., for certain mushrooms).

From radioactivity measurements and calculations performed in Germany (11) it was concluded that in the first year after the blast the average additional irradiation exposure for humans was 1 millisievert (mSv). This level is in the same magnitude as the average annual exposure from natural radiation sources (2 mSv). The additional average dose, accumulated over 50 years, which results from the Chernobyl accident is expected to be 3.5 mSv. This makes only 3.5% of the exposure of approximately 100 mSv from natural sources over the same time period.

From a retrospective point of view it can be concluded that the radioactive fallout from the Chernobyl blast, even though regionally of dramatic and globally of important consequences, did not seriously affect the purity of pharmaceutical excipients distributed on the world market. The importance of this example lies in the vast spread of the radioactive fallout and that it opened, for the first time in the field of pharmaceutical excipients, a scenario that required a comprehensive risk assessment for a large number of substances under consideration of various parameters, such as

Geographic distribution of the radioactive fallout
Radioactive nuclides involved
Absorption and accumulation of involved nuclides
Potential risk materials
Medical risks

B. Organic Impurities

1. Proteins

Among the potential impurities in excipients requiring special attention because of health reasons, the following two proteins need specific discussion: gluten and prion protein.

a. Gluten. Gluten is the protein fraction from certain cereals that causes celiac disease, a severe degeneration of the intestinal mucosa associated with malabsorption of food. The frequency of this disease is, on average, approximately 1 : 1500 individuals (12); however, the incidence can be considerably higher or lower in certain local regions (13,14).

Whereas certain cereals, such as wheat, barley, rye, and oats, contain gluten, others, such as maize and rice, do not. Among the excipients commonly used in the manufacture of pharmaceuticals the risk of a contamination with gluten is restricted to wheat starch. The first process step in the production of wheat starch from flour is the separation of starch and proteins. The protein content is normally reduced to a level of maximum 0.3% protein by this step. According to the definition of the Codex Alimentarius Standard 118-1981 for "Gluten-Free Foods" (15), food is now considered gluten free (i.e., no adverse reaction of celiac patients expected) if the protein content does not exceed 0.3%, but this limit is still under further discussion.

Gluten, as a fraction of the total protein content, has been determined and limited indirectly by the nonspecific nitrogen determination with the Kjeldahl method. The protein content is calculated from the nitrogen result by multiplication with a factor. This factor varies to some extent, depending on the amino acid composition of the specific cereal protein. The factor for wheat protein is 5.7, for maize 6.25. In practice most frequently a factor of 6.25 is applied. Since the development of methods for direct determination of gluten, direct limits for its active fraction, prolamin, also can be found in the literature. The following limits are considered equivalent: maximum 0.05% nitrogen, maximum 0.3% protein, maximum 200 ppm gluten, or maximum 100 ppm prolamin.

b. Prion Protein. No event has had a greater influence on the production, trade, and use of pharmaceutical excipients than the detection of the prions as the transmitting agent of bovine spongiform encephalopathy (BSE) and other transmissible spongiform encephalopathies (TSE), such as Scrapie in sheep. The main reasons for this development are as follows:

The absolute fatality of TSE diseases
The uncertainty about the potential transmission of BSE to humans causing Creutzfeldt–Jakob disease (CJD)
Detection of a new variant of CJD
The enormous expansion of BSE in the United Kingdom in the early 1990s
No analytical method available for the direct detection of prion proteins in excipients as well as in other substances
The enormous number of potential risk materials from the food, feed, pharmaceutical, and cosmetics sector that are traded on the global market

In view of the potentially serious consequences, in 1990, the authorities began to control the problem with the following series of directives and guidelines:

- Council Directive (92/118/ EEC of December 17, 1992) laying down animal health and public health requirements governing trade and imports into the European Community (EC) of products not subject to the said requirements laid down in specific Community rules referred to in Annex A (I) to Directive 89/662/EEC and, as regards pathogens, to Directive 90/425/EEC.
- Council Directive (90/667/EEC of November 27, 1990) laying down the veterinary rules for the disposal and processing of animal waste, for its placing on the market, and for prevention of pathogens in feedstuffs of animal or fish origin amending Directive 90/425/EEC.
- Commission Decision (92/562/EEC of November 17, 1992) on the approval of alternative heat treatment systems for processing high-risk material.
- Commission Decision (4/382/EC of June 27, 1994) on the approval of alternative heat treatment systems for processing animal waste of ruminant origin, with a view to the inactivation of spongiform encephalopathy agents.
- Commission Decision (96/449/EC of July 18, 1996) on the approval of alternative heat treatment systems for processing animal waste with a view to the inactivation of spongiform encephalopathy agents.
- Commission Decision (94/381/EC of June 27, 1994) concerning certain protection measures relative to BSE and the feeding of mammalian-derived protein.
- Commission Decision (95/60/EC of March 6, 1995) amending Decision 94/381/EC concerning certain protection measures relative to BSE and the feeding of mammalian-derived protein.
- Decision of the Swiss Intercantonal Control Office for Medicinal Products relative to BSE and medicinal products for human use (official from March 22, 1991).
- EC/CPMP "Guidelines for minimizing the risk of transmission of agents causing spongiform encephalopathies via medicinal products" (EC Dossier III/3298/91, adopted Dec. 11, 1991, official from May 1, 1992).
- FDA letter of Dec. 17, 1993 on BSE and FDA-regulated products.
- German Ministry of Health "Guidelines on safety measures in connection with medicinal products containing body materials obtained from cattle, sheep, or goats for minimizing the risk of transmission of BSE and scrapie" (official from February 16, 1994).

- FDA "Substances prohibited from use in animal food and feed; specified offal from adult sheep and goats prohibited in ruminant feed; scrapie" (21 CFR part 589/*Federal Register* Vol. 59, No. 166, Aug. 29, 1994).
- FDA "Bovine-derived materials; Agency letters to manufacturers of FDA-regulated products" (21 CFR part 589, *Federal Register* Vol. 59, No. 166, Aug. 29, 1994).
- Commission Decision (96/239/EC of March 27, 1996) on emergency measures to protect against BSE (total export ban of any kind of bovine material from United Kingdom origin, including also tallow-derived excipients).
- Commission Decision (96/ 362/ EC of June 11, 1996) amending Decision 96/239/EC on emergency measures to protect against bovine spongiform encephalopathy (this decision lays down the principles under which the export ban for bovine material from United Kingdom origin may be lifted).

By the end of 1997, several new directives and guidelines were under preparation. At the EC:

- Commission Decision (97/ 534/ EC of July 30, 1997) on the prohibition of the use of material presenting risks for transmission of spongiform encephalopathies
- Inclusion of Decision 97/ 534/ 97 as annex into EC Council Directive (75/318 EEC of May 20, 1975) on the approximation of the laws of the Member States relating to analytical, pharmatoxicological, and clinical standards and protocols for the testing of medicinal products.
- EMEA/CPMP: Revision of the "Note for Guidance on minimising the risk of transmitting animal spongiform encephalopathies via medicinal products" issued by the CPMP in 1991.

At the FDA:

- Guidance for Industry: The Sourcing and Processing of Gelatin to Reduce the Potential Risk Posed by Bovine Spongiform Encephalopathy (BSE) in FDA Regulated Products for Human Use (September 1997).

It seems that authorities will follow two different approaches in their corrective and preventive measures. In approach 1 (FDA), the authorities exclude use of bovine-derived materials originating from countries with BSE (relevant list: most recent version of the Office International des Epizooties [OIE] list on the worldwide incidence of BSE), with exemption of some excipients (e.g., tallow-derived excipients) still under discussion. In approach 2 (EU), the authorities require tracing of the origin (no use of material with United Kingdom origin) and absolutely excluding of the use of specified risk materials (SRM; brain, eye, spinal cord, or spleen) in the manufacture of substances to be used in humans.

Most pharmaceutical excipients affected by the governmental corrective and preventive measures belong to one of the following classes of materials:

Fat and fat derivatives (if manufactured from beef tallow and not from vegetable sources)

Gelatin, including hard and soft gelatin capsules (if manufactured from bovine and not from porcine source materials).

Excipients based on milk

Lanolin and lanolin derivatives.

By far the largest group affected is that of the fat derivatives. Considering the rigorous processing of such materials (see Fig. 1), they are exempted in regulatory directives and guidelines as a special case considered as safe.

The foregoing Guidelines of the German Federal Ministry of Health, dated February 16, 1994, provide a detailed model to evaluate the overall safety of a specific substance (active ingredient, excipient in a given product) on the basis of the following parameters and a numeric scoring system:

Country of origin (prevalence of BSE/scrapie) and animal environment (including herds and fodder)

Type of animal material used (organs, tissue, or body fluids)

Methods used to inactivate or remove potential spongiform encephalopathy (SE) agents

Quantities of animal raw material required to produce one daily dose of the raw material

Number of daily doses

Route of administration

The numeric score system to evaluate the overall safety on basis of the foregoing parameters is based on the following model:

- One gram of homogenate of native brain obtained from a hamster with manifest SE symptoms contains enough infectious material to theoretically infect a further 10^9 hamsters with 50% probability following a single injection. Thus, the 1 g of material contains $10^9 \times LD_{50}\%$. Classification of this material on basis of the foregoing parameter and the criteria defined in the guideline leads to a score of 5.

- Risk for transmission of SE agents must be less than $1:10^6$ (= natural incidence of Creutzfeldt–Jakob disease in humans). To reduce the 50% probability of transmission from 10^9 to 10^{-6} the risk needs to be reduced by 15 indices of 10 compared with the original raw material. This is equivalent to a score of $5 + 15 = 20$.

This model has been developed from the "CPMP Guidelines for minimizing the risk of transmission of agents causing SE via medicinal products," valid since May 1, 1992. It is in our opinion a practical and sound tool for a reliable risk assessment in this difficult field.

2. Additives

Frequently, starting materials for technical applications contain additives to ensure optimal processability, or to prevent degradation or microbial growth. For the same reasons, additives are also used in several pharmaceutical excipients: for example,

> Silicon dioxide (anticaking agent, approximately 0.5%) in cellulose ethers
> Butylated hydroxytoluene (antioxidant, 200 ppm) in anhydrous lanolin and certain fat derivatives
> alpha-Tocopherol (antioxidant, 0.1–0.2%) in soybean lecithin and certain fats and fat derivatives
> Hydrogen peroxide (preservative, ppm-range) in aqueous dispersions for tablet coating.

Occasionally additives can be found in excipients which, by declaration, do not contain additives owing to cross-contamination from other products (e.g., technical grades of the same material processed in the same equipment). A typical example of this is glyoxal in cellulose ethers. To prevent solutions of cellulose ethers from going lumpy during the dissolution process, grades for technical applications (e.g., hydroxyethylcellulose) frequently contain glyoxal. Its cross-linking effect delays the dissolving process of the substance, leading to homogeneous solutions. Use of such a grade for pharmaceutical application would also be desirable from a technological point of view. However, because of the high reactivity of glyoxal, a consequence of its chemical structure as a dialdehyde (Fig. 3) closely

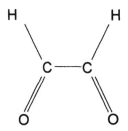

Figure 3 Chemical structure of a glyoxal.

related to the structure of formaldehyde, it requires adequate care for toxicological reasons and for stability reasons owing to potential interaction of the active
ingredient with the additive.

Whereas the *USP 23/NF 18* monograph for hydroxyethylcellulose only
says "... may contain a suitable anticaking agent ..." (16), the *EP* monograph
limits its concentration in hydroxyethylcellulose to a maximum of 200 ppm (17).
In an investigation of eight pharmaceutical-grade samples of hydroxyethylcellulose from different origins, the following results were obtained. In most of the
samples tested (origin: Egypt, France, Argentina, Germany, and Spain), the glyoxal levels were less than 20 ppm. Glyoxal levels were 90 ppm and 280 ppm
for India and Columbia, respectively. The highest glyoxal level was found in
hydroxyethylcellulose from Brazil (approximately 5000 ppm).

Another example of cross-contamination is the presence of ferrocyanide
used in some countries as an anticaking agent in food-grade qualities of sodium
chloride in a concentration of 5–10 ppm. The sodium chloride monograph of the
EP 2nd edition as well as that of the *USP 23* require the absence of this additive,
most likely because of the use of sodium chloride in injectables. In general the
possibility of a cross-contamination must be taken into consideration in cases
when the pharmaceutical industry is only a minor user.

Additives in excipients, even though usually added intentionally, must be
handled from the safety and efficacy standpoint as impurities. To reliably assess
the potential risk for an individual application, it is of high importance for the
pharmaceutical manufacturer to know whether an excipient contains an additive
and, if so, which one.

3. Degradation Products of Excipients

The stability of excipients must be discussed both from the physical and chemical
point of view. Changes in the physical properties (e.g., absorption of water, aggre-
gation of particles, decrease in viscosity, sedimentation of aqueous dispersions
from coprocessed excipients, or melting properties of fat derivatives owing to
changes in modification) may impinge on processability and quality of the finished drug product (e.g., instability of active ingredient owing to increased water
content of excipients, or different dissolution rates owing to physical changes in
dispersions for aqueous coating). In contrast with physical changes, chemical
degradation is relevant only for the stability of a drug product because of potential
interactions between the active ingredient and the excipient degradation product.

Whereas chemical degradation of excipients in powder form is rare (e.g.,
hydrolysis of cellulose acetate under influence of humidity), it is frequently found
in semisolid and liquid excipients, such as oils and highly unsaturated fat derivatives, or solvents, such as glycols and alcohols. The degradation pattern is usually
oxidation following a free radical mechanism. Products of degradation are mole-

cules with a high reactivity (mainly peroxides, aldehydes, and acids) that can easily interact with other substances in a formulation, including the active ingredient.

The excess of excipients in a formulation can cause significant stability problems (as outlined in Sec. III.B.1, of this chapter) even if the concentration of the degradation product of the excipient is low. The following example from industrial practice is typical for such problems. Benzyl alcohol, used as a preservative and, to some extent, also as a solubilizer in topical and occasionally in injectable products, decomposes by a free radical mechanism to form simultaneously benzaldehyde and hydrogen peroxide. As shown in Fig. 4, benzyl alcohol is extremely unstable in the presence of oxygen and light. When stored in daylight it decomposes within days to form approximately 600 ppm benzaldehyde, equivalent to 190 ppm hydrogen peroxide. In investigations of various samples, up to 1.5% of benzaldehyde, equivalent to 0.5% hydrogen peroxide, were found. In aqueous peptide preparations, hydrogen peroxide can rapidly oxidize the sulfhydryl group of amino acids, such as cysteine, to the corresponding sulfoxide and inactivate the biological activity of the molecule. As demonstrated in the following example, even very low amounts of hydrogen peroxide can cause serious stability problems:

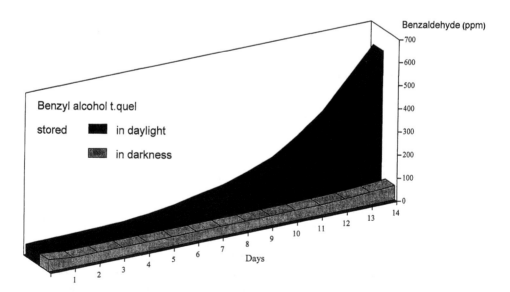

Figure 4   Decomposition of benzyl alcohol.

Formulation:

    0.5 mg peptide (one amino acid with a sulfhydryl group)

    5 mg benzyl alcohol

    0.5 mL total amount of formulation per ampoule

Molecular weights:

    Peptide: 2900

    Benzyl alcohol: 108

    Hydrogen peroxide: 18

    Hydrogen peroxide content of benzyl alcohol: 100 ppm equivalent to 0.5 µg

Amount of peptide inactivated by 0.5 µg of hydrogen peroxide: 80.5 µg, equivalent to 16.1% of the total peptide in the formulation.

Whereas 2-phenylethanol undergoes the same degradation mechanism, 2-phenoxyethanol does not. In other instances stability problems were observed caused by free radical degradation of isopropanol, resulting in peroxide levels of up to 0.1%.

4. *Monomers and Processing Aids in Synthetic Polymers*

Synthetic polymers are widely used as excipients in a variety of pharmaceutical formulations including oral, parenteral, ophthalmic and topical dosage forms. Because of their chemical reactivity, the starting materials in the polymerization reactions are generally hazardous. It is, therefore, a prerequisite for the use of synthetic polymers as pharmaceutical excipients to demonstrate the absence of such compounds by suitable analytical methods. Pharmacopeial testing monographs contain test methods, including specifications, that often represent the limit of quantitation of the monomer by the method described. The excipient manufacturer has to guarantee by the choice of the polymerization reaction parameters that predominantly all starting materials have completely reacted. Unreacted monomers, dimeric and undesired oligomeric products, catalysts, and processing aids have to be removed subsequently in a most efficient way by suitable methods. If stabilizers are used by the manufacturers, it is a matter of declaration and labeling on their products.

Despite the extremely high toxicity of the monomer ethylene oxide, ethoxylation products are widely used as excipients in pharmaceutical products, such as

    Polyethylene glycols

    Polyoxyethylene derivatives of natural or processed oils, fatty acids, fatty acid esters, or fat alcohols

    Copolymers (e.g., with propylene oxide)

The dimerization product of ethylene oxide, 1,4-dioxane, has a potential carcinogenic potency. Dioxane can be removed from the reaction products by a water-stripping process. The absence of ethylene oxide and dioxane in the final product can be demonstrated by gas chromatographic methods with limits of 1 ppm for ethylene oxide and 10 ppm for 1,4-dioxane.

Another group of synthetic polymers widely used as pharmaceutical excipients is based on the polymerization of vinyl derivatives. Povidones are polyvinyl pyrrolidones with pharmacopeial specifications for the monomer vinyl pyrrolidone. The monomer limit was, until recently, 0.2% (determined by a titrimetric method) and, in the framework of international harmonization, has been changed for toxicological reasons to maximum 10 ppm, determined by high-performance liquid chromatography (HPLC). Besides residual monomer, povidones can also contain hydrazine as a side product from the chemical reaction; its formation and concentration depend on the process conditions. Pharmacopeias limit this toxic impurity at maximum 1 ppm. Cross-linked variants (crospovidone) and variants manufactured by copolymerization with vinyl acetate (copolyvidone) are also widely used as excipients. Polyvinyl alcohol is prepared by hydrolyzation of polyvinyl acetate.

Polymethacrylates are used as film-coating agents for tablets. They are produced by copolymerization of methacrylic acid and its esters. The *USP/NF* specifies a maximum monomer content of 0.3% which seems rather high compared to other monomer limits (18).

However, not only polymers, but also low molecular weight substances, with simple structure, can contain highly toxic substances as side products. An example of this is triethanolamine, which can contain *N*-nitrosodiethanolamine, known for its carcinogenic properties. Until now this impurity has not been limited by pharmacopeial monographs, but a recent draft from the *EP* has proposed a limit of 10 ppb (19). Reliable determination of such extremely low levels requires sophisticated equipment and well-trained technicians in addition to special care in sample handling.

5. Residual Solvents

The release of an international guideline for residual solvents is one of the targets of the International Conference on Harmonization (ICH) of Technical Requirements for Registration of Pharmaceuticals for Human Use (20). In this guideline a classification of residual solvents into three classes is proposed by risk assessment. Class 1 comprises solvents that are known or strongly suspected human carcinogens or environmental hazards. Any use of these solvents should generally be avoided. Class 2 contains solvents that have to be limited in pharmaceutical products because of their inherent toxicity. Concentration limits are defined for these solvents as well as limits for permitted daily exposure. Solvents with low toxic potential are listed in class 3.

The guideline does not contain analytical procedures, but refers to gas chromatographic methods described in the pharmacopeias. Nonspecific methods, such as loss on drying, may be used if only class 3 solvents are present.

The guideline clearly states that manufacturers of pharmaceutical excipients have to submit information about the content of residual solvents. If solvents of class 3 are present in amounts greater than 0.5%, they must be identified and quantified. The same is true for solvents of class 2 in amounts exceeding the specified limits. Solvents of class 1 must be identified and quantified whenever they are likely to be present.

The ICH guideline does not yet apply to already-marketed drug products. For excipients that contain residual solvents of class 1 or 2 in amounts higher than those of the guideline, manufacturers and end-users of such excipients will face increasing pressure to take actions to reduce the amount of these solvents or to replace them completely, even if a higher level is specified in a pharmacopeial monograph. Modifications in the manufacturing procedure of these excipients may have an effect on their processability and on the specifications and behavior of the final drug products. Reformulation and reregistration of the pharmaceutical products may be necessary. A long-term approach for the replacement of such excipients therefore will be necessary.

Carbomers are widely used excipients and, unfortunately, examples of the aforementioned case. Carbomers are acrylic acid polymers, cross-linked with allyl derivatives. Benzene, a class 1 solvent, is used for their production. The pharmacopeial specifications for benzene in various carbomer grades now vary from 0.01 to 0.5. Programs to replace these carbomers by benzene-free grades have already been initiated.

6. Pesticides

A high percentage of pharmaceutical excipients is manufactured from vegetal or animal source material. To prevent or to cure infestation with parasitic organisms, plants and animals frequently undergo pest control with a great variety of highly active substances, generally known as pesticides. Most of the plants and animals that serve as source materials for pharmaceutical excipients are also used in the preparation of food and feed. Therefore, their treatment with pesticides is subject to strict regulation (e.g., waiting time before harvesting of fruits) to avoid a transfer into the food chain. As described under Sec. II.A, most pharmaceutical excipients of vegetal and animal origin undergo comprehensive and rigorous processing, which further reduces the risk of contamination with pesticides.

Nevertheless, contamination of pharmaceutical excipients cannot be ruled out completely as shown in Table 11, pesticide residues in lanolin. In the example shown in Table 11, the source of pesticide contamination is the treatment of sheep against ectoparasites in dip baths. Because most of the pesticides are lipophilic,

Table 11 Pesticide Residues in Lanolin

Origin of samples	Diazinon (ppm)	Bromphosethyl (ppm)	Lindan (ppm)	Dieldrin (ppm)	p,p'-DDT (ppm)
Germany	0.25	0.03	<0.02	0.02	—
France	7.2	1.7	46.9	0.3	0.3
Spain	9.9	0.7	5.9	1.3	5.9
Mexico	14.8	59.1	0.01	0.2	0.2
Argentina	23.5	0.9	21.6	0.9	1.0
Taiwan	13.2	89.7	0.02	0.1	0.4
India	53.9	66.0	—	0.02	—

they migrate into the wool fat. As evident from the data in Table 11, the concentrations can be considerable, and a limitation for substances that might be critical relative to pesticide residues is obviously necessary. Normally the acceptable daily intake (ADI) serves as a basis to calculate the limit. Because the main source for pesticide intake by humans is food, contribution from pharmaceuticals should be kept low (e.g., maximum of 25% of the total ADI).

If one takes diazinon, one of the pesticides frequently found in considerable concentrations in lanolin, as a model to calculate the limit on the basis of the ADI the result is 15 ppm (21)

Acceptable daily intake for diazinon	0.002 mg/kg (22)
Maximum amount of lanolin applied per day	2 g
Maximum absorption rate from lanolin	100%
Standard body weight	60 kg
Standard body surface	2 m²
Acceptable contribution from lanolin	25%

The foregoing conditions have been derived from practical experience.

Frequently residues of several pesticides are found (see Table 11). The *USP* lists, for instance, 34 possible pesticides (23). Limitation of the total residual amount of pesticides, is therefore, also necessary. This can be done either as absolute value or by transforming the results by means of the ADI in equivalents of one pesticide (21). It is possible to reduce the overall pesticide content in lanolin considerably by selection of the starting materials and special-processing conditions during the refining process. Frequently, these processes simultaneously also reduce residual detergents, used in the wool-washing process, and free lanolin alcohols. In the *USP* this grade of material is found under the monograph "Modified Lanolin." Today the *USP* limits pesticides in lanolin at the following levels (24):

Lanolin	Maximum 10 ppm of any individual substance specified in Table 1 of the mongraph Total amount maximum 40 ppm
Modified lanolin	Maximum 1 ppm of any individual substance specified in Table 1 of the monograph Total amount maximum 3 ppm

The *EP* limits, under chapter 2.8.13, pesticides in herbal drugs on the basis of ADI values for 34 individual pesticides (25).

7. Microbial Contamination, Mycotoxins, Residues from Antimicrobial Treatment

Despite the rigorous processing conditions most excipients undergo during their manufacture, microbial contamination of excipients plays an important role and requires adequate evaluation and control. Microbial contamination of excipients is most frequently caused by contaminated source materials (e.g., grain), processing aids, or insufficient cleaning of equipment. Among the processing aids, water is by far the most important source for contamination. Adequate design and close monitoring of the water supply units is, therefore, of utmost importance, not only when water is used as an excipient in pharmaceutical preparations, but also for water used as process water in the manufacture of excipients.

The spectrum of microorganisms observed in excipients covers a broad range of species, including pathogenic microorganisms such as *Salmonella* and *Staphylococcus aureus*, fungi, and yeasts. Risks associated with the microbial contamination of excipients are sixfold:

1. Infection of patients
2. Decomposition of formulations
3. Pyrogens
4. Mycotoxins
5. Residues from antimicrobial treatment
6. Contamination of pharmaceutical-processing equipment

Depending on the kind and origin of the source material and the extent of its processing, excipients occasionally contain, in addition to a high total count of microorganisms, pathogenic species such as *Salmonella, Escherichia coli,* and other coliforms, *S. aureus, Pseudomonas aeruginosa,* or *Aspergillus flavus,* which can lead to serious infections or intoxications. Examples of these are *Salmonella* species or *E. coli* in tragacanth or acacia; *E. coli/*coliforms in starch, milk derivatives, or gelatin; and *P. aeruginosa* in excipients requiring large quantities of water during the manufacturing process.

Whereas contamination with *P. aeruginosa* is, for instance, extremely critical for ophthalmical and topical applications, contamination with *Salmonella* species and coliforms is of main relevance for oral applications. For patients with reduced capacity of the immune system, high microbial counts can be critical, even if no pathogenic species are involved. High counts of microorganisms can also lead to a destruction of the formulation or the active ingredient, especially in liquid preparations.

Requirements for the microbiological purity of excipients depend on the use category of drug product. Even though product-specific requirements are found in only a few pharmacopeial monographs for excipients known to be critical and needing control of each batch (e.g., starches and gelatin), all excipients must comply with general specifications to ensure that the drug product or pharmaceutical preparation meets the requirements for microbiological purity as defined, e.g., in the *European Pharmacopoeia* Chapter 5.1.4: Microbiological Quality of Pharmaceutical Preparations (26). Considering the great variety of materials, processes, and plants involved in the manufacture of excipients, it is recommended to carefully evaluate excipients from new supply sources for microbiological purity and to perform rechecks at certain time intervals. Harmonization of microbiological requirements is further discussed in Chapter 14.

Microbiological purity of excipients intended for the manufacture of injectable preparations is of extreme importance to ensure sterility. For such excipients a low bioburden (e.g., maximum 100 colony-forming units; CFU) is required today. In addition to having a low germ count, excipients used in injectable preparations must meet the requirements for pyrogens/bacterial endotoxins, determined by the *limulus amebocyte lysate* (LAL) test. Pyrogens, or bacterial endotoxins, are metabolic products of predominantly gram-negative bacteria and can induce severe fever attacks in patients. A material with a high bioburden at an intermediate stage may contain pyrogens, even though the bacterial count of the final product is low. Pyrogens/bacterial endotoxins need special attention for water and carbohydrates manufactured by enzymatic processes (e.g., dextrose, mannitol, and citric acid). The *EP* limits pyrogens/bacterial endotoxins in such substances as follows (27):

Mannitol (bacterial endotoxins): maximum 4 IU/g for concentrations < 100 g/L, maximum 2.5 IU/g for concentrations > 100 g/L
Sodium chloride maximum 5 IU/g

Mycotoxins, similar to penicillins are metabolic products of certain fungi species, however, with a higher toxicity. Among the mycotoxins the metabolic products of some *Aspergillus* species, especially *A. flavus*, called aflatoxins, are of main concern because of their extreme toxicity, carcinogenicity, and heat resistance. Potential sources for a contamination of excipients with mycotoxins are oil seeds and grain with a high fungi count. As a result of high fungi counts in animal feed, aflatoxins have also been found in milk and milk products.

Whereas strict limits for aflatoxins are applied in the food sector, we are not aware of similar requirements for pharmaceuticals, with the exception of a draft from the German Health Authorities entitled *Verordnung ueber das Verbot der Verwendung von mit Aflatoxinen kontaminierter Stoffe bei der Herstellung von Arzneimitteln*, from 1997 (28). This draft document limits aflatoxins in starting materials for pharmaceutical products as follows:

Aflatoxin M_1	Maximum 0.05 µg/kg
Aflatoxin B_1	Maximum 2 µg/kg
Aflatoxin B_1, B_2, G_1, G_2, in sum	Maximum 4 µg/kg

To keep the microbial count under control, in the late 1960s–early 1970s, excipient manufacturers started antimicrobial treatment of excipients, frequently without informing pharmaceutical manufacturers of this step. The highly reactive and toxic formaldehyde was, for instance, used in the manufacture of injectable grade mannitol, ion-exchange resins for the manufacture of purified water, and for the treatment of starches. Up to 950 ppm formaldehyde was found at that time in supplies of maize starch. This uncoordinated and uncontrolled antimicrobial treatment resulted in a series of serious stability problems.

Antimicrobial treatment of excipients is still of considerable importance. The following techniques have to be distinguished: irradiation, gas treatment (e.g., ethylene oxide), and addition of chemical agents. As ethylene oxide has been abandoned, at least in Europe, for toxicological reasons, sulfur dioxide–sulfites, hydrogen peroxide (e.g., in starches), and chlorine are today the most widely applied agents in the antimicrobial treatment of excipients. Complete decomposition of residues of antimicrobial agents can take months. For example, starting from a concentration of 440 ppm hydrogen peroxide immediately after treatment, the residue found in wheat starch was still 75 ppm after 3 months.

Considering the high reactivity of antimicrobial agents as well as their usually low molecular weight, even small traces of such agents can cause serious stability problems. The use of excipients that have undergone antimicrobial treatment requires a specific risk assessment; declaration of such treatment by the supplier is indispensable.

V. CONCLUSION

Excipients are manufactured from a great variety of source materials by application of numerous different process technologies and production methods. Depending on the kind and quality of the source material and the process technology or production method applied, excipients can contain various impurities, ranging from practically nontoxic substances, such as sodium sulfate, to highly toxic or dangerous substances, such as ethylene oxide and aflatoxins, or dangerous microorganisms, such as *Salmonella* or *Staphylococcus* species. A reliable purity assessment, based on meaningful testing, requires detailed information about origin, production method, and intended application of the excipient.

Frequently, the pharmaceutical industry uses only a very low percentage of the total production output of materials predominantly designed and used for technical applications. Use of technical-grade material can have dramatic consequences, as in the Haiti glycerin case, the Nigeria–Bangladesh propylene glycol incident, and the Spanish olive oil case (when many people died because of intoxication with diethylene glycol or fatty acid anilides present as impurities or additives). Differentiation of technical- and pharmaceutical-grade materials is of special importance in an environment of a growing global trade.

Impurities in excipients can interact with active ingredients in drug products, leading to degradation and loss of efficacy, which can affect the overall evaluation of safety of excipients. Hence, purity of excipients must be considered not only from the toxicological, but also from the interaction point of view. Also, in this context, nontoxic impurities can be critical (e.g., precipitations in injectable solutions caused by interaction of the active ingredient with ions such as calcium or sulfate).

Purity characterization of excipients by pharmacopeial monographs has made considerable progress in recent years despite some still-existing shortcomings (e.g., discrepancies between different pharmacopeias, unreliable heavy metal test). Therefore, pharmacopeial monographs can be applied for a reliable assessment of the purity of excipients from established supply sources, with the clear understanding that

1. These are minimum requirements that do not necessarily cover the needs of the individual application (e.g., requirements for oral or parenteral application are different)
2. Evaluation and standardization of excipients from new supply sources cannot be based exclusively on testing against the pharmacopeial monograph.

REFERENCES

1. Pan American Health Organization (PAHO). Washington, DC, News Release June 27, 1996 and July 3, 1996.
2. World Health Organization (WHO). Geneva, Switzerland, Alert No. 50, June 28, 1996.
3. U.S. Department of Health and Human Services. Morbid and Mortal Wkly Rept (MMWR). 45(30):649–650, 1996.
4. The safety of tallow derivatives with respect to spongiform encephalopathy. The European Oleochemicals and Allied Products Group, Bruxelles, Belgium, May 4, 1997.
5. P Woditsch, A Westerhaus. In: Ullmann's Encyclopedia of Industrial Chemistry. vol A 20: Pigments, Inorganic. Weinheim: VCH Verlag, 1992, pp. 243–290.
6. ZT Chowhan. The formidable task of setting meaningful standards for excipients: case study—magnesium stearate. USP Pharm Forum 14:4621–4623, 1988.
7. European Pharmacopoeia. Prepublication of texts adopted during the June 1997 session of the European Pharmacopoeia Commission.
8. R Ciciarelli, D Jäkel, E König, R Müller-Käfer, J Pavel, M Röck, M Thevenin, H Ludwig. Determination of metal traces—a critical review of the pharmacopeial heavy metal test. USP Pharm Forum 21:1638–1640, 1995.
9. Adapted from: K Blake. Harmonization of the USP, EP, and JP heavy metals testing procedure. USP Pharm Forum 21:1632–1635, 1995.
10. EC Directive 95/2 on food additives other than colours and sweeteners. Feb 20, 1995.
11. HD Roedler. Strahlenexposition und Strahlenrisiko in der Bundesrepublik Deutschland durch den Reaktorunfall von Tschernobyl. Dtsch Apoth Zt (DAZ) 127(7):299–302, 1987.
12. PJ Ciclitira, HJ Ellis. Determination of gluten content of foods. Panminer Med 33(2): 75–82, 1991.
13. K Mitt, O Uibo. Low cereal intake in Estonian infants: the possible explanation for the low frequency of coeliac disease in Estonia. Eur J Clin Nutr 52(2):85–88, 1998.
14. G Fluge, JH Dybdahl, J Ek, A Lovik, R Rohme. Guidelines for diagnosis and follow-up of patients with celiac disease. Norwegian Coeliac Association. Tidskr Nor Laegeforen 117:672–674, 1997.
15. Codex Alimentarius. 4:100–103, 1994.
16. The United States Pharmacopeia (USP 23)/The National Formulary (NF 18). Monograph: Hydroxyethyl Cellulose. 1995, p 2253.
17. European Pharmacopoeia, 3rd ed. Monograph: Hydroxyethylcellulose, 1996, pp 987–989.
18. The United States Pharmacopeia (USP 23)/The National Formulary (NF 18). Monograph: Methacrylic Acid Copolymer. 1996. pp 2265–2266.
19. Triethanolamine. Pharmeuropea 10(1):46–47, 1998.
20. JC Connelly, R Hasegawa, JV McArdle, L Tucker. ICH guideline residual solvents. Pharmeuropa 9(1 suppl), pp. S1-S68, 1997.
21. D. Jäkel. Unpublished. Correspondence with USP, 1991.

22. Pesticides in vegetable drugs. Pharmeuropa 4(2):143, 1992.

23. The United States Pharmacopeia (USP 23)/The National Formulary (NF 18). Monograph: Lanolin. 1995, pp 869–871.

24. The United States Pharmacopeia (USP 23)/The National Formulary (NF 18). Monograph Modified Lanolin. 1995, pp 871–872.

25. Pesticide Residues. European Pharmacopoeia. 3rd ed. Chapter 2.8.13. 1996, pp 122-125.

26. Microbiological quality of pharmaceutical preparations. European Pharmacopoeia, 3rd ed. Chapter 5.1.4. 1996, pp 287–288.

27. European Pharmacopoeia. 3rd ed. Monographs: Mannitol, pp 1143–1144; Sodium Chloride, pp 1481–1482, 1996.

28. German Ministry of Health. Draft Verordnung über Höchstmengen an Aflatoxinen bei Arzneimitteln. April 30, 1997.

3
History of Excipient Safety and Toxicity

Charles L. Winek
Duquesne University, Pittsburgh, Pennsylvania

I. INTRODUCTION

The goal of this chapter is to review the history of the safe use of excipients and reported toxicities. This chapter does not deal with microbial contamination of foods or drugs. It deals specifically with excipients. The word is derived from the Latin word *excipiens* the present perfect of *ex-cipio*, meaning to "take out." According to *Black's Medical Dictionary* (1), *excipient* means any more-or-less inert substance added to a prescription to make the remedy as prescribed more suitable in bulk, consistency, or form for administration.

Excipients are defined in *Dorland's Medical Dictionary* (2) as "any more or less inert substance added to a prescription in order to confer a suitable consistency or form to the drug; a vehicle." And finally, *Morten's The Nurses' Dictionary* (3) defines excipients as the substance used as a medium for giving a medicament.

Excipients have been referred to as inert or inactive ingredients or inert additions because they do not have an active role in treating human ailments. For the purpose of this chapter, *pharmaceutical excipients* are defined as the substance(s), other than the active pharmacological agent(s), used in the formulation of a product or drug-delivery system.

A. Early History

Historically, the word excipient was used specifically to identify those substances used in the preparation of pills, which were a very popular dosage form for several

59

hundred years and were essentially replaced by tablets and capsules. In 1945 *American Pharmacy* (4) defined the ideal excipient as ''one that imparts the requisite degree of plasticity by its adhesive character, by a slight solvent action on the pill ingredients, or by a combination of these qualities. It should be biologically and chemically inert and should be capable of producing a mass which retains its plasticity for long periods of time. It should not retard disintegration of the pill in the alimentary tract.'' Liquid glucose and honey were the two most common excipients used in pill making up until the 1940s, and they were both on the generally recognized as safe (GRAS) list before it existed.

Historically, the first excipients used by humans were probably flavoring or sweetening agents. We know that initially most drugs were natural products. Virtually all natural products have a bitter or unpleasant taste, so it can be assumed that the first excipients were natural products, such as honey and molasses. Certainly, availability played a role in the use of early excipients, both geographically and locally. For example, sugarcane certainly was not available in frigid climates; however, its use was common in warm climates as a flavor enhancer. Bitter tastes of natural medicinals were probably avoided by some form of encapsulation, such as placing the medicinal substance on a leaf and rolling it and twisting it closed. It is also conceivable that medications could have been sealed by natural substances such as tree saps, which eventually would harden. Shellac is a natural insect excretion and certainly was available to primitive humans.

The history of medicine and pharmacy progressed simultaneously with advancements in science and technology. Most of the earlier references to medications do not specifically address excipients. It was only with the development of recorded and standard formulas and formularies that we began to see the use of specific excipients. Preservatives were probably not used as preparations and were either brewed from dried plant material or were alcoholic. Part of the history of pharmacy and medicine include the use of freshly made decoctions and infusions as well as tinctures and fluid extracts. These types of preparations were referred to as galenicals, named after Galen.

B. History of Regulation

At first there were no specific guidelines, rules, regulations, or such, for drug preparations, including excipients. Information on excipients used in the preparation of medicaments for the treatment of human ailments were passed down from caveman to caveman, from witch doctor to witch doctor, from tribe to tribe, from medicine man to medicine man. This passing of information and training ultimately became the apprentice system, which dealt with the art of dosage form preparation. Availability and common knowledge were most likely the key to excipient use. As mentioned earlier, encapsulation and coating were probably both used. Every nationality has something edible that is wrapped or rolled in some leaf or dough. We know from written history, including the Bible, that items such as honey, tree saps, and various alcoholic beverages were available. In fact, it is most likely that an alcoholic beverage of some type was used as the vehicle for both internal and external medicinals. Vallee (5) indicates: "For nearly 10,000 years of known Western history, beer and wine not water were the major daily thirst quenchers consumed by all ages." It was not until the 19th century that humans were able to produce water suitable for consumption. This historical fact, along with folklore, seems to indicate that alcoholic medicinals were popular and that hydroalcoholic vehicles were a major excipient for drug delivery. This is also evident by examining the early pharmacopeia.

Before official formularies and compendia, such as the *United States Pharmacopeia* (*USP*) and the *National Formulary* (*NF*) and the corresponding compendia of other nations, there were no specific regulations on pharmaceuticals whether prescription or over-the-counter (OTC). People relied on the reputation of the manufacturer or supplier. Pharmaceuticals generally fall under pharmacopeial guidelines and not regulatory statutes. The guidelines are enforceable under the Food, Drug, and Cosmetic Act. Voluntary guidelines have been established by the Pharmaceutical Manufacturers Association (PMA) and the Nonprescription Drug Manufacturers Association (NDMA). Table 1 lists some of the historical regulatory events and toxic episodes related to excipients.

II. SPECIFIC USES AND NEEDS OF EXCIPIENTS

It appears that the word excipient was exclusively pharmaceutical in origin beginning with the preparation of the dosage form "pills." At one point in the history of pharmacy, excipients were referred to as pharmaceutical necessities. For example, althea (marshmallow root) was official in the 10th edition of the *NF*. Its category was pharmaceutical necessity; ingredient of ferrous carbonate pills. Other examples of pharmaceuticals listed in *NF 10* are lard, lycopodium, mastic, white pine, and taraxacum. These items were official because they were necessary for the formulation of other official preparations. By today's definition they were excipients, but were called pharmaceutical necessities because of their inclusion in the formula of an officially recognized product. The word excipient does not appear in *NF 10* (published in 1955). The *USP*, 10th edition, published in 1920 does not mention either pharmaceutical necessity or excipient. It does not give a therapeutic category or indicated use, but it is obvious that excipients were official items with official monographs because of the need for them in other official formulas. Examples of excipient items listed in *USP 10* are: mel (honey), lycopodium, lactose, glucose (liquid glucose), glycerin, gelatin and dextrose. The word excipient is mentioned only once in the *United States Dispensatory* (*USD*) 1870 edition (6). Detailed descriptions are given for the preparation of formulas

Table 1   Some of the Historical Events Related to Excipient Toxicity

1820	First edition of the *United States Pharmacopeia* (originally referred to as the *National Pharmacopeia*).
1833	First edition of the *United States Dispensatory*.
1888	First edition of the *National Formulary*.
1902	USDA-Dr. Harvey W. Wiley's "Poison Squad" begins study of food preservatives.
1906	Pure Food and Drug Act—the first safe food law.
	Upton Sinclair's book, *The Jungle*—the meat inspection act of 1906.
1930	*100,000,000 Guinea Pigs*—a best selling book by Consumer's Research.
1937	Elixir of sulfanilamide tragedy—diethylene glycol used as the solvent for sulfanilamide.
1958	FDA—The Food Additive Amendment.
1959	GRAS additives and regulated additives.
1959	Report of adverse reaction to tartrazine
1960	The Delaney Clause—Food/Color Additive amendment.
1966	Hexylene glycol and coma in patients treated with burn dressing.
1969	Cyclamate is banned.
1980	Labeling requirements for pharmaceutical preparations containing tartrazine.
1982	Benzyl alcohol toxicity in neonates.
1982	Renal failure linked to polyethylene glycol.
1983	Propylene glycol hyperosmolality reported.
1984	Infant deaths due to E-Ferol.
1986	FDA adverse reaction reporting regulations are changed.
1986	Revocation of sulfite GRAS status.
1988	Fourteen deaths in India from ingestion of glycerin contaminated with diethylene glycol.
1989	Reformulation of Alupent to delete soya lecithin.
1992	Diethylene glycol poisoning in Nigerian children—glycerin contaminated with diethylene glycol.
1996	Haitian diethylene glycol acute renal failure in children caused by contaminated glycerin used in the manufacture of acetaminiphen syrup.

and the various medicinals used at the time, but no mention is made of excipients, inert ingredients, or pharmaceutical necessities. There are descriptions of substances used at the time as vehicles, solvents, flavors, and such, but they were not called or referred to as excipients. Amaranth, cudbear, and cochineal all are listed and were used as red-coloring agents. There is also an entry on the preservation of medicinals, but no specific chemical preservative is mentioned. The single mention of excipients is on page 1318 and refers to confection of roses and molasses as the best excipients when the pills are to be kept long. There is also an indication that some pills were coated with egg white, lycopodium, or mastic. Others were coated with glycyrrhiza (licorice powder).

A. Pills

Pills were made from *masses*, which were the boli of material prepared for various quantities of pills. The most popular and frequently prescribed pills required larger masses of starting material that were stored in reserve for the extemporaneous preparation of pills. All pills were hand-rolled until late 1883 (pharmacists were referred to as pill rollers) when Brochedon invented and received a patent for a machine for "shaping pills, lozenges, and black lead by pressure in dies" (7). Brochedon called the pills "compressed pills." These were actually tablets and led to the development of rotary tablet machines in 1874. At the time of compressed pills it was indicated that pills had been a popular dosage form for several hundred years. That would date pills to the 14th century. I do not know who made the first pill or who coined the term excipient. The first pills were likely made by a woman, but a man took credit for it.

B. Tablets

The dosage form tablets evolved as indicated and compressed tablets were prepared in the United States in 1871. Tablet formulas were divided into medicaments and excipients. The word excipient began with pills and became more popular with tablets. The excipients used in tablets were classified as follows: (a) liquids (i.e., water and alcohol); (b) adhesives (i.e., sugar, acacia, pectin, sodium alginate, and methylcellulose) (c) diluents (i.e., starch, lactose, talcum, kaolin, and dextrins); (d) disintegrators (i.e., starches, agar, and bentonite); (e) absorbents (i.e., milk sugar and starch); (f) lubricants (i.e., arrowroot, carnauba wax, cocoa butter, lycopodium, spermaceti, syrup and egg albumen, and talcum).

The use of the word excipient has been expanded to include all those items used in the preparation and formulation of a pharmaceutical product, except the active pharmacological agent. I have even seen the phrase "eye excipient" used to refer to the antimicrobial–preservative in the ophthalmic preparations. Excipients are used in the formulation and preparation of foods, drugs, and cosmetics. Some excipients are used in all three categories (i.e., acacia, propylene glycol, and glycerin). etc. Some are used only in drugs and cosmetics (i.e., phenol, phenylmercuric salts, isopropyl myristate).

The use of excipients in pharmaceutical products is a necessity for the formulation and preparation of drug delivery systems. Colorants were used historically and presently for product identification and acceptability. Color aids the patient in distinguishing among medications and avoids confusion. People associate colors and flavors, and the public expects that something mint-flavored would be colored green and that something cherry- or raspberry-flavored would be colored red.

III. TOXICITY VERSUS SAFETY

A. GRAS List

Almost all pharmaceutical excipients were originally used in foods and are generally recognized as safe (GRAS) and many have been reaffirmed GRAS. Their use in foods generally indicates that the amount of oral exposure is higher and that humans have used these items in foods for many years. So the safety of GRAS food ingredients used as excipients in drugs is established, provided there is no interference with the drug's availability. Some of the substances used in pharmaceuticals that are GRAS are listed in Table 2.

IV. SPECIFIC TOXICITIES

Historically, there have been some events of toxicity and lethality related to the excipient of a product. The most noteworthy is the 1937 report (8) of death caused by diethylene glycol. It was contained in elixir of sulfanilamide–massengil at a concentration of 72%. Deaths were due to acidosis and renal failure. The vehicle does play a role in acute oral toxicity studies: either increasing or decreasing toxicity. The sulfanilamide event served as a lesson that the toxicity of the final product must be determined. Today it is routine practice to establish the safety or toxicity of the active constituents alone and in combination with the excipient(s) used in the final formulation (see Chapter 4).

      Diethylene glycol toxicity was reported again in 1987, when a limited number of burn patients were reported to suffer from acidosis and oliguria following topical antibacterial treatment. Diethylene glycol was the excipient in the product (9).

      There were additional diethylene glycol fatalities reported in Haiti from November 1995 through June 1996. A total of 88 children were diagnosed with acute anuric renal failure. The outbreak was associated with diethylene glycol-contaminated glycerin used to manufacture acetaminophen syrup. The syrup was locally manufactured and sold under the names of Afebril and Valodon. Both products were removed from the market. The manufacturer announced a recall of the products and other syrups that it produced. A public information campaign was instituted. The glycerin was imported to Haiti from another country. Ten children were transferred to medical centers in the United States for intensive care and dialysis; 9 are still living. Of the 76 children who remained in Haiti,

Table 2 List of Excipients Included in Monograph[a] and Affirmed as GRAS[b]

Acacia[b]	Ethylparaben	Potassium citrate[b]
Alcohol[b]	Fumaric acid	Maize starch, sterilizable
Alginic acid[b]	Gelatin	Pregelinated starch
Ascorbic acid[b]	Liquid glucose	Potassium sorbate[b]
Bentonite[b]	Glycerin[b]	Povidone
Benzalkonium chloride	Glyceryl monostearate[b]	Propane[b]
Benzoic acid[b]	Glycofurol	Propylene glycol
Benzyl alcohol	Guar gum[b]	Propylene glycol alginate
Butane[b]	Hydrochloric acid	Propylparaben
Butylated hydroxyanisol[b]	Hydroxyethylcellulose	Saccharin
Butylated hydroxytoluene[b]	Hydroxypropylcellulose	Saccharin sodium
Butylparaben	Hydroxypropylmethylcellulose	Sesame oil
Precipitated calcium	Hydroxypropylmethylcellulose	Shellac
carbonate[b]	phthalate	Colloidal silicon dioxide
Dibasic calcium phosphate[b]	Isobutane	Sodium alginate[b]
Tribasic calcium phosphate[b]	Isopropyl alcohol	Sodium ascorbate
Calcium stearate	Isopropyl myristate	Sodium benzoate[b]
Calcium sulfate	Isopropyl palmitate	Sodium bicarbonate[b]
Carbomer	Kaolin	Sodium chloride
Carbon dioxide[b]	Lanolin	Sodium citrate dihydrate and
Carboxymethylcellulose	Lanolin alcohols	anhydrous[b]
calcium	Lecithin[b]	Sodium laurel sulfate
Carboxymethylcellulose	Magnesium aluminum silicate	Sodium metabisulfite
sodium[b]	Magnesium carbonate[b]	Sodium starch glycolate
Hydrogenated castor oil	Magnesium stearate[b]	Sorbic acid[b]
Cellulose acetate phthalate	Malic acid	Sorbitan esters (sorbitan fatty
Microcrystalline cellulose	Mannitol	acid ester)
Powdered cellulose	Methylcellulose[b]	Sorbitol[b]
Cetomacrogol emulsifying	Methylparaben	Starch
wax	Mineral oil	Stearic acid
Cetostearyl alcohol	Mineral oil and lanolin	Stearyl alcohol
Cetrimide	alcohols	Sucrose
Cetyl alcohol	Monoethanolamine	Sugar, compressible
Cetyl esters wax	Paraffin	Sugar, confectioner's
Chlorhexidine	Peanut oil	Suppository bases
Chlorobutanol	Petrolatum	(semisynthetic glycerides)
Chlorocresol	Petrolatum and lanolin	Talc
Citric acid[b]	alcohols	thimerosal
Coloring agents,	Phenylmercuric acetate	Titanium dioxide
pharmaceutical	Phenylmercuric borate	Tragacanth[b]
Corn oil	Phenylmercuric nitrate	Tricholoromonofluoromethane
Cottonseed oil	Polacrilin potassium	Triethanolamine
Dextrin	Poloxamer	Pharmaceutical waters
Dextrose	Polyethylene glycol	Carnauba wax[b]
Dichlorodifluoromethane	Polymethylacrylates	Emulsifying wax
Dichlorotetrafluoroethane	Polyoxyethylene alkyl ethers	Microcrystalline wax
Diethanolamine	Polyoxyethylene castor oil	White wax
Diethyl phthalate	derivatives	Yellow wax
Docusate sodium	Polyoxyethylene sorbitan fatty	Xylitol
Edetic acid and edetates	acid esters	Zein
Ethyl oleate	Polyoxyethylene stearates	Zinc stearate[b]
Ethylcellulose	Polyvinyl alcohol	

Sources: [a] Ref. 49; [b] Ref. 50.

only 1 is known to have survived. The Haitian outbreak is the most recent in the history of the glycols (10). This situation could have been avoided if Good Manufacturing Practices (GMPs) were followed and quality control and quality assurance procedures were in place.

There are about 1300 excipients used in the manufacture of pharmaceuticals. As already indicated, many of these are on the GRAS list. Only a few will be considered here because of specific problems that have been identified with their inclusion in pharmaceuticals.

A. Toxicity from Intentional Overdose

Abuse of some prescription and OTC products has caused toxicity and lethality, in part owing to the excipients (11,12). Most notably is the excipient ethyl alcohol. Its use in prescription and OTC cough syrups has contributed to toxicity in abuse situations when consumed in large quantities. It is not a major problem when used as directed. Table 3 lists some OTC pharmaceutical products and their ethanol content. Some of these have been involved in sporadic "fads" of abuse nationwide because of their ease of availability. Alcohol is one of the oldest "pleasure poisons," and its abuse in the form of pharmaceutical products will be as lasting as the alcohol problem itself. Since 1992, manufacturers have reformulated some OTC products to make them alcohol-free, especially pediatric products, or have reduced the alcohol content (e.g., cough syrups).

1. Alcohol and Asthma Products

In a recent court case in which I testified concerning the intoxication of the driver of a motor vehicle, the defense attorney showed me an OTC asthma product and asked that I read to the jury the alcohol content of the product. I could not, because I did not have my glasses. I tried several pairs from the attorneys and court reporter and finally was able to read the label with the help of the judge's glasses. The product was an epinephrine inhalation aerosol that contained no sulfites, but did have 33% ethanol (w/w). The defense attorney's suggestion was that his client had used the entire 1/2-ounce bottle, causing his intoxication. I calculated that at his weight, the use of one bottle all at once, and assuming complete absorption, that his blood alcohol content (BAC) would be 0.0066% and that to reach the determined BAC in the case he would have had to use forty-two bottles of the product.

There are alcohol-free products on the market and as a forensic toxicologist, if all asthma inhalation aerosols were alcohol-free, there would be one less group of products used as excuses for driving under the influence.

2. Toxicity Owing to Ethanol Content

Ethanol in intravenous pharmaceuticals is generally at a level of 10%, and the dose is such that alcohol intoxication does not occur. However, cases have been

Table 3 Some Over-the-Counter Products Containing Ethanol

Cough and cold preparations	Size	Ethanol (%)
Ambenyl-D	4 oz	9.5
Cotylenol Liquid Cold Formula	4 oz	7.5
Cheracol-D Cough Syrup	4 oz	4.75
Comtrex Cough Formula	4 oz	20
Comtrex Liquid	6 oz	20
Daycare Liquid	6 oz	10
Demazin Syrup	4 oz	7.5
Geritol Tonic Liquid	12 oz	12
Hall's Metho-Lyptus Decongestant Cough Mixture	3 oz	22
Novahistine DMX	4 oz	10
Novahistine Elixir	4 oz	5
Pertussin Syrup	3 oz	9.5
Pertussin ES	4 oz	9.5
Pertussin All-Night PM	8 oz	25
Primatene Mist Solution	½ oz	34
Primatuss Cough Mixture 4	6 oz	10
Robitussin CF	4 oz	4.75
Robitussin PE	4 oz	25
Triaminic Expectorant	4 oz	5
Trind	5 oz	5
Vick's Formula 44 Cough and Decongestant	4 oz	10
Vick's 44 M Cough, Cold and Flu	4 oz	10
Vick's Nyquil Liquid Multisymptom	6 oz	10

Source: Ref. 51.

reported in which the patient became clinically and legally intoxicated, reaching BACs as high as 0.267% (13–16). Ethanol does not always fit the definition of an excipient because it does have pharmacological properties and is considered a drug and one of the oldest drugs used by humans.

B. Diarrhea Associated with Sugar Alcohols

One case report (17) indicates that a 5-year-old child developed diarrhea after receiving valproic acid syrup. The excipients of the product included sucrose, glycerin, and sorbitol; each of which can cause osmotic diarrhea. Valproic acid itself is known to cause diarrhea. Another report (18) indicates diarrhea from sorbitol solution used in the preparation of a liquid hydralazine. Sorbitol caused osmotic diarrhea in a series of 12 patients receiving sugar-free theophylline elixir containing sorbitol (19). Sorbitol intolerance has been demonstrated and reported (20,21). It may be more common than previously thought. Sorbitol is metabolized to fructose and can be dangerous when administered to fructose intolerant patients (22).

C. Hypersensitivity

By far, the greatest number of reported adverse or toxic effects caused by excipients are those broadly classified as hypersensitivity reactions. These are allergic reactions, intolerance reactions, skin reactions, and others. Table 4 lists some excipients that have been implicated in hypersensitivity reactions. Several excipients fall into the intolerance group in which select humans react to an excipient substance such as gluten or lactose.

Acute hypersensitivity reactions have been attributed to sulfites used as antioxidants. A common effect is bronchoconstriction, and the route of administration plays a role, particularly in asthmatics (23). Sulfite is still contained in

Table 4   Some Excipients Causing Hypersensitivity-"Type" Reactions

Acacia		Ethylenediamine
Annatto		Phenylmercuric nitrate
Aspartame		Benzalkonium chloride
Cetyl alcohol		Tartrazine
Benzoic acid/benzoate		Sulfites
Peruvian balsam		Thimerosol
Lanolin		Soybean oil
Isopropyl myristate		Sorbic acid
Intolerance	Gluten	
	Lactose	
Sensitization	Propyl galate	
	Thimersal	
	Parabens	
Contact dermatitis/allergy	Propylene glycol	
	Cetyl alcohol	
	Benzyl alcohol	
	Chloroacetamide	
	Hathon CG	
	Polyethylene glycol	
	Benzalkonium Cl	
	Lanolin	
Photosensitivity	Cinnamon oil	

injectable epinephrine with U.S. Food and Drug Administration (FDA) approval. The product must carry the warning about potential allergic reactions. The FDA has estimated that about 5% of asthmatics are sulfite-sensitive (24). Chlorobutanol has caused anaphylactic shock (25). Benzoic acid and benzoates have been implicated as causing hypersensitivity reactions in asthmatics and aspirin-intolerant individuals.

Tartrazine (FD&C yellow number 5) continues to be implicated as causing urticaria (26). Tartrazine also has been implicated in aspirin-induced asthma (27,28). The use of tartrazine in pharmaceuticals has been reduced since 1990. It has been replaced by the safer FD&C yellow number 6. Contact allergy has been related to both lanolin and benzalkonium chloride. Sensitization reactions have been reported for the parabens (29), thimerosal (30), and propyl gallate (31). Transdermal drug delivery systems (patches) have been reported to cause skin irritation, allergic contact dermatitis, and adverse dermatological reactions (32–36).

V. SOME MAJOR EXCIPIENTS AND REPORTED EFFECTS

Some excipients have been implicated as causing, contributing to, or playing some role in certain human reactions and illnesses. The *Handbook of Food, Drug, and Cosmetic Excipients* (37) has reviewed many of these.

Benzyl alcohol has gained attention in adverse reactions in low birth weight, premature infants (38,39). Headache and migraine have been associated with aspartame (40,41). Propylene glycol has been implicated in cardiotoxicity (42,43), hyperosmolality in premature infants (44), and thrombophlebitis when used as a vehicle in injectables (45,46). The polysorbate event associated with the vitamin E product E-Ferol, and its effects in premature infants is still not fully understood (47). Many reports exist and much has been written about inert ingredients resulting in removal of products from the market, as well as changes in excipients used in products. For example, pediatric suspensions of Augmentin (amoxicillin–clavulanate) no longer contain mannitol as a sweetener, for it may have contributed to the diarrhea seen with the product.

Gelatin is widely used in the manufacture of capsules for pharmaceutical products. Gelatin has been considered safe and reaffirmed as safe. With the occurrence of mad cow disease in the United Kingdom, questions concerning the safety have been raised. The World Health Organization (WHO) has indicated that gelatin is considered safe for human consumption. Gelatin is derived from animal skin, cartilage and bones. WHO has indicated that the chemical extraction procedure for making gelatin destroys bovine spongiform encephalopathy (48). The mad cow disease demonstrates the concern for safety that can arise when an excipient substance is obtained from a natural source that may become diseased or contaminated (see Chapter 2).

VI. CONCLUSIONS

The toxicity of excipients has been extremely low. Morbidity, mortality, and human discomfort caused by excipients is minor when compared with the total pharmaceutical history of drug toxicity. Much of the undesired history of excipients could be avoided if patients were aware of their sensitivities and products were labeled to alert such patients. Much progress has been made in this area of patient education in the industrialized world. Excipients have had some unfortunate effects in premature infants and asthmatics. I learned early in my toxicology career that absolute safety is an impossibility. This may be true, but the pharmaceutical industry continues to strive for safety. The use of GMPs and quality control-quality assurance programs by pharmaceutical manufacturers worldwide would end many of the serious consequences of excipient-related morbidity and toxicity.

REFERENCES

1. Black's Medical Dictionary. 23rd ed. London: Adam and Charles Black, 1958.
2. Dorland's Medical Dictionary. 25th ed. Philadelphia: WB Saunders, 1974.
3. Morten's, The Nurse's Dictionary. 24th ed. London: Faber & Faber, 1957.
4. RA Lyman. American Pharmacy. 2nd ed. Philadelphia: JB Lippincott, 1948.
5. BL Vallee. Alcohol in human history. In: B Jansson, H Jornvali, U Rydberg, L Terenius, BL Vallee, eds. Toward a Molecular Basis of Alcohol Use and Abuse. Basel: Birkhauser Verlag, 1994, pp 1–8.
6. GB Wood. The Dispensatory of the United States of America. 13th ed. Philadelphia: JB Lippincott & Co., 1876.
7. LP Kebler. Tablet industry—its evolution and present status. The composition of tablets and methods of analysis. J Am Pharm Assoc 3:820, 937, 1062, 1914.
8. EMK Geiling, PR Cannon. Pathologic effects of elixir of sulfanilamide (diethylene glycol) poisoning: a clinical and experimental correlation: final report. JAMA 111: 919–926, 1938.
9. MC Cantarell, J Fort, J Camps, M Sans, L Piera, M Rodamilans. Acute intoxication due to topical application of diethylene glycol. Ann Intern Med 106:473–479, 1987.
10. MMWR, Morbid Mortal Wkly Rep, Aug, 85(30): 1–2, 1996
11. CL Winek, WD Collom. Codeine fatality from cough syrup. Clin Toxicol 3:97–100, 1970.
12. CL Winek. Dangerous cough syrups. N Eng J Med 280:840–841, 1969.
13. P Andrien, L Lemberg. An unusual complication of intravenous nitroglycerin. Heart Lung 15:534–536, 1986.

14. TL Shook, JM Kirshenbaum, RF Hundley, JM Shorey, GA Lamas. Ethanol intoxication complicating intravenous nitroglycerin therapy. Ann Intern Med 101:498–499, 1984.

15. J Shorey, N Bhardwaj, J Loscalzo. Acute Wernicke's encephalopathy after intravenous infusion of high-dose nitroglycerin. Ann Intern Med 101:500, 1984.

16. SH Korn, JB Comer. Intravenous nitroglycerin and ethanol intoxication. Ann Intern Med 102:274, 1985.

17. MW Veerman. Excipients in valproic acid syrup may cause diarrhea: a case report. Ann Pharmacother 24:832–833, 1990.

18. EB Charney, JN Bodurtha. Brief clinical and laboratory observations: intractable diarrhea associated with the use of sorbitol. J Pediatr 98:157–158, 1981.

19. TE Edes, BE Walk. Nosocomial diarrhea: beware the medicinal elixir. South Med J 82:1497, 1989.

20. NK Jain, DB Rosenberg, MJ Ulahannan, MJ Glasser, CS Pitchumoni. Sorbitol intolerance in adults. Am J Gastroenterol 80:678, 1985.

21. NK Jain, PP Vijaykumar, CS Pitchumoni. Sorbitol intolerance in adults. Am J Gastroenterol 9:317, 1987.

22. MJ Schulte, W Lenz. Fatal sorbitol infusion in patients with fructose-sorbitol intolerance. Lancet 2:188, 1977.

23. G Goldfarb, R Simon. Provocation of sulfite sensitive asthma. J Allergy Clin Immunol 73:135, 1984.

24. FDA Drug Bull 14:24, 1984.

25. H Hoffman, G Goerz, G Plewig. Anaphylactic shock from chlorobutanol-preserved oxytocin. Contact Dermatitis 15:241, 1986.

26. DD Stevenson, RA Simon, WR Lumry, DA Mathison. Adverse reactions to tartrazine. J Allergy and Clin Immunol 78:182–191, 1986.

27. M Samter, RF Beers. Intolerance to aspirin: clinical studies and consideration of its pathogenesis. Ann Intern Med 68:975–983, 1968.

28. O Michel, N Naeije, M Bracamonte, J Duchateau, R Sergysels. Decreased sensitivity to tartrazine after aspirin desensitization in an asthmatic patient tolerant to both aspirin and tartrazine. Ann of Allergy 52:368–370, 1984.

29. WP Schorr, AH Mohajerin. Paraben sensitivity. Arch Dermatol 93:721, 1966.

30. P Lisi, P Perno, M Ottaviani, P Morelli. Minimum eliciting patch test concentration of thimerosal. Contact Dermatitis 24:22, 1991.

31. G Kahn, P Phanuphak, HN Claman. Infantile methemoglobinemia caused by food additives. Clin Toxicol 15:273, 1979.

32. JFGM Hurkmans, HE Bodde, LMJ Van Driel, H Van Doorne, HE Junginger. Skin irritation caused by transdermal drug delivery systems during long-term (5 days) application. Br J Dermatol 112:461, 1985.

33. PW Letendre, C Barr, K Wilkens. Adverse dermatologic reaction to transdermal nitroglycerin. DICP 18:69, 1984.

34. RG Fischer, M Tyler. Severe contact dermatitis due to nitroglycerin patches. South Med J 78: 1523, 1985.

35. CEH Grattan, CTC Kennedy. Allergic contact dermatitis to transdermal clonidine. Contact Dermatitis 12:225, 1985.

36. BK Schwartz, WE Clendenning. Allergic contact dermatitis hydroxypropyl cellulose in a transdermal estradiol patch. Contact Dermatitis 18:106, 1988.

37. SC Smolinske, Handbook of Food, Drug, and Cosmetic Excipients. Boca Raton, FL CRC Press, 1992.
38. WJ Brown, NRM Buist, HTC Gipson, RK Huston, NG Kennaway. Fatal benzyl alcohol poisoning in a neonatal intensive care unit. Lancet 1:1250, 1982.
39. J Gershanik, B Boecler, H Ensley, S McCloskey, W George. The gasping syndrome and benzyl alcohol poisoning. N Engl J Med 307:1384, 1982.
40. DR Johns. Migraine provoked by aspartame. N Engl J Med 315:456, 1986.
41. SS Schiffman, CE Buckley III, HA Sampson, EW Massey, JN Baraniuk, JV Follett, ZS Warwick. Aspartame and susceptibility to headache. N Engl J Med 317:1181, 1987.
42. GL Gellerman, C Martinez. Fatal ventricular fibrillation following intravenous sodium diphenylhydantoin therapy. JAMA 200:161, 1967.
43. AH Unger, HJ Sklaroff. Fatalities following intravenous use of sodium diphenylhydantoin for cardiac arrhythmias. JAMA 200:159, 1967
44. AM Glasgow, RL Boeckx, MK Miller, MG MacDonald, GP August. Hyperosmolality in small infants due to propylene glycol. Pediatrics 72:353, 1983.
45. A Schou-Olesen, MS Huttel. Local reactions to IV diazepam in three different formulations. Br J Anaesth 52:609, 1980.
46. MAK Mattila, M Ruoppi, M Korhonen, HM Larni, L Valtonen, H Keikkinen. Prevention of diazepam-induced thrombophlebitis with cremophor as a solvent. Br J Anaesth 51:891, 1979.
47. JB Arrowsmith, GA Faich, DK Tomita, JN Kuritsky, FW Rosa. Morbidity and mortality among low birth weight infants exposed to an intravenous vitamin E product, E-Ferol. Pediatrics 83:244, 1989.
48. Executive Newsletter NDMA, May 10, 1996, p 7.
49. American Pharmaceutical Association. Handbook of Pharmaceutical Excipients. Washington, DC: APA Press, 1986.
50. JM Concon. Food Toxicology, Part B. New York: Marcel Dekker, 1988.
51. NF Billups, SM Billops, eds. American Drug Index. 42 ed. Facts and Comparisons, 1998.

4
Regulation of Pharmaceutical Excipients

Christopher C. DeMerlis
FMC Corporation, Philadelphia, Pennsylvania

I. INTRODUCTION

The goal of the chapter is to review the regulatory status and requirements for excipients (inactive ingredients) for drug products in the United States and medicinal products in the European Union. European Union requirements are discussed; however, individual Member State requirements are not included.

Table 1 lists the references to excipients in the U.S. Code of Federal Register. Table 2 lists the references to excipients in the European Union.

II. REGULATION OF EXCIPIENTS IN THE UNITED STATES FOR PRESCRIPTION AND OVER-THE-COUNTER DRUG PRODUCTS

A. Excipients in the Registration of a Drug Product

1. Definition and General Requirements for Excipients

Under Section 201(g)(1) of the Federal Food, Drug, and Cosmetic Act (FD&C Act; 1), the term *drug* is defined as:

> (A) articles recognized in the official *United States Pharmacopeia*, official *Homeopathic Pharmacopeia of the United States*, or official *National Formulary*, or any supplement to any of them; and (B) Articles intended for use in the diagnosis, cure, mitigation, treatment, or prevention of disease in man or other animals; and (C) Articles (other than food) intended to affect the

Table 1 U.S. Code of Federal Register References to Excipients

Subject	Reference	Content
General	21 CFR § 210.3(b)(8)	Definitions
	21 CFR § 201.117	Inactive ingredients
	21 CFR § 210.3(b)(3)	Definitions
Over-the-counter drug products	21 CFR § 330.1(e)	General conditions for general recognition as safe, effective, and not misbranded
	21 CFR § 328	Over-the-counter drug products intended for oral ingestion that contain alcohol
Drug Master Files	21 CFR § 314.420	Drug master files
Investigational New Drug Application	21 CFR § 312.23(a)(7)	IND content and format
New Drug Application	21 CFR § 312.31	Information amendments
	21 CFR § 314.50(d)(1)(ii)(a)	Content and format of an application
	21 CFR § 314.70	Supplements and other changes to an approved application
Abbreviated New Drug Application	21 CFR § 314.94(a)(9)	Content and format of an abbreviated application
	21 CFR § 314.127	Refusal to approve an abbreviated new drug application
	21 CFR § 314.127(a)(8)	Refusal to approve an abbreviated new drug application
Current Good Manufacturing Practice	21 CFR § 211.84(d)	Testing an approval or rejection of components, drug product containers and closures
	21 CFR § 211.165	Testing and release for distribution
	21 CFR § 211.180(b)	General requirements
	21 CFR § 211.80	General requirements
	21 CFR § 211.137	Expiration dating
Listing of drugs	21 CFR § 207	Registration of procedures of drugs and listing of drugs in commercial distribution
	21 CFR § 207.31(b)	Additional drug listing information
	21 CFR § 207.10(e)	Exemptions for domestic establishments
Labeling	21 CFR § 201.100(b)(5)	Prescription drugs for human use
	21 CFR § 201.20	Declaration of presence of FD&C Yellow No. 5 and/or FD&C Yellow No. 6 in certain drugs for human use
	21 CFR § 201.21	Declaration of presence of phenylalanine as a component of aspartame in over-the-counter and prescription drugs for human use
	21 CFR § 201.22	Prescription drugs containing sulfites; required warning statements

structure of any function of the body of man or other animals; and (D) Articles intended for use as a component of any articles specified in clause (A), (B), or (C).

An excipient meets the definitions as listed in (A) and (D) above.

In 21 CFR § 210.3(b)(8) (2), an "inactive ingredient means any component other than an active ingredient." According to the CFR, the term inactive ingredient includes materials in addition to excipients. 21 CFR § 201.117 states the following:

> Inactive ingredients: A harmless drug that is ordinarily used as an inactive ingredient, such as a coloring, emulsifier, excipient, flavoring, lubricant, preservative, or solvent in the preparation of other drugs shall be exempt from Section 502(f)(1) of the Act. This exemption shall not apply to any substance intended for a use which results in the preparation of a new drug, unless an approved new-drug application provides for such use.

Excipients also meet the definition of component in the Good Manufacturing Practice (GMP) regulations in 21 CFR § 210.3(b)(3): "Component means any ingredient intended for use in the manufacture of a drug product, including those that may not appear in such drug product."

The *NF* Admissions Policy in the *United States Pharmacopeia 23/National Formulary 18* defines the word *excipient* (3): "An excipient is any component other than the active substance(s), intentionally added to the formulation of a dosage form. It is not defined as an inert commodity or an inert component of a dosage form."

Similar to all other drugs, excipients must comply with the adulteration and misbranding provisions of the FD&C Act. Under Section 501(a), an excipient shall be deemed to be adulterated if it consists in whole or in part of any filthy, putrid, or decomposed substance, or if it has been prepared, packed, or held under insanitary conditions whereby it may have been contaminated with filth, or whereby it may have been rendered injurious to health. An excipient is adulterated if the methods used in, or the facilities or controls used for its manufacture, processing, packing, or holding do not conform to or are not operated or administered in conformity with current Good Manufacturing Practice to assure that such drug meets the requirements of the act as to safety and has the identity and strength, and meets the quality and purity characteristics which it purports or is represented to possess. In addition, under Section 501(b), an excipient shall be deemed to be adulterated if it purports to be or is represented as a drug the name of which is recognized in an official compendium, and its strength differs from, or its quality or purity falls below, the standards set forth in such compendium.

Under Section 502 of the FD&C Act, an excipient shall be deemed to be misbranded if its labeling is false or misleading in any particular, or if in a package form unless it bears a label containing the name and place of business of the

Table 2 European Union References to Excipients

Subject	Reference	Content
Regulation	EC No 541/95 of 10 March 1995	Examination of variations to the terms of a marketing authorization granted by a competent authority of a member state
	EC No 542/95 of 10 March 1995	Examination of variations to the terms of a marketing authorization falling within the scope of Council Regulation (EEC) No 2309/93
	Notice to Applicants for Marketing Authorizations, Volume II. 1998	Medicinal Products for Human Use in the Member States of the European Community: Rules Governing Medicinal Products in the European Community
Directive	65/65/EEC of 26 January 1965	Approximation of provisions laid down by law, regulation or administrative action relating to medicinal products
	75/319/EC of 20 May 1975	Approximation of provisions laid down by law, regulation or administrative action relating to proprietary medicinal products
	75/318/EC of 20 May 1975	Approximation of the laws of the Member States relating to analytical, pharmacotoxicological and clinical standards and protocols in respect of the testing of medicinal products
	91/507/EEC of 19 July 1991	Analytical pharmacotoxicological and clinical standards and protocols for testing of medicinal products

	Reference	Description
	91/356/EEC of June 13 1991	Laying down the principles and guidelines of good manufacturing practice for medicinal products for human use
	92/27/EEC of 31 March 1992	Labeling of medicinal products for human use and on package leaflets
	Amendment to 75/319/EEC and 81/851/EEC	First draft of a proposal for a European Parliament and Council Directive on GMP for starting materials and inspection of manufacturers of both medicinal products and their starting materials
Note for Guidance	EU-CPMP-GDL 3196/91	Excipients in the Dossier for Application for Marketing Authorization of Medicinal Products
	EU-CPMP/CVMP-QWP 297/97	Inclusion of Antioxidants and Antimicrobial Preservatives in Medicinal Products
	EU-CPMP-QWP 155/96	Development Pharmaceutics and Process Validation
	EU-CPMP-GDL 3324/89	Specifications and Control Tests on the Finished Product
	EU-CPMP-ICH 367/96	Specifications: Test Procedures and Acceptance Criteria for New Drug Substances and New Drug Products: Chemical Substances
	EU-CPMP-ICH 297/97	Summary of Requirements for Active Substances in Part II of the Dossier
Guideline	1997	Guideline on Excipients in the Label and Package Leaflet of Medicinal Products for Human Use
Rule	1992	Good Manufacturing Practice for Medicinal Products. Volume IV

manufacturer, packer, or distributor and an accurate statement of the quantity of the contents in terms of weight, measure, or numerical count. The label must bear the established name, under 502(e), which means the applicable official name designated pursuant to Section 508, or if there is no such name and such drug, or such ingredient, is an article recognized in an official compendium, then the official title thereof in such compendium or if neither of the above apply the common or usual name, if any, of such drug or ingredient.

The U.S. Food and Drug Administration (FDA) compliance officials strongly encourage the use of inactive ingredients that meet compendial standards when standards exist. The FDA Center for Drug Evaluation and Research maintains an Inactive Ingredient Committee whose charter includes the evaluation of the safety of inactive ingredients on an as-needed basis, preparation of recommendations concerning the types of data needed for excipients to be declared safe for inclusion in a drug product, and other related functions (4).

From a regulatory standpoint, the FDA's concern regarding safety involves the toxicity, degradants, and impurities of excipients, as discussed in other chapters in this book. In addition, other chapters of this book address types of toxicity concerns, toxicity testing strategies, and exposure and risk assessment of excipients.

Excipients must be safe for their intended use. Under 21 CFR § 330.1(e), over-the-counter (OTC) human drugs that are generally recognized as safe and effective and not misbranded, may only contain inactive ingredients if they are suitable and if the amounts administered are safe and do not interfere with the effectiveness of the drug or with required tests or assays. Color additives may be used in accordance with the provisions of the FD&C Act and the regulations of 21 CFR Parts 70–82. The FDA proposed that to make it clear that, to be considered as suitable within the meaning of 21 CFR § 330.1(e), each inactive ingredient in an OTC human drug product should perform a specific function (5). The proposed regulation defined *safe and suitable* to mean that the inactive ingredient meets various conditions as mentioned in the foregoing. OTC drug manufacturers are responsible for assuring that these conditions are met. There is no formal approval mechanism.

In the United States, the safety and suitability of excipients used in new drugs are considered as part of the New Drug Application (NDA) process. There is no separate and independent review and approval system for excipients. There are no specific regulations or guidelines that specify the requirements needed to gain approval of a new drug that contains a new excipient. Generally, pharmaceutical companies choose excipients that previously have been approved for commercial use in other NDAs. The FDA's Inactive Ingredient Guide (6), discussed later in this chapter, contains a listing of inactive ingredients present in approved drug products. There is currently no way of gaining a listing for an excipient in the guide independent of the NDA route. The FDA reviews the status of an excipient in food as information to support its use in drug products. Factors relative to the use of an excipient, such as dosing regimen and route of administration, are also reviewed. Advances in excipient technology and drug dosage form technology have created a need for a separate regulatory approval process for new excipients. The *USP* published IPEC's Excipient Safety Evaluation Guidelines as Information Chapter ⟨1074⟩ Excipient Biological Safety Evaluation Guideline (7).

Information on existing or new excipients can be described and provided to the FDA in an NDA directly. Alternatively, the manufacturers of excipients may prepare and submit type IV Drug Master Files (DMF) to support the use of an excipient in one or more NDAs. The DMFs are discussed in FDA's regulations under 21 CFR § 314.420 and the FDA-issued Guidance for Drug Master Files (8). When authorized by the DMF submitter (i.e., the excipient manufacturer) and cross-referenced by an NDA submitter, the FDA reviews the DMF to make determinations on the safety, manufacture, and quality of the excipient use in the new drug that is the subject of the then pending NDA. The DMF becomes active when reviewed in conjunction with the review and approval of an NDA.

The FD&C Act designates the *USP/NF* as an official compendia. The *NF* section contains official standards for excipients for strength, quality, and purity. Generally, pharmaceutical manufacturers use excipients listed in the *USP/NF*. The *USP* gives priority for admission of a monograph to excipients used in formulations for which there is an approved NDA (3).

The *USP/NF* provides a listing of excipients by categories in a table according to the function of the excipient in a dosage form, such as tablet binder, disintegrant, and such (9). A recent published article categorized excipients for use in injectable products (10). An excellent reference for excipient information is the *Handbook of Pharmaceutical Excipients* (11).

2. Drug Product Application Requirements for Excipients (Components and Composition)

Under Section 505(b)(1) of the FD&C Act, a person filing a drug application shall submit to the FDA a full list of the articles used as components of a drug product, and samples of the articles used as components as the FDA may require.

A sponsor of an NDA must submit specific information on excipients. Requirements for an Investigational New Drug Application (IND) include the information listed in 21 CFR § 312.23(a)(7) "chemistry, manufacturing, and control information." Information for the drug product should include a list of all components, which may include reasonable alternatives for inactive compounds, used in the manufacture of the investigational drug product, including both those components intended to appear in the drug product and those that may not appear, but that are used in the manufacturing process and, where applicable, the quantitative composition of the investigational drug product, including any reasonable variations that may be expected during the investigational stage. Reference to the current edition of the *United States Pharmacopeia/National Formulary* may satisfy certain requirements by provision of the required IND information. Under 21 CFR § 312.31, information amendments may include new chemistry, manufacturing, and controls information concerning the use of excipients. Under 21 CFR § 312.33, IND annual reports may contain a summary of any significant manufacturing changes made during the past year which may include excipient changes.

Requirements for a drug product in an NDA include the information listed in 21 CFR § 314.50(d)(1)(ii)(a) "chemistry, manufacturing, and control information." Information for the drug product should include a list of all components used in the manufacture of the drug product regardless of whether they appear in the drug product, a statement of composition of the drug product, and a statement of the specifications and analytical methods for each component. Reference to the current edition of the *United States Pharmacopeia/National Formulary* may satisfy relevant requirements. For each batch of drug product used to conduct a bioavailability or bioequivalence study or used to conduct a primary stability study, the following information must be included: the batch production record; the specifications and test procedures for each component and for the drug product; the names and addresses of the sources of the active and noncompendial inactive components; the results of any test performed on the components used in the manufacture of the drug product, as required by 21 CFR § 211.84(d), and on the drug product, as required by 21 CFR § 211.165.

The chemistry, manufacturing, and controls information requirements for inactive ingredients in abbreviated new drug applications (ANDA) are more specific and are listed in 21 CFR § 314.94(a)(9). Applicants must include the same information required by NDAs under 21 CFR § 314.50(d)(1). In addition, the applicant shall identify and characterize the inactive ingredients in the proposed drug product and provide information demonstrating that such inactive ingredients do not affect the safety of the proposed drug product.

a. Parenteral.   A drug product intended for parenteral use shall contain the same inactive ingredients and the same concentration as the reference listed drug (RLD) identified by the applicant. An applicant may seek approval of a drug product that differs from the reference listed drug in preservative, buffer, or antioxidant, provided that the applicant identifies and characterizes the differences and provides information demonstrating that the differences do not affect the safety of the proposed drug product.

b. Ophthalmic.   A drug product intended for ophthalmic or otic use shall contain the same inactive ingredients and in the same concentration as the reference listed drug identified by the applicant. An applicant may seek approval of a drug product that differs from the reference listed drug in preservative, buffer, substance to adjust tonicity, or thickening agent provided that the applicant identifies and characterizes the differences and provides information demonstrating that the differences do not affect the safety of the proposed drug product, except that, in a product intended for ophthalmic use, an applicant may not change a buffer or substance to adjust tonicity for the purpose of claiming a therapeutic advantage over a difference from the listed drug, for example, by using a balanced salt solution as a diluent as opposed to an isotonic saline solution, or by making a significant change in the pH or other change that may raise questions of irritability.

c. Topical. A drug product intended for topical use shall contain the same inactive ingredients as the reference listed drug identified by the applicant. An applicant may seek approval of a drug product that differs from the reference listed drug provided that the applicant identifies and characterizes the differences and provides information demonstrating that the differences do not affect the safety of the proposed drug product.

The FDA issued various guidances for industry which include requirements for excipients for drug product applications. The Guideline for Submitting Supporting Documentation in Drug Applications for the Manufacture of Drug Products (12) lists specific requirements for components and composition, and the guideline should be consulted for more information. If any proprietary preparations or other mixtures are used as components, their identification should include a complete statement of composition and other information that will properly describe and identify these materials. Proposed alternatives must be justified for any listed substances by demonstrating that the use of these alternatives does not significantly alter the stability and bioavailability of the drug product and the suitability of manufacturing controls. The guideline additionally requires a statement of the quantitative composition specifying, by unit dose, a definite weight or measure or appropriate range for all excipients contained in the drug product.

Various other guidelines include requirements for excipients (13–15). One inspectional guideline discusses the importance of control of the physical characteristics of the excipient, stating that variations in such characteristics may affect the performance of the dosage form (14). Inspectional guidelines instruct the FDA investigator during inspection of the validation of the manufacturing process of the drug product to review the firm's data to ensure that the physical and chemical characterization of the drug substance and other raw materials conform to application specifications (15).

The FDA's regulations contain several provisions about inactive ingredients for parenteral, ophthalmic and otic, and topical generic drug products. Additional regulations codified under 21 CFR § 314.127 direct that FDA will refuse to approve ANDAs that contain inactive ingredients that are not permitted to be first reviewed in an ANDA, as described in more detail later. On November 17,

1994, the FDA issued the Interim Inactive Ingredients Policy (16) which divides excipients into two classes:

1. Exception excipients: changes are allowed from the reference listed drug (RLD).
2. Nonexception excipients: no differences are allowed from the RLD.

Generally, excipients in the proposed generic drug product, other than an oral dosage form, should be qualitatively identical and quantitatively essentially the same as the excipients in the RLD. The applicant must identify and characterize any difference in inactive ingredients between the proposed drug product and the RLD. If an exception excipient is different, either qualitatively or quantitatively, information must be submitted to demonstrate that the difference does not affect the safety of the proposed drug product. The policy discusses differences for exceptions and nonexceptions that will be permitted.

Under 21 CFR § 314.127(a)(8), the FDA will refuse to approve an ANDA if the inactive ingredients are unsafe for use under the conditions prescribed, recommended, or suggested in the labeling proposed for the drug product. In addition, the FDA will refuse to approve an ANDA if the composition of the drug product is unsafe, under the conditions prescribed, recommended, or suggested in the proposed labeling because of the type or quantity of inactive ingredients included or the manner in which the inactive ingredients are included. The inactive ingredients or composition of a proposed drug product will be considered to raise serious questions of safety if the product incorporates one or more of the following changes:

1. A change in an inactive ingredient such that the product does not comply with an official compendium.
2. A change in composition to include an inactive ingredient that has not been previously approved in a drug product for human use by the same route of administration.
3. A change in the composition of a parenteral drug product to include an inactive ingredient that has not been previously approved in a parenteral drug product.
4. A change in composition of a drug product for ophthalmic use to include an inactive ingredient that has not been previously approved in a drug for ophthalmic use.
5. The use of a delivery or a modified-release mechanism never before approved for the drug.
6. A change in composition to include a significantly greater content of one or more inactive ingredients than previously used in the drug product.

7. If the drug product is intended for topical administration, a change in the properties of the vehicle or base that might increase absorption of certain potentially toxic active ingredients, thereby affecting the safety of the drug product, or a change in the lipophilic properties of a vehicle or base (e.g., a change from an oleaginous to a water-soluble vehicle or base).

FDA's Current Good Manufacturing Practice (CGMP) Regulations for finished dosage forms outline the requirements for the control of components of drug products. For example, under 21 CFR § 211.180(b), firms must establish written specifications for all raw materials. All records must be readily available for inspection by the FDA. The records must include the identity and quality of each shipment of each lot of components, the name of the supplier, the supplier's lot number if known, the receiving code, as specified under 21 CFR § 211.80, and the date of receipt. Records must include the results of any test or examination performed and the conclusions, an individual inventory record of each component, and a reconciliation of the use of each lot of each component. Master production and control records must include a complete list of components.

The Division of Drug Information Resources of the FDA compiles the *Inactive Ingredient Guide*. The guide contains all inactive ingredients present in approved drug products or conditionally approved drug products currently marketed for human use (6). The guide provides FDA reviewers with information on inactive ingredients in products that have been approved by the FDA. Once an inactive ingredient appears in a currently approved drug product for a particular route of administration, the inactive ingredient would not usually be considered new and may require a less extensive review. The guide contains the following information for an inactive ingredient:

1. *Chemical Abstract Service* (CAS) Registry Number
2. NDA count: total number of NDAs in which a particular inactive ingredient currently appears.
3. Approval date and division: specifies the approval date and review division responsible for evaluating this most recent NDA.
4. Potency range: specifies the minimum and maximum amounts of inactive ingredients for each route of administration and dosage form.

The Certification Branch of the Division of Color Technology of the FDA has designated permanently listed, provisionally listed, and delisted color additives; this listing appears in the *Inactive Ingredient Guide* (6). Detailed information on color additives uses, restrictions, and tolerances are listed in 21 CFR parts 70–82.

21 CFR Part 207 requires the registration of producers of drugs and listing of drugs in commercial distribution. 21 CFR § 207.31(b) requests a qualitative

listing of the inactive ingredients be provided to the FDA for each initial drug listing form (Form FDA 2657) (17). The applicable official name designated pursuant to Sec. 508 of the FD&C Act should be used for the listing. If there is no such name or drug, or the ingredient is an article recognized in an official compendium, then the official title in the compendium will be used. If there is no official title in the compendium, then the common or usual name, if any, will be used.

The FDA's *Compliance Program Guidance Manual* 7346.832 on preapproval inspections/investigations (18) discusses the district's objectives and responsibilities in conducting the inspections. Under this program, FDA's Center for Drug Evaluation and Research (CDER) may request that a noncompendial excipient manufacturing facility be inspected. These excipients are typically used in specialized dosage forms and drug delivery systems. Data submitted for excipients for tests, methods, and specifications is reviewed by CDER chemists and audited by the district office under the preapproval inspection program. To facilitate the pre-approval inspection, certain information on the development of the dosage form is required. This data must include physical and chemical specifications for the excipients.

Under 21 CFR § 314.70 supplements and other changes to an approved application, to add or delete an ingredient, or otherwise to change the composition of the drug product, other than deletion of an ingredient intended only to affect the color of the drug product, require a supplement for FDA approval before the change is made. Changes described in an annual report include the deletion of an ingredient intended only to affect the color of the drug product.

In November 1995, the FDA issued a guidance (19) to provide recommendations to sponsors of new drug applications, abbreviated new drug applications, and abbreviated antibiotic applications who intend, during the postapproval period, to change components or composition. The guidance defines the levels of change, recommended chemistry, manufacturing, and controls tests for each level of change, in vitro dissolution tests or in vivo bioequivalence tests, or both, for each level of change, and documentation that should support the change. The FDA issued another guidance for changes to components or composition for nonsterile semisolid dosage forms (20) and a guidance for modified-release solid oral dosage forms (21); both guidelines discuss components and composition changes.

3. Labeling and Nomenclature Requirements for Excipients in Drug Products

21 CFR § 201.100(b)(5) requires that prescription drugs for human use be labeled with the name of all inactive ingredients if it is for other than oral use. Flavorings and perfume may be designated as such without naming their components, and

color additives may be designated as coloring without naming specific color components unless the naming of such components is required by a color additive regulation prescribed in Subchapter A of the regulations. Trace amounts of harmless substances added solely for individual product identification need not be named. If it is for parenteral injection administration, the quantity or proportion of all inactive ingredients must be listed, except that ingredients added to adjust the pH or to make the drug isotonic, may be declared by name and a statement of their effect. If the vehicle is water for injection it need not be named.

The General Labeling Provisions of Part 201 of the regulations require specific declarations or warnings. Under 21 CFR § 201.20 the presence of FD& C Yellow No. 5 or FD&C Yellow No. 6 in certain drugs for human use must be declared. Under 21 CFR § 201.21 the presence of phenylalanine as a component of aspartame in over-the-counter and prescription drugs for human use must be declared. Under 21 CFR § 201.22 prescription drugs containing sulfites are required to be labeled with warning statement. Under 21 CFR Part 328, any over-the-counter drug product intended for oral ingestion shall not contain alcohol as an inactive ingredient in concentrations that exceed those established in Part 328; specific-labeling requirements are also included in the regulation.

The *United States Pharmacopeia 23/National Formulary 18* General Chapter ⟨1091⟩, Labeling of Inactive Ingredients, provides guidelines for dosage forms. The name of an inactive ingredient should be taken from the current edition of one of the following reference works in the following order of precedence: (a) *USP/NF*; (b) *USP Dictionary of USAN and International Drug Names*; (c) *CTFA Cosmetic Ingredient Dictionary*; (d) *Food Chemicals Codex*. The general chapter outlines other requirements. The Nonprescription Drug Manufacturers Association (NDMA) has published *Voluntary Codes and Guidelines of the OTC Medicines Industry*; the guideline contains a section for disclosure of inactive ingredients (22). The voluntary identification of excipients on the label of an OTC drug product allows consumers with known allergies to select products with confidence of safe use.

In February 1997, FDA published proposed changes to the labeling regulations for OTC human drugs (23). This proposal would establish a standardized-labeling format for all OTC drug products marketed under a drug monograph or a marketing application. The revised format would require that OTC drug products include a heading on the label designated "Other Ingredients" or "Inactive Ingredients" followed by a listing of the inactive ingredients contained in the product. In November 1997, the Food and Drug Modernization Act of 1997 was passed by Congress (24). Section 412 of this act requires that OTC drugs have the "established name of each inactive ingredient listed in alphabetical order on the outside of the container of the retail package and, if determined to be appropriate by the Secretary, on the immediate container." FDA is required to promulgate regulations pertaining to this law.

B. Official Pharmacopeial Standards for Excipients

The Federal Food, Drug, and Cosmetic Act recognizes the *United States Pharmacopeia/National Formulary* as an official compendia. The statute empowers the FDA to enforce the law using certain defined aspects of the compendia. Most commonly recognized are *USP/NF* standards for determining the identity, strength, quality, and purity of the articles, and specifications for packaging and labeling. The *United States Pharmacopeia* describes drug substances and dosage forms, whereas the *National Formulary* is limited to excipients. The General Notices and Preface to the *USP* and *NF* contain requirements for official substances. Where an article is used as both a therapeutic agent and an excipient, it is included in the *USP* with a cross-reference in the *NF* to that *USP* monograph. The *USP* General Notices and Requirements and *USP* General Chapters and Reagents, Indicators, and Solutions and Reference Tables should be referenced for *NF* articles.

The designation "NF" in conjunction with the official title on the label of an article means that the article purports to comply with *NF* standards. Where an article differs from the standards of strength, quality, and purity determined by the application of the assays and tests in the *NF* monograph, its difference shall be plainly stated on its label. Articles in the *NF* are official, and the standards in the monographs apply only when the articles are intended or labeled for use as drugs and when bought, sold, or dispensed for these purposes, or when labeled as conforming to the *NF*.

Official substances must be prepared according to recognized principles of good manufacturing practice and from ingredients complying with specifications designed to assure that the resultant substances meet the requirements of the compendial monographs. An official substance must contain no added substances except where specifically permitted in the individual monograph. Where such addition is permitted, the label indicates the name(s) and amount(s) of any added substance(s). The *USP* General Notices address added substances to official preparations (9).

A discussion of requirements for foreign substances and impurities is contained in Chapter 2 of this book.

The Preface to the *NF* (3) outlines the Admissions Policy for the established order of priorities for the inclusion of excipients in the *NF*:

1. The standards and the test methods for the most widely used and critical excipients will be reviewed thoroughly and updated. Priority will be given to improving and harmonizing the standards and the test methods with those of other compendia.
2. Monographs for new excipients present in dosage forms marketed in the United States will be considered for inclusion in the *NF* provided the excipient suppliers submit adequate data and information. Excipi-

ents utilized in articles marketed in the United States will be considered in the following order:

Multisource, single-component excipients
Single-source, single-component excipients
Mixtures of excipients that have been altered by processing

3. Single-component or multicomponent mixtures that do not appear in articles marketed in the United States, but are found in other compendia or are involved in widespread studies for use in dosage forms, will be considered on a lower priority for inclusion in the *NF*.
4. For excipients not appearing in articles marketed in the United States, a draft monograph will be considered for publication in the *Pharmacopeial Previews* section of *Pharmacopeial Forum* to solicit public comment. The Preface also includes information necessary for the review for inclusion in the *NF*.

C. Manufacturing and Quality Requirements for Excipients

Excipients must be manufactured under current Good Manufacturing Practices (cGMPs). Although the GMP regulations under 21 CFR Parts 210 and 211 apply specifically to drug products, Sec. 501(a)(2)(B) of the act requires that all drugs be manufactured, processed, packed, and held in accordance with GMPs. The FDA states that no distinction is made between bulk pharmaceutical chemicals and finished pharmaceuticals, and failure of either to comply with cGMPs constitutes a failure to comply with the requirements of the act (25).

Bulk pharmaceutical chemicals (BPCs) include both active and inactive ingredients. The requirements under part 211 are used by the FDA as guidelines for the inspection of BPC manufacturers as interpreted in the FDA's Guide to Inspection of Bulk Pharmaceutical Chemicals (September 1991). The guide states that it is neither feasible nor required to apply rigid controls during the early-processing steps of a BPC; the requirements should be increasingly tightened according to some reasonable rationale. At some logical step, such as where the BPC can be identified and quantified for those processes in which the molecule is produced, appropriate GMPs should be imposed and maintained throughout the remainder of the process. Various articles have been written discussing the requirements for bulk pharmaceutical chemicals (26,27).

In March 1998, the FDA issued a draft guidance for Industry—Manufacturing, Processing, or Holding Active Pharmaceutical Ingredients (28). Although the guidance focuses on the manufacture of active pharmaceutical ingredients (APIs), the FDA states that much of the guidance provided may be useful for the manufacture of excipients.

The International Pharmaceutical Excipients Council (IPEC) developed an industry GMP Guide for Bulk Pharmaceutical Excipients (29). Although the IPEC guide does not have official status with the FDA, drug product manufacturers use the guide to audit excipient manufacturers. The IPEC guide discusses general guidance, excipient quality systems, and auditing considerations. The U.S. Pharmacopeial Convention published IPEC's GMP guide (30).

Formal written procedures should be established for the review and approval of changes for the production process and other relevant changes for BPCs. Appropriate technical evaluation should be conducted. When necessary, the changes should be communicated to users of the excipients, and Drug Master Files (DMFs) should be amended.

Under 21 CFR § 207.10(e), manufacturers of harmless inactive ingredients that are excipients, colorings, flavorings, emulsifiers, lubricants, preservatives, or solvents that become components of drugs and who otherwise would not be required to register under part 207 are exempt from registration and drug listing in accordance with Part 207. FDA states in the Guide to Inspection of Bulk Pharmaceutical Chemicals (25) that whereas manufacturers of inactive ingredients may not be required to register with the FDA, they are not exempt from complying with GMP concepts, and they are not exempt from inspection. Whether or not manufacturers of inactive ingredients will be inspected on a surveillance basis is generally discretionary. However, an excipient manufacturer is always subject to "for cause" inspection. FDA has authority to inspect the manufacturing, packaging, and holding sites for excipients under Section 704(a)(1) of the FD&C Act. The Prohibited Acts and Penalties under Chapter III of the FD&C Act apply to excipients.

III. REGULATION OF EXCIPIENTS IN THE EUROPEAN UNION FOR PRESCRIPTION MEDICINAL PRODUCTS

A. Excipients in the Registration of a Medicinal Product

1. The European Union Regulation of Medicinal Products

Consistent with its goal of establishing a "common market" where goods can freely circulate, the European Union (EU) is in the process of harmonizing the legislation of its Member States in various areas, including medicinal products.

The harmonization of Member States legislation on medicinal products is nearly complete. It is realized mainly through the adoption of "Directives." Directives are legislative acts that set forth objectives for the Member States to achieve. Member States must achieve these objectives by passing national legislation implementing the directives. Matters that are not yet fully harmonized remain subject to the national legislation of the Member States.

There are two "foundation directives" applicable to medicinal products in the EU. These are Directives 65/65/EEC (31) and 75/319/EEC (32). These directives have been amended several times and supplemented by other directives.

Directive 65/65/EEC defines *medicinal products* as "any substance or combination or substances presented for treating or preventing disease in human beings or animals" and "any substance or combination of substances which may be administered to human beings or animals with a view to making a medical diagnosis or to restoring, correcting or modifying physiological functions in human beings or in animals" (31).

Council Directive 75/319/EEC (32), as amended, requires that Member States take appropriate measures to ensure that the manufacturer of medicinal products is subject to the holding of an authorization. Council Directive 75/318/EEC (33) provides several requirements in order to obtain a marketing authorization. These requirements apply not only to finished medicinal products, but also to active substances, excipients, and packaging. Council Directives 65/65/EEC and 75/319/EEC, as amended, require that medicinal products be subject to a marketing authorization before they may be placed on a European market.

The EU directives on medicinal products have to be "implemented" to be effective in the Member States. As EU regulations on medicinal products are embodied in directives, national implementing legislation may differ from one Member State to the other. However, generally speaking, EU regulations on medicinal products have been properly "implemented" in the Member States.

In addition to the foregoing directives, reference must be made to the "Notes for Guidance" elaborated by the EU Committee for Proprietary Medicinal Products (CPMP) in consultation with the European Commission and the Member State authorities. These documents have no legal force in the EU, but contain relevant information, including information to assist applicants in the development of a marketing authorization for a medicinal product.

2. Regulatory Status of Excipients

The European Pharmacopoeia Commission published a general monograph for excipients that includes the following definition for an *excipient*. "Excipients are all substances contained in a dosage form other than the active substance" (34).

Council Directive 65/65/EEC as amended does not contain a definition for excipient (31). EU legislation has no specific approval procedure for excipients. The EU does not have a drug master file system for excipients similar to that of the United States. Excipients are reviewed as part of the procedure leading to the marketing authorization of the finished medicinal product. Therefore, relevant information on excipients must be included in the Marketing Authorization Application. More specifically, a Marketing Authorization Application must contain qualitative and quantitative particulars of all the constituents of the proprietary product in usual terminology. A description of the control methods by the manufacturer must include qualitative and quantitative analysis of the constituents, including excipients.

3. Data Requirements for Excipients in the Dossier for Application for Marketing Authorization of a Medicinal Product

The requirements for obtaining a Marketing Authorization for Medicinal Products in the EU are described in Part II, Sec. A, C, E, F of the Annex to the Directive 75/318/EEC (33). Directive 91/507/EEC of 19 July 1991 (35) modified the Annex to Council Directive 75/318/EEC. The standard format for the data is described in the "Notice to Applicants" published by the European Commission, in the sections referring to Parts II A, C, E, and F (36). Also, the Note for Guidance: Excipients in the Dossier for Application for Marketing Authorization of a Medicinal Product (37) ("Excipients Guidance") discusses the requirements pertaining to excipients for purposes of the application for a marketing authorization. In addition, a Note for Guidance covers the inclusion of antioxidants and antimicrobial preservatives in medicinal products (38).

As described more fully in the following, the requirements vary depending on whether or not the excipient is listed in the *European Pharmacopoeia* (*Ph. Eur.*) or a Member State pharmacopeia. The Excipients Guidance provides that excipients must be listed in the composition of the medicinal product part, with their common name, quantity, use, and reference standard. When the common name is not sufficient to indicate functional specifications, the brand name with commercial grade should be specified. Qualitative and quantitative information should be provided for mixtures. Only the qualitative composition is required for flavoring agents and aromatic substances.

Required information for development pharmaceutics for a medicinal product is discussed in another Note for Guidance on developmental pharmaceutics (39). It specifies that the results of compatibility studies of the active ingredient(s) with the excipients should be provided where appropriate. The choice and the characteristics of excipients should be appropriate for the intended use. An explanation should be provided relative to the function of all constituents in the formulation, with justification for their inclusion. In some cases, experimental data may be necessary to justify the inclusion of an excipient: for example, a preservative. The choice of the quality of the excipient should be guided by its role in the formulation and manufacturing process. In some cases, it may be necessary to address and justify the quantity of certain excipients in the formulation. Compatibility of excipients with other excipients, where relevant, (e.g., combination of preservatives in a dual preservative system) should be established and supporting stability data may be sufficient. If novel constituents are used in the manufacture of the product (e.g., a new matrix of a prolonged-release preparation, a new propellant, or a permeability enhancer), full information on the composition and function of the constituent in the formulation of the product should be furnished with safety documentation.

The Guidance on Development Pharmaceutics discusses new substances. A new substance introduced as a constituent will be regarded in the same way as that of a new active ingredient and full supporting data is required in accordance with the Note for Guidance on Excipients, unless it is already approved for use in food for orally administered products, or in cosmetics for topical administration. Additional data may still be required where an excipient is administered by an unconventional route, or in high doses.

The Excipient Guidance Annex lists eight examples of requirements concerning different kinds of excipients, such as mixtures of chemically related components (37). Routine tests that are to be carried out on each batch of starting materials must be stated in the application for excipients described in the *Ph. Eur.* or, if not in the *Ph. Eur.*, in a Member State pharmacopoeia. In addition, and when necessary, the test used to determine the quality of the excipient should be shown to be in relation to the function that it fulfills in the medicinal product. Data on microbiological contamination of excipients used in the manufacture of sterile products should always be given where membrane filtration is used to achieve sterility. For excipients not described in the *Ph. Eur.* or in the pharmacopoeia of a Member State, the guidance lists test procedures and storage conditions. Various tests are listed that must be followed to establish the specifications for an excipient not listed in the *Ph. Eur.* or a Member State pharmacopoeia.

Documentation should be presented in the scientific data section of the dossier under control of the starting materials (Part IIC) to justify the choice of the excipient. The data determines the properties that must be checked during routine tests and that will be subject to certain specifications in connection with the bioavailability of the product. The Note for Guidance: Specifications and Control Tests on the Finished Product (40) requires that excipients that affect the availability of an active substance must have a quantitative determination in each batch of drug product unless bioavailability is guaranteed by other appropriate tests. The determination must be established on a case-by-case basis as a function of development studies. The guidance also addresses the requirements for preservatives (40). The Guidance for Specifications has other requirements for excipients: namely, "if necessary, identification and assay of the constituents of the excipients in the medicinal product such as identification of colourants used and identification and assay of antimicrobial agents or antioxidant preservatives (with acceptance limits)." The International Conference on Harmonization (ICH) is developing a guideline to assist in the establishment of a single set of global specifications for new drug substances and new drug products (41).

Scientific data are not systematically required for well-known excipients that have been used in similar medicinal products for a long period and when their characteristics and properties have not changed significantly. For solid and semisolid dosage forms, the scientific data should, if necessary, provide information on the relevant characteristics of the excipient. Special tests are often necessary: for example, to verify the capacity of the excipients to emulsify and disperse, or to measure the viscosity. Appropriate data are needed for excipients used in a new route of administration.

The Excipient Guidance also discusses the requirements for a dossier that must be developed for new excipients. The data are the same as those required for a new active ingredient. Table 3 outlines the requirements.

The documentation on chemistry should be based on the Note for Guidance: Chemistry of Active Ingredients (42). The routine test procedures and limits should be established based on the documentation in the dossier. Apart from those situations discussed in the Note for Guidance: Specifications and Control Tests on the Finished Product, it is not usually necessary to carry out identity testing and an assay of the excipients in the finished product at release. For new excipients, stability data should be provided as required for new active sub-

Table 3   Required New Excipient Information

Definition	Existing use data	Chemistry documentation
Function	Chemistry and toxicology and field of existing uses	Name, address of manufacturer
Condition of use	Food additive, toxicology and quality specifications	Synthesis outline
Composition of mixtures	International specifications (FAO/WHO/JECFA)	Structure
	Cosmetic starting material data for topical use	Physical, chemical properties, identification (ID) and purity tests
	Toxicology data for the specific dosage form and route of administration	Validated analytical methods with batch results
		Microbiological tests, etc.
		Contamination, presence of foreign substances, residual solvents
		Quality of components for mixture and physicochemical tests for mixture

stances. Part II of this book should be referenced for safety evaluation studies for new excipients.

Council Directive 75/318/EEC states that the toxicology and pharmacokinetics of an excipient used for the first time in the pharmaceutical field shall be investigated; the directive makes reference in Part 3 Toxicological and Pharmacological Tests. The IPEC Europe Safety Committee published the Proposed Guideline for the Safety Evaluation of New Excipients (43).

The maintenance of the physicochemical properties of the finished product are dependent on the properties and the stability of the excipients (see Note for Guidance: Specifications and Control Tests on the Finished Product).

Directive 91/356/EEC describes the principles and guidelines of GMPs for medicinal products for human use (44). Detailed guidelines in accordance with Directive 91/356/EEC are published in the Guide to Good Manufacturing Practice which is used in assessing applications for manufacturing authorizations and as a basis for inspection (45). The guide describes various documentation requirements for starting materials including excipients. Table 4 outlines these requirements.

Other detailed documentation is required for excipients as starting materials, such as written procedures and records for the receipt of each delivery, sampling, testing, release, and rejection. Production Standard Operating Procedures (SOPs) should describe requirements for starting materials, such as purchasing from approved suppliers, checking of deliveries, appropriate labeling, and other requirements. The Quality Control GMPs describe specific requirements for sampling and testing of starting materials. Annex 8 in the Guide to Good Manufacturing Practice describes requirements for sampling of starting materials.

There are two regulations EC No 541/95 (46) and 542/95 (47) that describe procedures for approval of a variation or change to an approved marketing authorization. Type I variations are minor and Commission Regulations EC No 542/95 lists 33 types of changes in Annex I. The conditions that must be satisfied are also detailed and a notification procedure is required. If the authority is not satisfied with the data, they must advise the applicant within 30 days, otherwise

Table 4 Specifications for Starting Materials

Description: Name, internal code, pharmacopeial
reference, approved suppliers
Sampling and testing directions
Qualitative and quantitative requirements with
acceptance limits
Storage conditions and precautions
Maximum period of storage before reexamination

the variation can be considered as approved. Type II variations are major and are subject to an approval procedure. The regulation also describes changes to a marketing authorization requiring an entire new application as referenced to in Article II.

For the types of changes listed in Annex I, there are several examples of excipient changes as type I variations in Annex I of Regulations No 542/95 with the condition to be fulfilled. Table 5 outlines these changes and conditions.

4. Labeling and Nomenclature Requirements for Excipients in Medicinal Products

The legal provisions for the labeling and package leaflets for medicinal products for human use are in Council Directive 92/27/EEC (48). On June 12, 1997, a Guideline on the Excipients in the Label and Package Leaflet of Medicinal Products for Human Use was adopted by the EU Pharmaceutical Committee (49). The guideline is applicable to all applications for a marketing authorization and to all renewals of a marketing authorization made after September 1, 1997.

For parenteral products, topical products (includes inhaled medicines), and ophthalmic products, all excipients must appear on the label. For all other medici-

Table 5 Excipient Changes Type I Variations

Change	Condition to be fulfilled
Replacement of excipient with comparable excipient	Same functional characteristics No change in dissolution profile
Deletion of colorant or replacement with another	—
Addition, deletion, or replacement of flavor	Flavor must be in accordance with Directive 88/388/EEC
Change in coating tablet weight or change in capsule shell weight	No change in dissolution profile
Synthesis or recovery of nonpharmacopeial excipient	Specification not adversely affected, no new impurities or change in level of impurities requiring further safety study qualification, no change in physicochemical properties
Change in specification of excipients in medicinal product	Specifications must be tightened or addition or new test and limits
Changes to comply with pharmacopeial supplements	Change is made exclusively to implement new provisions of the supplement
Change in test procedures of nonpharmacopeial excipients	Method validation for equivalence to former test procedure

nal products, only those excipients known to have a "recognized action or effect" need to be declared on the label. The excipients appearing in the Annex of the Guidelines have a recognized action or effect and, therefore, when a medicinal product contains any of these it must be stated on the label, together with a statement, such as "see leaflet for further information." The label statement of these excipients should be phrased so that it does not imply that these are the only excipients present in the product; therefore, "includes . . ." would be preferable to "contains." Certain excipients in the Annex only need to be declared on the label in certain circumstances, related to the dose form or the quantity, as specified in the third column of the Annex.

According to Article 7.1(a) of Directive 92/27/EEC, all of the excipients must be stated on the package leaflet by name. Even those excipients that are present in very small amounts should be stated in the leaflet, including the constituents of ingested capsule shells or the constituents of a compound excipient preparation used, for example, in direct compression or in a film coat or a polish for an ingested dose form. Additional examples are given in the directive.

The fourth column in the Annex provides information corresponding to each excipient. The text of this information should be put into consumer understandable language. The excipients appearing in the Annex of the Guideline are referred to by their international nonproprietary name (INN), as recommended by the World Health Organization, or failing this, their usual common name. The INN lists nomenclature for pharmaceutical substances (50). The E number alone may be used for an excipient on the label, provided that the full name (INN where it exists, or usual common name), and the E number are stated in the user package leaflet in the section where the full qualitative composition is given. Where the full composition of a flavor or fragrance is not known to the marketing authorization holder, it should be declared in general terms (e.g., "orange flavor," "citrus fragrance/perfume"), and any components that are known should be stated (e.g., "orange flavor including orange oil and maltodextrin"). Chemically modified excipients should be declared in such a way as to avoid confusion with the unmodified excipient (e.g., modified starch).

B. Official Pharmacopeial Standards for Excipients

The purpose of the *European Pharmacopoeia* (*Ph. Eur.*) is to promote public health by the provision of recognized common standards for use by health care professionals and others concerned with the quality of medicines (51). Such standards are to be of appropriate quality as a basis for the safe use of medicines by patients and consumers.

Council Directives specify that starting materials used in medicinal products comply with monographs appearing in the *Ph. Eur.* Starting materials mean all the constituents of a medicinal product and include the active ingredient(s), and the excipients. The Annex of Directive 91/507/EEC (35) contains substantive requirements for the particulars and documents that must accompany applications for authorization of a medicinal product. A requirement for excipients (starting materials) is given in the Annex that "the monographs of the *European Pharmacopoeia* shall be an applicable to substances appearing in it."

The use of the title or the subtitle of a monograph implies that the article complies with the requirements of the relevant monograph. Unless otherwise indicated in the General Notice or in the monographs, statements in the monographs constitute mandatory requirements. An article is not of pharmacopeia quality unless it complies with all the requirements stated in the monograph.

Certain materials that are the subject of a pharmacopeial monograph may exist in different grades suitable for different purposes. Unless otherwise indicated in the monograph, the requirements apply to all grades of the material. In excipient monographs, a list of critical properties that are important for the use of the substance may be appended to the monograph for information and guidance. Test methods for determination of one or more of these critical properties may also be given for information and guidance.

The manufacturer of a substance in the *European Pharmacopoeia* may provide proof that the purity of the substance is suitably controlled by the monograph by means of a certificate of suitability granted by the Secretariat of the European Pharmacopoeia. The manufacturer must submit a detailed dossier, which may contain confidential data (52).

C. Manufacturing and Quality Requirements for Excipients

The first draft of a proposal for a European Parliament and Council Directive on GMP for starting materials and inspection of manufacturers of both medicinal products and their starting materials (amending Directives 75/319/EEC and 81/851/EEC) was issued (53). A detailed GMP guidance will be published by the European Commission.

The International Pharmaceutical Excipients Council (IPEC) of Europe developed an industry GMP Guide for Bulk Pharmaceutical Excipients (54). Although the IPEC guide does not have official status, medicinal product manufacturers use the guide to audit excipient manufacturers. The IPEC guide discusses general guidance, excipient quality systems, and auditing considerations.

IV. CONCLUSION

The regulations and guidances for the United States and the regulations, directives, and notes for guidances for the European Union were reviewed for the use of the excipients in drug and medicinal products. The development of an independent approval system for new excipients would encourage and advance the use of new excipient technology in drug products.

REFERENCES

1. Federal Food, Drug, and Cosmetic Act. Public Law Number 75-717, 52 Stat. 1040 (1938), 21 U.S.C. §§ 301 et seq.
2. Office of the Federal Register, National Archives and Records Administration. Code of Federal Regulations, Section 21 Food and Drugs. April 1, 1996.
3. United States Pharmacopeial Convention. United States Pharmacopeia 23/National Formulary 18, Suppl 6. Rockville, MD: USP Convention, 1997.
4. RE Osterberg. Excipient safety concerns does CDER have any?. Presentation by Co-Chairperson, CDER Inactive Ingredients Committee.
5. Office of the Federal Register, National Archives and Records Administration. Proposed rules. Fed Reg 42(70), Tuesday, April 12, 1977.
6. Division of Drug Information Resources, Food and Drug Administration, Center for Drug Evaluation and Research, Office of Management. Inactive Ingredient Guide. January 1996.
7. Pharmacopeial Forum 24(3) Rockville, MD: United States Pharmacopeial Convention, 1997.
8. U.S. Department of Health and Human Services, Food and Drug Administration, Center for Drug Evaluation and Research (CDER). Guidance for Drug Master Files. September 1989.
9. United States Pharmacopeia 23/National Formulary 18. Rockville, MD: The United States Pharmacopeial Convention, 1995.
10. S Nema, RJ Washkuhn, RJ Brendel. Excipients and their use in injectable products. PDA J Pharm Sci Technol, July–August 1997.
11. American Pharmaceutical Association. Handbook of Pharmaceutical Excipients. 2nd ed. Washington, DC: The Pharmaceutical Press, 1994.
12. U.S. Department of Health and Human Services, Food and Drug Administration, Center for Drug Evaluation and Research. Guideline for submitting supporting documentation in drug applications for the manufacture of drug products. February 1, 1987.
13. U.S. Department of Health and Human Services, Food and Drug Administration, Center for Drug Evaluation and Research. Guideline for submitting documentation for the manufacture of and controls for drug products. February 1, 1987.
14. U.S. Department of Health and Human Services, Food and Drug Administration, Division of Field Investigations, Office of Regional Operations, Office of Regulatory Affairs. Guide to inspections of oral solid dosage forms pre/post approval issues for development and validation. January 1994.
15. U.S. Department of Health and Human Services, Food and Drug Administration, Division of Field Investigations, Office of Regional Operations, Office of Regulatory Affairs and Division of Manufacturing and Product Quality, Office of Compliance. Guide to inspection of solid oral dosage form validation. 1993.

16. U.S. Department of Health and Human Services, Food and Drug Administration, Center for Drug Evaluation and Research, Office of Generic Drugs. Interim inactive ingredients policy. November 1994.

17. U.S. Department of Health and Human Services, Food and Drug Administration, Center for Drug Evaluation and Research. Drug registration and listing instruction booklet. May 1996.

18. Food and Drug Administration. New drug evaluation. Program 7346.832 pre-approval inspections/investigations. Compliance Program Guidance Manual, Chap 46. August 1994.

19. U.S. Department of Health and Human Services, Food and Drug Administration, Center for Drug Evaluation and Research. Guidance for industry, immediate release solid oral dosage forms, scale-up and post-approval changes: chemistry, manufacturing and controls, in vitro dissolution testing and in vivo bioequivalence documentation. November 1995.

20. U.S. Department of Health and Human Services, Food and Drug Administration, Center for Drug Evaluation and Research. Guidance for industry, nonsterile semi-solid dosage forms, scale-up and postapproval changes: chemistry, manufacturing and controls, in vitro release testing and in vivo bioequivalence documentation. May 1997.

21. U.S. Department of Health and Human Services, Food and Drug Administration, Center for Drug Evaluation and Research. Guidance for industry, modified release solid oral dosage forms, scale-up and post approval changes: chemistry, manufacturing, and controls: in vitro dissolution testing and in vivo bioequivalence documentation. September 1997.

22. Nonprescription Drug Manufacturers Association. Voluntary Codes and Guidelines of the OTC Medicines Industry. Washington, DC: NDMA.

23. Office of the Federal Register, National Archives and Records Administration. Proposed Rules. Federal Register 62(39), Thursday, February 27, 1977.

24. Food and Drug Administration Modernization Act of 1997. S.830. Passed by the 105th Congress, November 9, 1997.

25. U.S. Department of Health and Human Services, Food and Drug Administration, Center for Drug Evaluation and Research. Guide to inspection of bulk pharmaceutical chemicals. September 1991.

26. DB Barr, WC Crabbs, D Cooper. FDA regulation of bulk pharmaceutical chemical production. Pharm Technol, September 1993.

27. D Harpaz. Bulk pharmaceutical chemicals (BPCs). Regul Affairs J, January 1996.

28. U.S. Department of Health and Human Services, Food and Drug Administration, Center for Drug Evaluation and Research, Center for Biologics Evaluation and Research, Center for Veterinary Medicine. Draft guidance for industry—manufacturing, processing, or holding active pharmaceutical ingredients. March 1998.

29. The International Pharmaceutical Excipient Council (IPEC). Good Manufacturing Practices Guide for Bulk Pharmaceutical Excipients. 1995.

30. Pharmacopeial Forum 24(2). Rockville, MD: United States Pharmacopeial Convention, 1997.

31. Council Directive 65/65/EEC of 26 January 1965 on the approximation of provis-

ions laid down by law, regulation or administrative action relating to medicinal products.

32. Second Council Directive 75/319/EEC of 20 May 1975 on the approximation of provisions laid down by law, regulation or administrative action relating to proprietary medicinal products.

33. Council Directive 75/318/EEC of 20 May 1975 on the approximation of the laws of the Member States relating to analytical, pharmacotoxicological and clinical standards and protocols in respect of the testing of medicinal products.

34. Council of Europe. Pharmeuropa, Vol. 10, No. 2. Strasbourg, June 1998.

35. Council Directive 91/507/EEC of 19 July 1991 on the analytical pharmacotoxicological and clinical standards and protocols for testing of medicinal products.

36. Commission of the European Communities. Notice to Applicants for Marketing Authorizations for Medicinal Products for Human Use in the Member States of the European Community: The Rules Governing Medicinal Products in the European Community. Vol II. 1998.

37. Commission of the European Communities, Committee for Proprietary Medicinal Products, Working Party on Quality of Medicinal Products. Note for Guidance EU-CPMP-GDL 3196/91. Excipients in the dossier for application for marketing authorization of medicinal products.

38. Commission of the European Communities, Committee for Proprietary Medicinal Products and Committee for Veterinary Medicinal Products, Working Party on Quality of Medicinal Products. Note for Guidance EU-CPMP/CVMP-QWP 297/97. Inclusion of antioxidants and antimicrobial preservatives in medicinal products.

39. Commission of the European Communities, Committee for Proprietary Medicinal Products, Working Party on Quality of Medicinal Products. Note for Guidance EU-CPMP-QWP 155/96. Development pharmaceutics and process validation.

40. Commission of the European Communities, Committee for Proprietary Medicinal Products, Working Party on Quality of Medicinal Products. Note for Guidance EU-CPMP-GDL 3324/89. Specifications and control tests on the finished product.

41. Commission of the European Communities, Committee for Proprietary Medicinal Products, Working Party on Quality of Medicinal Products. Note for Guidance EU-CPMP-ICH 367/96 Specifications: Test procedures and acceptance criteria for new drug substances and new drug products: chemical substances.

42. Commission of the European Communities, Committee for Proprietary Medicinal Products, Working Party on Quality of Medicinal Products. Note for Guidance EU-CPMP-QWP 297/97: summary of requirements for active substances; in Part II of the Dossier.

43. IPEC Europe Safety Committee. The proposed guidelines for the safety evaluation of new excipients. Eur Pharm Rev, November 1997.

44. Council Directive 91/356/EEC of 13 June 1991 laying down the principles and guidelines of good manufacturing practice for medicinal products for human use.

45. Commission of the European Communities. Products in the European Community. Vol IV. Good Manufacturing Practice for Medicinal Products. Brussels, 1992.

46. Commission Regulation (EC) No 541/95 of 10 March 1995 concerning the examination of variations to the terms of a marketing authorization granted by a competent authority of a Member State.

47. Commission Regulation (EC) No 542/95 of 10 March 1995 concerning the examination of variations to the terms of a marketing authorization falling within the scope of Council Regulation (EEC) No 2309/93.
48. Council Directive 92/27/EEC of 31 March 1992 on the labeling of medicinal products for human use and on package leaflets.
49. Commission of the European Communities, Pharmaceutical Committee. A Guideline on the Excipients in the Label and Package Leaflet of Medicinal Products for Human Use. 1997.
50. World Health Organization. International Nonproprietary Names (INN) for Pharmaceutical Substances. 1992.
51. Council of Europe. European Pharmacopoeia, 3rd ed. Strasbourg, 1997.
52. Council of Europe. Public Health Committee (Partial Agreement). Certification of suitability to the monographs of the European Pharmacopoeia, Resolution AP-CSP (98) 2.
53. First draft of a proposal for a European Parliament and Council Directive on GMP for starting materials and inspection of manufacturers of both medicinal products and their starting materials (amending Directives 75/319/EEC and 81/851/EEC).
54. P Rafidison, P Maillere. Good Manufacturing Practices guide for bulk pharmaceutical excipients. Pharmeuropa, June 1997.

5

Development of Safety Evaluation Guidelines

Joseph F. Borzelleca
Medical College of Virginia, Richmond, Virginia

Myra L. Weiner
FMC Corporation, Princeton, New Jersey

I. INTRODUCTION

Excipients have not been subjected to extensive safety testing because they have been considered a priori to be biologically inactive, therefore, nontoxic. Many, if not most, excipients used are approved food ingredients, the safety of which has been assured by a documented history of safe use or appropriate animal testing. Some of the excipients are Generally Recognized As Safe (GRAS) food ingredients (see Chapter 3). The excipient is an integral component of the finished drug preparation and, in most countries, is evaluated as part of this preparation. There has been no apparent need to develop specific guidelines for the safety evaluation of excipients, and most developed countries do not have specific guidelines. However, as drug development has become more complex and/or new dosage forms have developed, improved drug bioavailability has became more important. It was noted that the available excipients were often inadequate, new pharmaceutical excipients specifically designed to meet the challenges of delivering new drugs were needed, and these are being developed. The proper safety evaluation of these excipients has now become an integral part of drug safety evaluation.

In the absence of official regulatory guidelines, the Safety Committees of the International Pharmaceutical Excipients Council (IPEC) in the United States, Europe, and Japan developed guidelines for the proper safety evaluation of new pharmaceutical excipients (1–3). The Committees critically evaluated guidelines for the safety evaluation of food ingredients, cosmetics, and other products, as well as, textbooks, and other appropriate materials. Guidelines consulted included the U.S. Food and Drug Administration's Red Book (4), EEC methods (5), OECD Guidelines for the Testing of Chemicals (6) and criteria documents from the International Program on Chemical Safety FAO/WHO Joint Expert Committee on Food Additives (7). Several other texts were also consulted, including Hayes' *Principles and Methods of Toxicology* (8) and The National Academy of Sciences' *Toxicity Testing* (9). Review articles and other appropriate materials were also critically considered. Because there are no guidelines specifically for pharmaceutical excipients as there are for food ingredients, the Committee developed a unique tiered approach that incorporates sound, scientific principles; conserves resources; simulates human exposure in animal testing; and involves human testing early in the safety evaluation program (1–3).

II. BACKGROUND INFORMATION

Before initiating a safety evaluation program for a new pharmaceutical excipient, it is advisable to address the following (1–3):

1. Chemical and physical properties and functional characterization of the test material.
2. Analytical methods that are sensitive and specific for the test material and that can be used to analyze for the test material in animal food used in the feeding studies or in the vehicle used for other studies.
3. Available biological, toxicological, and pharmacological information on the test material and related materials (which involves a thorough search of the scientific literature).
4. Intended conditions of use, including reasonable estimates of exposure.
5. Potentially sensitive segments of the population.

A. Literature Search

A comprehensive and critical search of the scientific literature on the test material and related materials is essential before the start of any testing program for the following reasons:

1. To conserve resources by preventing duplication.
2. To determine structure–activity relations. For example, if chemicals with similar structures have been reported to be genotoxic, an Ames test should be conducted before investing further resources in the test material.

3. To aid in the design of experimental investigations. Although studies identified in the literature search may not satisfy Good Laboratory Practice (GLP) guidelines, they can provide data on toxicity that can be considered in the design of definitive GLP studies.

The focus of the literature evaluation should be biological effects, including pharmacological, toxicological, and nutritional. The data collected from the literature search and from the characterization of the test material are critically evaluated by competent scientists to develop the appropriate studies necessary to establish the safety of the test material (experimental protocols).

B. Characterization of the Test Material

1. Chemical and Physical Characterization and Analysis

Characterization involves the identity, including structure; the Chemical Abstracts Service (CAS) Registry Number and other identifiers or designations; chemical and physical properties, including the melting point, boiling point, vapor pressure, solubility, and the form of the excipient to be used (and evaluated in animal studies). The method of preparation (synthesis or extraction) should be critically reviewed to identify starting materials, intermediates, and impurities that are potentially toxic and that could be present in the finished product. The stability of the test material during preparation of the final product and during storage should be assessed. Specifications for the test material must be developed and the material to be evaluated must meet these specifications. Trace metals and bacterial count may be necessary for current Good Manufacturing Practice (GMP) specifications.

Because the physical form of the test material may influence the design of the safety evaluation program, information on the physical state, melting point, vapor pressure, and other standard physical variables should be determined. An analytical method that is sensitive, specific, and reproducible must be developed to ensure compliance with specifications, but also to aid in the design of the biological studies, including biodisposition. Methods may include gas chromatography (GC), liquid chromatography (LC), gas chromatography–mass spectrometry (GC–MS) and molecular weight. The product to be tested, the test material, is the final commercial product; any ''significant'' change in the production will require further evaluation and/or testing.

2. Biological Characterization

Pharmacological activity should be assessed using standard pharmacological screens (e.g., isolated tissues and receptor-binding assays). In vitro biotransformation studies could provide information useful in designing the appropriate safety evaluation studies.

III. GENERAL PRINCIPLES

A. Principles

The proper evaluation of the safety of excipients, drugs, food ingredients, and other materials to which the public is exposed, involves rigorous adherence to sound, scientific principles in the design, execution, and interpretation of all studies, in vitro and in vivo, including animal and human. The following are principles to which one should adhere:

1. The test material is the exact chemical entity that will be used in the preparation of the final product. It is the material to which the consumer will be exposed; it is the item of commerce, and it meets established specifications.
2. The test material must be characterized as thoroughly as possible (including chemical, physical, and biological properties).
3. Experimental exposure to the test material must simulate human exposure conditions; for example, if it is to be taken orally, then animals should be exposed by gavage, as a dietary admixture, or in the drinking water.
4. Resources must be conserved by obtaining the maximum amount of useful information from the minimal numbers of subjects (animals or humans), but adequate for appropriate statistical analyses.
5. The selection of animal species must be scientifically defensible; for example, the biodisposition of the test material in the animal model and human should be identical or very similar.
6. Biological and statistical significance must be thoroughly substantiated; for example, by reference to standard texts, handbooks, or appropriate literature for biological significance and by using only currently accepted statistical methods.
7. The extent of testing is a function of the chemical and physical nature of the test material and conditions of exposure, including the extent of human exposure (who will be exposed), the frequency and duration of exposure because biological activity is a function of the dose. For example, if the biodisposition study indicates that the test material is not absorbed, fewer studies may be required.
8. Studies will be designed to show a dose–response relation. The effects produced must be compound-related and dose-dependent to be considered biologically significant.

9. Interspecies extrapolation should be cautionary and conservative and made by qualified scientists only (judgment and not merely the use of safety factors is encouraged).

The single most important component of a safety evaluation program is strict adherence to accepted sound scientific principles of experimentation. Deviation from these principles is totally unacceptable. Resources must be conserved, and this involves using the minimum number of animals consistent with accepted scientific procedures, current animal welfare regulations, and statistical requirements. Procedures are designed to obtain maximum information possible from each experimental subject, animal or human. All studies must satisfy GLP requirements.

B. Animal Models: Selection of Animal Species

Only appropriate validated animals models should be used. The selection of appropriate species is determined by several factors, including similarities in the biodisposition of the test material to that in humans; functional and/or morphological similarities; and applicable existing regulations. Whether one or both sexes and young or old animals will be used is dependent on the intended use of the final product and appropriate regulations. However, the use of young adult animals of both sexes is recommended. The number of animals used must be sufficient to permit proper statistical analysis and to satisfy relevant regulations.

C. Conditions of Exposure

1. Route of Exposure

The route of exposure must be the intended route for humans; that is, the experimental animal studies must simulate human exposure conditions. Appropriate, validated animal models for specific routes include, for example, the pig for dermal and oral studies, the rat for inhalation studies, and the guinea pig for skin or inhalation sensitization studies.

2. Dose Selection

In all studies involving chemical–biological interactions, such as pharmacological and toxicological ones, it is essential to establish dose–response relations. Usually only those responses that are dose-dependent are considered biologically significant. A no-observed adverse effect level (NOAEL), a subthreshold dose, must be identified for purposes of risk assessment and extrapolation to humans. The results observed at the NOAEL dose should be similar to those observed in the control group.

As pharmaceutical excipients are assumed to be biologically nonreactive, dose–response relations cannot always be established. An acceptable alternative is to use a maximum attainable or maximum feasible dose. This is the highest dose possible that will not compromise the nutritional or health status of the animal. Table 1 summarizes the maximum or limit doses for various types of studies by different routes of exposure. For example, 2000 mg/kg body weight of an orally administered test material is the maximum dose recommended for

Table 1 Limit Doses for Toxicological Studies

Nature of test	Species	Limit dose[a]	Ref.
Acute oral	Rodent	2000 mg/kg bw	6
Acute dermal	Rabbit Rat	2000 mg/kg bw	6
Acute inhalation[b]	Rat	5 mg/L air for 4 h or maximum attainable level under conditions of study	6
Dermal irritation	Rabbit	0.5 mL liquid 0.5 g solid	6
Eye irritation	Rabbit	0.1 mL liquid 100 mg solid	6
14-day/28-day oral repeated dosing; 90-day subchronic	Rodent, Non-rodent	1000 mg/kg bw/day	6
14-day/28-day repeated dermal; 90-day subchronic	Rat, rabbit	1000 mg/kg bw/day	6
Chronic toxicity, carcinogenicity	Rats, mice	5% maximum dietary concentration for nonnutrients	6
Reproduction	Rats	1000 mg/kg bw/day	6
Developmental toxicity (teratology)	Mice, rats, rabbits	1000 mg/kg bw/day	6

[a] mg/kg bw, milligrams of test material dosed per kilogram of body weight to the test species.
[b] Acute inhalation guidelines that indicate this limit dose are U.S. Environmental Protection Agency Toxic Substance Health Effect Test Guidelines, Oct, 1984; (PB82-232984) Acute Inhalation Toxicity Study; the OECD Guidelines for the Testing of Chemicals, Vol 2; Section 4; Health Effects, 403, Acute Inhalation Toxicity Study, May 12, 1982, and the Official Journal of the European Communities, L383A, Vol 35, Dec 29, 1992; Part B.2.

Table 2 Summary of Parameters Evaluated in Toxicology Studies

Type of study	Cage-side observations	Food and water intake	Body weight/gain	Gross pathology	Organ weight/ratio	Histopathology	Clinical chemistry/ hematology
Acute toxicity	Y	Y	Y	Y	N	N	N
Repeated-dose toxicity	Y	Y	Y	Y	Y	N[d]	Y
Subchronic toxicity	Y	Y	Y	Y	Y	Y	Y
Reproductive toxicity[b]	Y	Y	Y	Y	N[e]	N	N
Developmental toxicity[a]	Y	N	Y	Y	N	N	N
Chronic toxicity	Y	Y	Y	Y	Y	Y	Y
Carcinogenicity[c]	Y	Y	Y	Y	Y	Y	Y

Special parameters include the following:

[a] Evaluation of external malformations per treatment group and per litter, structural, skeletal and soft tissue anomalies, number of corpora lutea/litter, number of implantations per litter, percentage preimplantation loss, number of resorption sites, percentage postimplantation death, number of live fetuses per litter, sex distribution, fetal weight for males and females.

[b] Evaluation of the fate of all females (percentage mated, percentage pregnant, survival, litters totally resorbed or aborted, litters surviving through lactation); number of estrous cycles through dosing period; number of days needed for mating, natural delivery data (length of pregnancy), number of pups at birth, sex distribution at birth and day 7, live birth index, viability index, growth rate of offspring during lactation, body weight of male and female pups during lactation, gross malformations and internal malformations, implantation sites at autopsy for determination of postimplantation death.

[c] Tumor types and historical background incidence for each type of tumor that is elevated compared with the control; time to tumor analysis.

[d] Histopathology can be included, based on the anticipated toxicity to target organs and the need to choose doses for the subchronic study.

[e] Sex organ weight ratios only are included for organ weight determinations.

acute oral toxicity studies by the EEC and OECD (5,6). If this dose does not elicit an adverse effect, it is not necessary to administer higher doses. This is consistent with sound scientific principles and concern for the experimental animal. It also recognizes chemical and physical limitations of the test material.

D. Parameters to Be Evaluated

Table 2 summarizes the various types of parameters evaluated in different types of toxicology studies. Animals should be examined at least once daily, although twice daily observations, 5 h between observations, is recommended. Cage-side observations should be made by trained observers and properly recorded. These observations include changes in behavior; somatomotor activity; changes in skin and fur, eyes, mucous membranes, respiration, cardiovascular activity, autonomic and central nervous system activity (lethargy, sleep, coma, tremors, convulsions); secretory activity; changes in gastrointestinal activity (condition of mouth and perianal area, consistency of stool, salivation, diarrhea, or constipation); urinary volume; other signs suggestive of intoxication; signs preceding death; and time of death. Food and water consumption should be measured weekly and food efficiency ratios determined. Body weights are determined immediately before initiating the study, at least weekly thereafter, and at the time of death or sacrifice. All animals that die during the study or are killed due to moribundity should be subjected to a full and detailed gross necropsy at the time of sacrifice. Selected tissues from all animals are weighed and preserved for histopathological evaluation. These tissues include, but are not limited to the following: adrenal glands, brain, epididymides, heart, liver, kidneys, ovaries, spleen, and testes. Any abnormal tissue (gross lesion) is also preserved for histopathology. If the test material is applied dermally, the exposed skin and surrounding nonexposed skin are preserved for histopathological evaluation. If the test material is inhaled, the respiratory tract is carefully dissected and preserved for histopathological evaluation.

For repeated-dosing, subchronic, and chronic studies, hematology, clinical chemistry, and, often, urinalysis studies, are conducted (4–7). The frequency of these tests is dependent on the duration of the exposure. Hematology studies include hematocrit, and hemoglobin levels; erythrocyte, total and differential white blood cell, platelets, and reticulocytes counts; blood-clotting time, and potential. Clinical chemistries include electrolytes, glucose, cholesterol, urea, creatinine, protein (total and albumin) levels, and enzyme activity indicative of hepatocellular and renal effects.

IV. BASE SET TESTS

A. Tiered-Testing Strategy

A testing strategy has been developed for new pharmaceutical excipients that takes into consideration the physical–chemical nature of the product and the potential route(s) and duration of exposures, both through its intended use as part of a drug product and through workplace exposure during manufacturing (1–3). The number and types of studies recommended in this tiered approach are based on the duration and routes of potential human exposure. Thus, the longer the exposure to the new pharmaceutical excipient, the more studies are necessary to assure safety. Table 3 summarizes the entire set of toxicological studies recommended for new pharmaceutical excipients (1–3).

Tests have been grouped into tiers based on duration of human exposure to drug products using the U.S. FDA guidelines for new drugs (10). Exposures to drug products in this approach fall into three categories:

1. From a single dose to up to 2-weeks exposure in humans
2. Limited repeated exposure in humans from 2 weeks to 6 weeks
3. Extended exposure in humans for longer than 6 weeks

Tests have been outlined for each exposure category to assure safe use for the time period designated. The tests for each exposure category assure the safe use of the new pharmaceutical excipient for the time frame specified for the specific exposure category. Additional tests are required for longer exposure times. The types of studies in each exposure category and the rationale for each study follow.

B. Base Set Tests for All New Excipients

1. Strategy for a Tiered Approach

As shown in Table 3, a minimum data set is required for all new pharmaceutical excipients, regardless of route of intended use and duration of use (1). This "base set" approach develops data at an early stage in product development, which assures safety to workers exposed to the new pharmaceutical excipient during manufacture and also will assure safe exposure to humans in the final drug product for a single exposure to a maximum of 2 weeks of exposure. Table 4 indicates the specific tests and the purpose for each test required in the Base Set, or Appendix 1.

2. Importance of Acute Toxicity Data for Worker Exposure by All Routes

Regardless of the route of intended use of a new pharmaceutical excipient in pharmaceutical applications, a base set of acute tests is recommended to determine potential hazards through brief exposure via oral, dermal, inhalation, or ocular routes to protect workers exposed during manufacture (see Table 4). These tests include acute oral toxicity, acute dermal toxicity, acute inhalation toxicity, eye irritation, skin irritation, and skin sensitization. These studies are used for hazard identification for worker safety and Material Safety Data Sheets (MSDSs)

Table 3 Summary of Toxicological Studies Recommended for New Pharmaceutical Excipients Based on Route of Exposure

Tests	Oral	Mucosal	Transdermal	Dermal/ topical	Parenteral	Inhalation/in tranasal	Ocular
Appendix 1-base set							
Acute oral toxicity	R	R	R	R	R	R	R
Acute dermal toxicity	R	R	R	R	R	R	R
Acute inhalation toxicity	C	C	C	C	C	R	C
Eye irritation	R	R	R	R	R	R	R
Skin irritation	R	R	R	R	R	R	R
Skin sensitization	R	R	R	R	R	R	R
Acute parenteral toxicity	—	—	—	—	R	—	—
Application site evaluation	—	—	R	R	R	R	—
Pulmonary sensitization	—	—	—	—	—	R	—
Phototoxicity/photoallergy	—	—	R	R	—	—	R
Ames test	R	R	R	R	R	R	R
Micronucleus test	R	R	R	R	R	R	R
ADME-intended route	R	R	R	R	R	R	R
28-day toxicity (2 species)-intended route	R	R	R	R	R	R	R
Appendix 2							
90-day toxicity (most appropriate species)	R	R	R	R	R	R	R
Developmental toxicity (rat and rabbit)	R	R	R	R	R	R	R
Additional assays	C	C	C	C	C	C	C
Genotoxicity assays	R	R	R	R	R	R	R
Appendix 3							
Chronic toxicity (rodent, nonrodent)	C	C	C	C	C	C	C
Photocarcinogenicity	—	—	C	C	—	—	—
Carcinogenicity	C	C	C	C	C	C	—

R, required; C, conditionally required
Source: Ref. 1.

Table 4 Appendix 1, Base Set Studies for a Single Dose up to 2-Weeks Exposure in Humans

Test	Purpose
Acute oral toxicity	To determine the potential acute toxicity–lethality following a single oral dose
Acute dermal toxicity	To determine the potential acute toxicity–lethality following a single dermal exposure
Acute inhalation toxicity	To determine the potential acute toxicity–lethality following a single 4-h inhalation exposure to a test atmosphere containing the new pharmaceutical excipient (aerosol, vapor, or particles)
Eye irritation	To determine the potential to produce acute irritation or damage to the eye
Skin irritation	To determine the potential to produce acute irritation or damage to the skin
Skin sensitization	To determine the potential to induce skin sensitization reactions
Ames test	To evaluate potential mutagenic activity in a bacterial reverse mutation system with and without metabolic activation
Micronucleus test	To evaluate the clastogenic activity in mice using polychromatic erythrocytes
ADME—intended route	To determine the extent of absorption, distribution, metabolism, and excretion by the intended route of exposure following a single dose and repeated doses
28-day toxicity—intended route	To assess the repeated-dose toxicity in male and female animals of two species following dosing for 28 days by the intended route of exposure

in compliance with the Chemical Manufacturers Association Responsible Care Program, OSHA Hazard Communication Standard (11), European Community Standard (12), and Canadian WHMIS (13). Although these standards (11–13) do not require the development of new test data, it is highly recommended that these data be developed for hazard classification, product stewardship, and product liability purposes. Because regulations often undergo revisions, the most recent regulatory guidelines should be consulted for the study design of all relevant tests (4–6, 13–15).

Although low toxicity or lack of toxicity is expected for most pharmaceutical excipients, particularly by the intended route(s) of pharmaceutical application, these tests ensure that any acute hazard likely to be encountered at high doses in the workplace or through accidental exposure are known. Such exposures are often orders of magnitude greater than the exposure to the excipient in its pharmaceutical applications. For example, a dermal excipient may be completely nontoxic by the dermal route, but may produce some toxicity when inhaled, ingested or instilled into the eyes.

The objective of the acute toxicity study is to evaluate the potential to cause acute toxicity, including lethality, after a single dose by the chosen route of exposure (see Ref. 16 for details on the test protocols and various regulatory guidelines for each type of study). Groups of animals, usually rats, are dosed with several doses or a single "limit" dose by the selected route of exposure and observed for 14 days for lethality and clinical observations for biological activity, including toxicity. From the dose–response data, a dose is calculated that estimates the dose causing lethality in half the test animals. This dose is known as the LD_{50} or lethal dose 50. It is a standard term used worldwide to express the degree of acute toxicity of a chemical by a given route. Even though an excipient is expected to be nontoxic by a specific route, a study can be conducted at a maximal or limit dose to conserve animal usage. One can conclude that a test substance is practically nontoxic or nonlethal at or above the limit dose (see Table 1). The limit doses or maximum doses for acute toxicity studies and repeated dose studies (discussed in the following) are shown in Table 1. For acute oral toxicity, the limit dose is generally considered to be 2000 mg/kg orally (4–6,14). For acute dermal toxicity, the limit dose is generally considered to be 2000 mg/kg dermally (4–6,14). For acute inhalation toxicity, a limit dose of 5 mg/L air, if achievable, or a maximally attainable dose is selected that approximates the highest level of test material that can physically be generated in a test atmosphere based on the substance's physical and chemical properties (6). If no deaths occur and no toxicity is observed at the limit dose or maximum attainable dose, then the test material is considered nontoxic by inhalation following brief exposure.

Primary eye irritation in rabbits, primary skin irritation in rabbits, and skin sensitization in guinea pigs are conducted to determine the potential hazards through contact with eyes and skin. These studies provide information on the irritation and sensitization potential of excipients when applied directly into eyes or onto skin for workplace or accidental exposure to excipients (see Ref. 16). For excipients intended to be used in pharmaceutical applications by ocular or dermal routes, these studies provide key information early in the product development cycle. Excessive irritation or sensitization reactions in these tests for an ocular or dermal excipient usually results in discontinuance of further testing and development.

The skin sensitization study evaluates the potential of a substance to produce an immune-mediated–type response. There are several different skin sensitization protocols from which to choose (see Chapter 8 for details). The choice should be based on the regulatory guidelines in the region where the product will be sold.

Acute toxicity tests by the intended pharmaceutical route(s) of exposure to the new pharmaceutical excipient which are not routes of worker or likely accidental exposure (parenteral, vaginal, rectal) are also necessary to understand the immediate hazard(s) through excess exposure. These tests will be discussed in the respective chapters in this section. However, if new applications are found for an existing oral excipient by intranasal or parenteral routes, for example, the conduct of additional toxicity studies should be considered to assure safety by the new routes of exposure. Special studies required for given routes (i.e., phototoxicity, photoallergy, and pulmonary sensitization) will be discussed in the chapters related to these routes.

3. The Importance of Genotoxicity Studies in the Base Set Tests

The base set includes evaluation of the potential for a chemical to induce genetic mutations and chromosomal damage or "genotoxic potential." *Genotoxic substances* can be defined as agents that induce alterations in the nucleic acids and associated cellular components, resulting in modified hereditary characteristics or DNA inactivation (see Ref. 17 for details). Assessment of genotoxic potential is an important part of any base set screening for potential hazards of new chemicals. Because damage to the genome by a chemical is potentially irreversible and heritable, it is critical to include tests that adequately evaluate this endpoint. The results of a single genotoxicity assay may not be adequate enough to make a determination of "genotoxicity" or lack thereof. The weight of the evidence approach is taken after several tests have been conducted on various endpoints of genotoxicity (i.e., mutagenicity, DNA damage, chromosomal damage; see Chapter 11 for details).

Various regulatory agencies throughout the world have different requirements for specific genotoxicity test batteries. The specific tests required often vary with the type of product marketed and the potential for human exposure. It is important to determine the specific requirements in the countries where one plans to commercialize the new pharmaceutical excipient. Nevertheless, some standard tests acceptable in many countries are included in the base set (18,19). A combination of both in vitro and in vivo tests allows one to make assessments of both "worse-case" estimates (direct exposure to cells in vitro) and more relevant exposures to humans using exposure in animals. These are the Ames test of bacterial mutagenicity and the mouse micronucleus test.

The Ames test, is a relatively inexpensive and predictive test performed in vitro using *Salmonella typhimurium*. This test has been widely used and validated and is excellent as a first screen. The Ames test has a high degree of concordance and accuracy in predicting known carcinogens (20). The Ames test evaluates the mutagenic potential in bacteria, a prokaryote, in vitro both in the absence and in the presence of exogenous metabolic activation by rat liver microsomes (S9 fraction) to mimic mammalian metabolic pathways. Because many substances are activated to genotoxic compounds by metabolism in the body, inclusion of rat liver microsome enhances the sensitivity of this test to detect potential mammalian mutagens (17).

To gain additional information on the genotoxic potential in the whole animal the mouse micronucleus test provides a good model for assessing damage to chromosomes, also known as clastogenic activity (17). Micronuclei are formed from chromosomes left behind during cell division. Test materials that interfere with normal cell division or affect spindle fiber function or formation increase the number of micronuclei in cells. Polychromatic erythrocytes taken from bone marrow are evaluated for micronuclei: 1,000 of these cells per animal are scored. Known positive and negative controls materials are included.

Positive results in any one of the genotoxicity assays must be followed up with additional studies, based on sound professional judgment and current international protocols. The weight of the evidence for genotoxic potential is determined following review of all of the test results, both positive and negative, and the models evaluated (see Chapter 11).

4.   The Importance of Toxicokinetics in the Base Set

In evaluating the safety of a new pharmaceutical excipient, it is important to understand its disposition in the body or biodisposition. The term *toxicokinetics* is used to encompass all the processes that describe the movement of a chemical throughout the body. These include absorption, distribution, metabolism, and excretion (ADME) and the kinetics of these processes over time (pharmacokinetics; PK). Thus, ADME–PK studies after a single dose and after multiple doses by the pharmaceutical route of exposure are required to gain an adequate understanding of a chemical's half-life in the body, absorption, or lack thereof, metabolism and excretion. The importance of the ADME–PK study is discussed in detail in Chapter 11.

The ADME–PK study is included in the base set for several reasons. First, the goal in a safety assessment program is to provide data from which one may estimate risk to humans. The ideal species for evaluating potential repeated-dose (28-day), subchronic, and chronic toxicity in animals should resemble that in humans, as much as possible, relative to biodisposition (7). Enhanced toxicity in a given species for the chemical may be due to a unique metabolic pathway or to an unusual excretory or absorption pattern. Realistically, the species available for standard laboratory studies are limited to those most frequently used because there is a large database of historical background information (i.e., histopathology, clinical chemistry, hematology, and spontaneous lesions). Within this framework of selection, it is best to test standard species for ADME–PK studies prospectively to select the best species for further toxicity tests (7). If there is reason to suspect that a certain standard laboratory species more closely approximates humans for a given route of exposure (i.e., minipigs for oral and dermal routes; hairless guinea pigs for dermal; rats for inhalation), then ADME–PK studies should be compared in the species known to exhibit the appropriate responses for extrapolation to humans for at least one of the two test species.

Second, because the base set of tests requires the conduct of 28-day repeated-dose studies in two species by the appropriate route(s) of intended use (see later), the ADME–PK studies are required to help select the two most appropriate species for the 28-day studies (1–3). The results of the ADME–PK are also used to design the longer-term studies (see later discussion).

The single-dose and repeated-dose ADME–PK studies each provide different types of information on the toxicokinetics of the excipient. The toxicokinetics of a compound may vary with the length of exposure (and other factors, such as dose level, i.e., single-dose versus multidoses). For example, the induction of liver enzymes to metabolize certain chemicals occurs more frequently and the gut microflora show adaptive changes more readily following repeated exposure than following single doses. Therefore, the requirement for a single-dose ADME–PK study provides data on the relative kinetics of these parameters without adaptive responses, and the requirement for a repeated-dose ADME–PK study provides data to evaluate adaptive responses to the chemical on these same parameters (7). Both single-dose and repeated-dose ADME–PK studies are important to the overall scientific evaluation of risk to humans.

5. The Rationale for Two Species for the 28-day Study

Different species often respond differently to xenobiotics. In designing an appropriate testing strategy to adequately evaluate the toxicity of a new pharmaceutical excipient, a requirement is included to test for repeated (28-day) exposure in two different species by the route of intended exposure, based on the ADME–PK profile. Inclusion of two species ensures that any major differences in target-organ toxicity across species will be detected and that decisions on the most appropriate species for longer-term and special studies (see following Appendix 2 and 3) are based on sound scientific data. The precedent for the use of two species comes, in part, from the U.S. Environmental Protection Agency guidelines for pesticide registration, which require two species for subchronic and chronic studies (21).

The purpose of the 28-day toxicity study is to determine the potential adverse effects of a new pharmaceutical excipient when dosed repeatedly for 28-days by the route of intended exposure. It provides information on toxicity over a range of doses that can be used to select doses for the 90-day subchronic toxicity study. It provides preliminary information on potential target-organ effects, dose-response, clinical signs of toxicity, body weight, survival, and potential cumulative toxicity.

The toxicity testing guidelines are similar, but not identical, across various regulatory agencies around the world. Depending on the area of the world in which the excipient and final drug product will be sold, different guidelines may be chosen. Harmonization of testing guidelines has started, but a "harmonized guideline" does not exist for every type of study (see Chapters 4 and 14). The guidelines referenced here include U.S. Environmental Protection Agency (14); U.S. Food and Drug Administration (4); European Economic Community (5); Organization of Economic Co-Operation and Development (6); and the Society of Agricultural Chemical Industry of Japan (15). Occasionally, it is possible to include multiple guidelines in a single, customized protocol when conducting certain studies. Toxicologists should be consulted for the specific study designs and protocols to meet the requirements for the guideline(s) for various world regions.

B. Review of the Base Set Tests Before Conduct of Additional Tests

1. Critical Scientific Evaluation of All Data to Allow Human Exposure of a Single Dose up to 2 Weeks of Repeated Exposure

Following the completion of the base set, all data are critically evaluated and compared with levels of anticipated human exposure by the route of intended use. The NOAEL in the 28-day studies will determine the safe dose administered to humans as a single dose or for a limited duration (see Chapter 13, for details). The safe dose for humans receiving a single dose or for several doses for a short, limited duration, will be determined based on biological activity and NOAELs in the 28-day studies, the lack of genotoxicity, and the ADME–PK profile. Depending on the results of the ADME–PK study, a decision on whether only a single dose or several doses of a limited duration of the new pharmaceutical excipient may be permitted in humans. However, the studies in the base set are intended to allow human exposure for no more than 2 weeks.

The results of the acute toxicity tests and genotoxicity tests are also evaluated for potential hazards to workers during the manufacture of the new pharmaceutical excipient. Appropriate hazard warnings should be prepared based on the regulatory guidelines in the country or countries where the product will be manufactured and sold. Material Safety Data Sheets and labels can be prepared using the results of these tests (11–13).

2. Potential Environmental Impact of the Product

In addition to the mammalian toxicity tests discussed in the foregoing, consideration should be given to conducting acute environmental studies to determine the environmental effect of the new pharmaceutical excipient from a spill or accident during manufacture or transportation. Such studies usually include acute toxicity to freshwater and saltwater invertebrates and vertebrates and to algae. Various regulatory agencies have specific guidelines on the conduct of these tests and the choice of appropriate environmental species. Depending on the results of these tests, chronic fish or invertebrate tests may be recommended after review by a qualified environmental toxicologist.

V. LEVEL I AND II TESTS

A. Duration of Human Exposure Triggers Longer-Term Toxicity Studies

If exposure to the new pharmaceutical excipient is expected to occur for longer than 2 but no more than 6 weeks, additional toxicological studies are required, as shown in Table 5, Appendix 2. The longer the expected human exposure, the more extensive will be the toxicological studies to assure safety. A tiered approach assures that those tests necessary to ensure safety for the expected duration of human exposure are conducted. Thus, to assure safe use for greater than 2 weeks, but no more than 6 weeks in humans, subchronic toxicity and develop-

Table 5 Appendix 2 Studies for Limited Repeated Exposure of 2–6 Weeks in Humans

Test	Purpose
90-day toxicity study	To assess the subchronic toxicity in male and female animals of two species following dosing for 90-days by the intended route of exposure
Developmental toxicity study	To assess the potential to induce birth defects in the fetuses of animals exposed during pregnancy in rats or rabbits, or both
Additional genotoxicity or other tests	As deemed appropriate

Table 6   Appendix 3 Studies for Repeated Chronic Exposure in Humans

Test	Purpose
Chronic toxicity	To assess the toxicity following chronic (lifetime) exposure by the route of intended exposure
Oncogenicity	To assess the potential to induce tumors by the intended route of exposure
One-generation reproduction	To assess the potential reproductive and developmental toxicity in males and females by the intended route of exposure

mental toxicity studies are required, (see under Appendix 2 in Tables 5 and 3). To assure safe use for greater than six continuous weeks, chronic or oncogenicity studies are conditionally required (See under Appendix 3 in Tables 6 and 3). This means long term studies should be considered for prolonged human exposures, but may not be absolutely required. A thorough, scientific review of the data generated in the base set and Appendix 2 studies should be undertaken. From a critical evaluation by a competent toxicologist, the results of the physical-chemical properties of the test material, the 28-day, and 90-day tests, the ADME–PK acute and repeated-dose tests, and the developmental toxicity test(s), a final determination can be made on the value of chronic toxicity or oncogenicity studies.

For example, if no toxicity is observed at a limit dose of 1000 mg/kg body weight per day following the 90-day toxicity study, no genotoxicity was found, and the ADME–PK profile indicates that the material is not absorbed and is completed excreted unchanged in the feces, then it is likely that a chronic study is not necessary. The decision to conduct chronic studies should be determined on a case-by-case basis using scientific judgment.

B.   Rationale for Each Study Type in the Tiered Approach

1.   Rationale for Additional Studies in Appendix 2

Appendix 2 in Table 5 summarizes the studies required for new pharmaceutical excipients to which humans will be exposed for 2–6 weeks. Studies required in addition to the base set tests include both a 90-day subchronic toxicity study in the most appropriate species and teratology studies in rats or rabbits by the intended route of human exposure.

The purpose of the 90-day toxicity study is to develop an understanding of the potential hazards of the new pharmaceutical excipient following subchronic exposure. The 90-day study represents exposure during a significant part of the animals' life cycle and can more clearly identify long-term effects on physiological functions. The 90-day study is also used to determine dose levels for the chronic study if longer-term human exposure is anticipated (see later discussion). The parameters evaluated in the standard 90-day study are essentially the same as those in the definitive 28-day study, including effects on body weight, food consumption, survival, clinical observations, organ weights, clinical chemistry and hematology, and gross and histopathological evaluation.

Compared with the 28-day study, the 90-day study provides additional data on toxicological effects from repeated exposure over a longer segment of the animal's life cycle and on the dose–response for adverse effects found in the 28-day study.

Appendix 2 tests (see Table 5) require the conduct of a teratology study to evaluate the potential of the new pharmaceutical excipient to affect development, including induction of skeletal, soft tissue, or other birth defects and anomalies. Rats and rabbits are recommended for the developmental toxicity study requirement because these species are widely used for this type of study, and there is a large historical database of background anomalies that can be used in evaluating the incidence of findings in treated and control groups. The route of exposure should be the intended route of pharmaceutical use. Because completion of the level 2 studies could permit use in humans for up to 6 weeks, it is important to know the potential teratogenicity, because a woman could be pregnant during this time period and not be aware of the pregnancy.

Evaluation of the data from the base set tests may trigger the need to conduct additional in vivo or in vitro genotoxicity studies. The need for these tests should be decided on a case-by-case basis using sound scientific judgment and knowledge of the specific area of expertise (i.e., genotoxicity). In addition, positive findings in the base set should be carefully reviewed to determine if additional tests are necessary to better understand the potential hazard for more-extended human use. For example, an equivocal or positive finding of skin sensitization in the base set test may have critical implications for a dermal, transdermal, or mucosal excipient, requiring extensive follow-up, but may not require additional studies for parenteral, oral, or ocular excipients. If there is any doubt about the significance of the finding, consultation should be made with an appropriate scientific expert and an individualized testing program developed. Additional studies should be designed to understand the potential toxicity and mechanism of the effect to define the importance of the findings for humans.

2. Rationale for Additional Studies in Appendix 3

Appendix 3 (see Table 6) includes additional studies that should be considered in determining the potential hazards from prolonged exposure to a new pharmaceutical excipient. These include chronic toxicity studies in two species, preferably rodent and nonrodent; one-generation reproduction study; and carcinogenicity. For topical and transdermal excipients, consideration should be given to conducting a photocarcinogenicity study (see Chapter 7).

The purpose of the chronic toxicity studies in two species is to obtain definitive information on the effects of the new pharmaceutical excipient when administered throughout the lifetime of the test species. The length of the chronic study depends on the species: 12 months for dogs, 18 months for mice, and 24 months for rats in standard guidelines. Principles for dose selection in chronic studies have been widely debated, and they often vary for the type of product and regulatory agency (see Ref. 22 for full discussion). Data are developed in two species to obtain the information for extrapolation to humans. The study design of the chronic study is similar to that of the 90-day study. More animals are included, but the parameters measured are generally the same. In contrast, the oncogenicity study includes the addition of complete tumor analyses (incidence of tumors, time-to-tumor analyses, significance of increased tumor incidence, based on historical control incidence rates). It is often cost-effective to combine the chronic toxicity and oncogenicity study in rats. This is more difficult for nonrodents. Generally, the chronic dog study is used for regulatory purposes for drug- and pesticide-active ingredients (10,21).

The one-generation reproduction study assesses the potential effects of the new pharmaceutical excipient on reproduction, including measurement of fertility, mating behavior, development and maturation of gametes, and preimplantation–implantation loss of the embryos. The results of this study are important in assessing the risks to male and female reproductive performance. Evaluation of sperm morphology and motility is now being included in the standard protocols for testing new pesticides, chemicals, and drugs. Developmental milestones for offspring can be included to obtain information on postnatal development.

VI. CONCLUSIONS

The guidelines presented herein provide a framework to be used with professional judgment and a sound, scientific, and tiered approach to the proper safety evaluation of excipients. These guidelines provide flexibility based on the nature of the excipient and anticipated human exposure; conserve resources; simulate human exposure conditions; recognize appropriate regulations; and require the expertise and judgment of a professional toxicologist. The choice of an animal model is dependent on the conditions of human exposure. The parameters to be evaluated are a function of the nature of the excipient and the anticipated conditions of human exposure. The studies proposed should be viewed as the minimum number needed to provide data to establish safe exposure conditions for humans. From the data generated, additional studies might be identified and executed.

The proper safety evaluation of an excipient involves adherence to rigorous scientific principles, appropriate experimental design, and recognition of regulatory guidelines or requirements. Route-specific issues of dosing including site, vehicle, concentration or dose of test material, and duration of exposure must be considered. These are discussed in detail in the following chapters of this section. Human exposure conditions must be simulated in the safety evaluation program. Attention to these issues will result in a flexible program, for fixed safety evaluation requirements adequate for all excipients is not possible. The guidelines presented herein should be viewed as general principles for the proper conduct of the safety evaluation of excipients. The definitive protocols for the proper safety evaluation of an excipient should be developed on a case-by-case or compound-by-compound basis. This will provide the flexibility needed. The route of exposure must simulate anticipated human exposure. For example, if the excipient is to be used in a dermatological preparation, the following issues need to be considered: which animal model is the most appropriate (e.g., pig vs. rabbit); which anatomical site is most appropriate; should the material be applied repeatedly to the same site (which would simulate human exposure) or to different sites.

The doses to be used should be multiples of anticipated human exposure. The duration of exposure in animals is dependent on the extent of anticipated human exposure. The endpoints evaluated should include both local and site-specific (e.g., if dermal, irritation; if oral, vomiting or diarrhea) and systemic (e.g., CNS depression, organ toxicity) effects. Evaluation of endpoints may be qualitative (e.g., behavioral effects, clinical observations) or quantitative (e.g., clinical chemistry, hematology). Variability is usually less with quantitative assessments. The results of the safety evaluation studies may suggest additional testing (e.g., to determine the mechanism of action).

The evaluation of the safety of excipients must involve sound science, appropriate experimental design, adequate data, professional judgment, and a case-by-case (compound-by-compound) approach to permit maximum flexibility. The following chapters of this section explore the design and conduct of appropriate studies for each of the major routes of exposure.

REFERENCES

1. M Steinberg, JF Borzelleca, EK Enters, FK Kinoshita, A Loper, DB Mitchell, CB Tamulinas, ML Weiner. A new approach to the safety assessment of pharmaceutical excipients. Regul Toxicol Pharmacol 24:149–154, 1996.
2. ML Weiner. Proposed guidelines for safe use of new excipients. PharmTech Conf '92 Proc, 1992, pp. 252–259.
3. IPEC Europe Safety Committee. The proposed guidelines for the safety evaluation of new excipients. Eur Pharm Rev, Nov:13–20, 1997.

4. U.S. Food and Drug Administration, Bureau of Foods. Toxicological principles for the safety assessment of direct food additives and color additives used in food. In Redbook II (Draft). Rockville, MD: U.S. Food and Drug Administration, 1993.
5. EEC. Methods for Determination of Toxicity and Other Health Effects. Part B of Annex V of Directive 67/548/EEC, 1997.
6. Organization for Economic Co-Operation and Development (OECD). OECD Guidelines for the Testing of Chemicals. Paris: OECD Publications, 1981.
7. International Programme on Chemical Safety (IPCS) in cooperation with the Joint FAO/WHO Expert Committee on Food Additives (JECFA). Environmental Health Criteria 70. Principles for the Safety Assessment of Food Additives and Contaminants in Food. Geneva: World Health Organization, 1987, pp 51-53.
8. AW Hayes. Principles and Methods of Toxicology. 3rd ed. New York: Raven Press, 1994
9. The National Academy of Sciences. Toxicity Testing: Strategies to Determine Needs and Priorities. Washington, DC: National Academy Press, 1984.
10. Food and Drug Administration. Proposed Implementation of International Conference on Harmonization Consensus Regarding New Drug Applications; Proposed Implementation Document; Availability. Fed. Reg. 57 (73): 13105-13106, Docket No. 92N-0136, 1992.
11. Code of Federal Regulations, OSHA Hazard Communication Standard, 29CFR 1910.1200.
12. Official Journal of the European Communities, 91/155/EEC, March 5, 1991.
13. Canada Gazette Part II, Vol 122, No. 2, Hazardous Products Act, December 31,1987.
14. U.S. EPA. Health Effects Test Guidelines. Arlington, VA: Office of Prevention, Pesticides and Toxic Substances, June 1996.
15. Society of Agricultural Chemical Industry: Agricultural Chemicals Laws and Regulations. Japan (II) (English translation), 1985.
16. PK Chan, AW Hayes. Acute toxicity and eye irritancy. In: AW Hayes, ed. Principles and Methods of Toxicology. 3rd ed. New York: Raven Press, 1994, pp 579–597.
17. D Brusick. Genetic toxicology. In: AW Hayes, ed. Principles and Methods of Toxicology. 3rd ed. New York: Raven Press, 1994, pp 545–577.
18. D Kirkland. New OECD genotoxicity guidelines: a critical review in relation to other international guidelines. Mutat Res 379(1 suppl. 1):S192, 1997.
19. D Marzin. The position of the in vitro micronucleus test within the battery of screening for genotoxic potential determination and regulatory guidelines. Mutation Res 392:175–181, 1997.
20. J Ashby, RW Tennant. Chemical structure, *Salmonella* mutagenicity and extent of carcinogenicity as indicators of genotoxic carcinogenesis among 222 chemicals tested in rodents by the U.S. NCI/NTP. Mutat Res 204:17–115, 1988.
21. Code of Federal Regulations, Protection of Environment, 40 CFR 158.340.
22. JA Foran. Principles for the Selection of Doses in Chronic Rodent Bioassays. Washington, DC: ILSI Press, 1997.

6
Routes of Exposure: Oral

Lois A. Kotkoskie
FMC Corporation, Princeton, New Jersey

I. INTRODUCTION

Pharmaceutical excipients are used in oral dosage forms of drug products, such as tablets, capsules, powders, liquids, suspensions, elixirs, and syrups. Pharmaceutical excipients, by definition, do not possess pharmacological activity, but they can provide a variety of functional characteristics that facilitate the oral administration of a pharmacologically active compound in a drug product. Some examples of functional categories are diluents, lubricants, coatings, flavors, dissolution agents, suspending agents, stabilizers, plasticizers, and preservatives. The extensive functional properties of some pharmaceutical excipients have led to their use as food additives and, likewise, some food additives have been used as pharmaceutical excipients. Many food additives and generally recognized as safe (GRAS) substances have applications as pharmaceutical excipients. Examples of GRAS substances used as oral pharmaceutical excipients include glycerin, acacia, and sodium chloride.

The chemical structures of existing pharmaceutical excipients are quite diverse. Oral pharmaceutical excipients can be inorganic or organic compounds, proteins, sugars, carbohydrates, surfactants, or fatty acid derivatives. Colors and flavoring agents are also pharmaceutical excipients used by the oral route of exposure. Given the diversity in chemical structures and functionalities of existing pharmaceutical excipients, any procedure designed to assess the safety of a new pharmaceutical excipient by the oral route of exposure should be very flexible.

The Safety Committee of the International Pharmaceutical Excipients Council (IPEC) has recommended a set of proposed guidelines for the safety

123

Table 1 Summary of Toxicology Tests Recommended for New
Pharmaceutical Excipients Administered by the Oral Route of Exposure

---

Appendix 1 (base set): single or limited (<2 wks) exposure in humans

Acute oral toxicity	Skin sensitization
Acute dermal toxicity	Bacterial gene mutation
Acute inhalation toxicity[a]	Chromosomal damage
Eye irritation	ADME-intended route
Skin irritation	28-day oral toxicity study (2 species)

Appendix 2: limited and repeated exposure (2–6 wks) in humans

90-day toxicity (1species)	Genotoxicity assays
Teratology (rat and/or rabbit)	Additional assays[a]

Appendix 3: long-term exposure (>6 wks) in humans
Chronic toxicity (rodent, nonrodent)[a]
Carcinogenicity[a]
One-generation reproduction

---

[a] Conditionally required
Source: Ref. 1.

assessment of new pharmaceutical excipients (1). Table 1 summarizes the required and conditionally required toxicology tests for new pharmaceutical excipients administered by the oral route of exposure. General principles for application of these guidelines have been discussed previously (see Chapter 5). It is beyond the scope of this chapter to fully discuss the assessment of new pharmaceutical excipients that are already used in food or by another route of exposure. The reader is encouraged to consult a recent IPEC Europe Safety Committee publication (2) for more information on this topic. The purpose of this chapter is to review the safety evaluation and testing strategies for new pharmaceutical excipients administered by the oral route of exposure.

II. BACKGROUND INFORMATION

New pharmaceutical excipients are often developed for other uses in addition to the excipient use in drug products. For example, pharmaceutical excipients may be used as food additives, industrial chemicals, cosmetic ingredients, and inert ingredients in pesticide formulations. When designing a toxicology testing program for a new pharmaceutical excipient for use by the oral route of exposure, it is important to consider the following questions: Will the excipient be used solely in drug products, or will it have other uses, such as food or industrial uses?

In what country (or countries) will the excipient be sold? What are the long-term marketing plans for this excipient? Will products similar to this one be designed and sold? The answers to these questions can assist the scientist in the development of a long-term toxicology testing program for a new pharmaceutical excipient.

A. Chemical and Physical Properties

The first and primary source of background information is the chemical and physical property information on the new pharmaceutical excipient. This is a broad category of information on such topics as raw materials, solvents, manufacturing, impurities, specifications, and other physicochemical properties.

Review of the supplier Material Safety Data Sheet (MSDS) for raw materials used in the manufacture of the new excipient is a logical first step in this process. The supplier MSDS may contain important information on impurities, chemical reactivity, or toxicological effects. The chemical composition section of the MSDS is particularly important. In the European Union, any chemical that is present at concentrations of 0.1% or higher and is classified as a carcinogen, developmental or reproductive toxicant, mutagen, toxic, or very toxic chemical must be listed on the MSDS (3,4). Under the U.S. Occupational Safety and Health Administration (OSHA) Hazard Communication Standard, any chemical that is present at levels of 0.1% or higher and is a carcinogen must be listed on the MSDS (5). Any hazardous chemicals identified from this first step should be evaluated relative to the final product.

The manufacturing process should also be reviewed in consultation with a chemist and/or chemical engineer to identify potential impurities in the final product. The solvents used during production and any chemical intermediates produced should also be identified. If the excipient has been produced in a pilot plant, then the manufacturing process should be reevaluated when production begins in the full-scale manufacturing plant.

The purity and composition of a new pharmaceutical excipient produced by a given manufacturing process should be determined, in addition to other relevant analyses such as pH, vapor pressure, particle size, specific gravity, stability, storage conditions, reactivity, and solubility in various solvents. The limits of impurities are then determined based on a review of all information and additional analyses, if necessary. Limits of impurities can also be obtained from the specifications of structurally similar compounds. The levels of residual solvents in a new pharmaceutical excipient can be evaluated relative to the International Conference on Harmonization (ICH) Guideline for Residual Solvents (6) or other standards.

At the conclusion of these analyses, it should be determined if any carcinogens, mutagens, developmental or reproductive toxicants, or toxic or very toxic chemicals are present in the final product at levels greater than or equal to 0.1%. If these types of chemicals are present in the final product, then the manufacturing process or raw materials should be reexamined to decrease the levels of hazardous chemicals in the new pharmaceutical excipient. It may be necessary to change the manufacturing process to improve the quality of the final product.

If the decision is made to test the new pharmaceutical excipient, then the product tested should be representative of the commercial product sold. An adequate amount of material from the same production lot should be made available for toxicology testing.

B. Review of Scientific Literature

Additional background information can be obtained by conducting a literature search on the new pharmaceutical excipient. Relevant scientific articles are reviewed, such as those on acute toxicity, absorption and metabolism, oral repeated-dose toxicity studies, genetic toxicity tests, and human clinical studies and case reports. A review of the literature for a new pharmaceutical excipient can be used to predict potential target organs of toxicity, nutritional concerns, interactions with active ingredients, or pharmacological effects. The literature review can also provide information on the potency of the new pharmaceutical excipient in animals or humans.

In some instances, a new pharmaceutical excipient for the oral route of exposure may have an extensive human history of food use. Appropriate scientific judgment should be used to determine if historical use in food can either fulfill the guidelines or preclude their application (1,2).

If no data are available for the new pharmaceutical excipient, then a literature search can be conducted on a structurally similar chemical to obtain the listed information. Based on the chemical similarity between the two materials and the amount and quality of the literature reviewed, it may be possible to "bridge" the toxicology information from one structurally similar compound to another. In this manner, fewer animals are used for toxicology testing of a new pharmaceutical excipient.

The information gathered from a review of the literature can be used to prepare an MSDS for the new pharmaceutical excipient and to conduct a safety assessment of the excipient based on the available data. The information obtained in the literature review is also important in the design of toxicology studies and in dose selection.

C. Exposure Assessment

The potential human exposure to a new pharmaceutical excipient used for the oral route of exposure is determined from several factors. The frequency and duration of use are two very important factors that determine the extent of toxicology testing needed for a new pharmaceutical excipient. The categories of human exposure listed in Table 1 are (a) single or limited exposure (less than 2 weeks), (b) limited and repeated exposure (2–6 weeks), and (c) long-term exposure (exposures longer than 6 weeks).

The user population can also be a factor in the determination of the extent and type of testing needed. For example, if an excipient will be used in children's drug products, then additional testing at higher dosages may be necessary because children may have a higher pharmaceutical excipient intake than adults, on a per kilogram body weight basis. The issue of children versus adult intake has previously been reviewed for food additives or contaminants (7).

The intended use level of the excipient in one or more oral dosage forms (such as tablet, capsule, liquid, or other) can be calculated as an average daily dose (see Chapter 12). The anticipated intake from other uses (e.g., food additive) should be added to the pharmaceutical excipient intake. This rough estimate of human intake can be used to set dose levels for repeated-dose toxicology studies that will be conducted on the new pharmaceutical excipient.

D. Assessment of Background Information

A critical review of the background information is necessary before proceeding with a toxicology-testing program for a new pharmaceutical excipient. The information should be thoroughly evaluated and additional information should be obtained, if necessary, before safety assessment of the new pharmaceutical excipient. If the existing data do not support the safe use of the excipient in an oral dosage form, then additional toxicology testing should be conducted (1). Relevant background information can be used in the design of studies, as discussed in the following section.

III. FACTORS TO CONSIDER IN STUDY DESIGN

Many new pharmaceutical excipients developed for use by the oral route of exposure may also have potential uses in food and industrial applications. The most appropriate testing guideline should be selected based on a thorough review of the background information and knowledge of potential additional applications of the excipient with time. For example, a new pharmaceutical excipient may be developed solely for use in oral drug formulations and, in this case, the ICH guidelines may be appropriate. If the new pharmaceutical excipient is being developed for both food and pharmaceutical applications in the United States, then it may be appropriate to use the Food and Drug Administration (FDA) Redbook guidelines (8,9); if it is being developed for these uses in multiple countries, then it may be appropriate to use OECD guidelines (10), which are accepted by most OECD member countries. Flavoring substances sold in the United States can be evaluated by the Flavor and Extract Manufacturers Association of the United States (FEMA) through the FEMA GRAS program (11) or by other means such as the Joint FAO/WHO Expert Committee on Food Additives (12).

A. Method of Oral Administration

The method of administration of a pharmaceutical excipient is dependent on several factors, the most basic of which is the type of toxicology study (in vitro or in vivo). For in vitro studies, the test material is usually diluted in an appropriate solvent and the dilutions are used in the test. The specific test guideline to be used for a particular in vitro test should be consulted for a list of approved solvents. The remainder of this section will be devoted to the selection of the method of oral administration of a pharmaceutical excipient for in vivo toxicology studies.

The most common methods of oral administration of a pharmaceutical excipient are (a) oral intubation (gavage), either undiluted or in a vehicle; or (b) incorporation of the test material into the animals' diet. Less frequent methods of oral administration include addition of the test material to drinking water, or administration by capsule. Administration by oral intubation or by capsule will result in accurate delivery of a specific dosage of test material to animals. Incorporation of a test material either in drinking water or diet is a less accurate way of administering a specific dose of test substance to animals owing to individual variability in the animals' body weights and food consumption values (13).

During the design of a toxicology-testing program for a new pharmaceutical excipient, it may be decided that one method of administration may be preferred over another, based on the physical properties of the pharmaceutical excipient and the human exposure pattern. After the testing program has begun, it may be necessary to change the method of administration for future studies, based on pharmacokinetic results, analytical considerations, or toxicological effects noted in the base set toxicology studies.

1. Human Exposure Pattern

The philosophy for determining the method of administration of a new pharmaceutical excipient to animals is to mimic the human exposure pattern as much as possible. It can be difficult to follow this philosophy when the human exposure pattern to oral pharmaceutical excipients can include multiple drug products such as tablets, capsules, gelcaps, liquids, and solid powders. Frequently, a new excipient is developed for use in only one drug product, such as a tablet, and then, as time passes, the excipient use is expanded to numerous drug products for both over-the-counter (OTC) and pharmaceutical drug formulations. Therefore, the method of administration that is chosen should not limit the future uses of the excipient in other oral drug products.

If the proposed use of a new pharmaceutical excipient is in tablet, capsule, or liquid drug products, then oral gavage administration of the excipient can be used to simulate the human exposure pattern in animal studies. Whatever method of administration is chosen, it must allow administration of an excipient at dosages that are orders of magnitude greater than human exposure and must also allow analytical verification of concentration, homogeneity, and stability of the excipient in the dosing solution or diet.

2. Physical State of the Test Material

A new pharmaceutical excipient developed for the oral route of exposure may be a solid, powder, wax, liquid, or semiliquid when manufactured and sold. However, there are many instances when the physical state of the excipient changes once it is incorporated into a drug product; consequently, humans are exposed to a different physical form of the excipient. For example, an excipient may be manufactured as a solid, but it may be used in drug products as a liquid or gel. Coatings may be sold as liquids which undergo polymerization to a solid (film) with the addition of a plasticizer during the drug formulation process.

Therefore, the question arises: Which physical state of the excipient should be used as the test material in toxicology studies—the manufactured form or the final dosage form in a drug product? It is generally advisable to test the excipient in the physical state in which it is manufactured, unless the physical properties of the test material preclude this approach. At the initiation of a toxicology-testing program for a new pharmaceutical excipient, it is not always possible to predict the ways an excipient will be used in drug products, and testing a new excipient as a final dosage form can limit the extent to which the data can be extrapolated to evaluate the safety in other drug products. In addition, if another chemical has been added to the excipient to simulate a final dosage form for a toxicology-testing program, (i.e., if a plasticizer is added to a coating to produce a film), then the results are no longer valid if the chemical (in this example, the plasticizer) is no longer used, either because of toxicity or regulatory concerns. For example, this concept was put into practice for the pharmaceutical excipient Aquacoat ECD ethylcellulose aqueous dispersion, a liquid that is commonly used in conjunction with plasticizers and annealing agents to form continuous, strong, and flexible film coatings for tablets, granules, and beads. The liquid form of the material (as manufactured) was used in a toxicology testing program (14) to maximize the usefulness of the results of the studies.

Table 2 Physicochemical Parameters of a
Pharmaceutical Excipient to Be Considered in
Determining the Method of Oral Administration

Vapor pressure
Particle size
Hygroscopic
Stability and storage conditions
Stability in vehicle (i.e., water or corn oil)
Interactions between test substance and vehicle

3. Physical Properties of the Test Material

The physicochemical properties of the test material should be considered when determining the method of administration and dose selection for toxicology studies (Table 2). Test materials that are volatile or hygroscopic should not be administered in diet unless it can be shown that the concentration and/or stability of the test material in the diet remains unaffected. Pharmaceutical excipients with very small particle sizes may not be homogeneous when admixed with diet. The stability and storage conditions of the test material should also be taken into consideration. For example, if the test material must be stored under refrigeration, then it would not be advisable to place the test material in the diet or drinking water at room temperature in an animal room for a long period. Some pharmaceutical excipients, such as the antioxidants butylated hydroxyanisole (BHA) and butylated hydroxytoluene (BHT), will lose their activity when exposed to light (15). The solubility of the test material in the vehicle (water, corn oil, or other) may dictate the maximum dose that can be administered to the animals. There can also be interactions between the test material and vehicle (water, corn oil, diet), that may need to be considered. This topic will be further discussed in Sec. III.D.

B. Choice of Animal Species

The decision to use a particular animal species in a specific toxicology study is based on several factors. For the acute toxicity studies outlined in Appendix 1 (base set) of the tests recommended for new pharmaceutical excipients (see Table 1), certain species are commonly used. The choice of species for repeated-dose toxicity studies requires evaluation of additional factors (Table 3). The results of the adsorption, distribution, metabolism, and excretion (ADME) studies can greatly influence the choice of species for the remainder of the toxicology-testing program for a new pharmaceutical excipient. Ideally, the species chosen for

Table 3 Selection Criteria for Animal Species and Strain in
Repeated-Dose Toxicity Studies

Requirements by regulatory agencies
Metabolism of test material in a manner similar to humans
Availability of historical control data
Most sensitive species and strain
Responsiveness of particular organs and tissues to toxic chemicals
Availability of species and strain
Capability and experience of laboratory in use of the species

Source: Ref. 13.

repeated-dose toxicity studies should metabolize the test substance in a manner
similar to humans (13).

1. Rodent

The most common rodent species used in toxicology studies are the rat and
mouse; less commonly used rodent species include the guinea pig and hamster.
Rodents can be used in the following required studies listed in Table 1: acute oral
toxicity, acute dermal toxicity, skin sensitization, ADME, 28-day oral toxicity,
subchronic (90-day) toxicity, teratology, in vivo genotoxicity, and one-generation
reproduction. Rodents can also be the species of choice for the following conditionally required studies listed in Table 1: acute inhalation toxicity, chronic toxicity, and carcinogenicity.

2. Nonrodent

The most common nonrodent species used in toxicology studies are the dog and
rabbit; primates and miniature swine are less commonly used. The rabbit is typically used in the following required studies listed in Table 1: eye irritation, skin
irritation, and teratology. The dog can be used in the following studies listed in
Table 1: 28-day oral toxicity study (required), and chronic toxicity study (conditionally required).

C. Dose Selection

The selection of doses for toxicology studies is influenced by two factors: study
objective and practical (biological or chemical) considerations (16). For the
repeated-dose toxicology studies recommended for testing of new pharmaceutical
excipients by the oral route of exposure, the purpose of the studies is to identify
the health hazard of a test substance and the dose–response curve for a particular

adverse effect. The dose range tested in a study should include a dose at which no adverse effects occur; this dose is termed the no-observed adverse effect level (NOAEL). The NOAELs obtained from a number of studies conducted with the same test substance may differ based on the type of study and species tested. Therefore, the number of dose groups and dose selection should be decided in a case-by-case basis for each type of study and species in a toxicology-testing program for a new pharmaceutical excipient. The results of the ADME and range-finding studies can be useful in the selection of the doses for a repeated-dose toxicology study.

1. Number of Dose Groups

The number of dose groups tested in a repeated-dose toxicology study is typically three (low, middle, and high). A separate group of animals acts as a concurrent control group and receives the vehicle only. Alternatively, if the test substance is expected to have low toxicity, only one dose group may be tested at the limit dose in addition to a concurrent control group.

If inadequate data are available to choose doses for a repeated-dose toxicity study, then a short repeated-dose range-finding study may be conducted using more than three dose groups to determine the dose–response curve for mortality and other serious toxicological effects. The results of the range-finding study can then be used to set doses for repeated-dose toxicology studies such as the 28-day toxicity study or teratology studies.

2. Dose Selection

For toxicology studies conducted with three dose groups, the highest dose selected should be sufficiently high to produce a toxicological response in the test animals and should not cause more than 10% mortality (9). The high dose should not exceed the limit dose for a particular method of administration. For chronic toxicity or carcinogenicity studies, the high dose should be the maximum-tolerated dose (MTD); note that the definition of MTD varies between international regulatory agencies (16). The intermediate dose should be sufficiently high to induce minimal toxic effects. The low dose should not induce measurable toxic responses in the test animals. Adequate dose selection should permit clear determination of a NOAEL in repeated-dose toxicology studies.

For toxicology studies conducted at the limit dose, the limit dose for the oral route of exposure differs between toxicology-testing guidelines. In the current OECD guidelines the limit dose is 1000 mg/kg per day (or the equivalent in the diet or drinking water) for 28-day and 90-day studies (10); the limit dose is 5% of the test substance in the diet for all studies in the current FDA guidelines and the OECD chronic toxicity or carcinogenicity study guideline (8–10). There is no published limit dose for the administration of a test substance in drinking water, although on a practical basis, the concentration of test substance should neither interfere with normal water intake (owing to palatability or osmotic problems) nor should it produce an electrolyte imbalance in animals. It is not appropriate to use a limit dose if the potential human exposure is expected to be high and there is an inadequate margin of exposure between the limit dose and the potential human exposure.

Dose selection should also be based on a number of other practical considerations, such as nutritional effects, physicochemical properties, palatability, and potential human exposure (16). The doses tested in a repeated-dose toxicity study must be much greater than potential human exposure.

D. Analytical Considerations

Analytical concerns are of paramount importance in toxicology study design. If the analytical aspect of the study is not well planned, it can result in administering less than the expected dose to the animals owing to test substance instability, degradation, or homogeneity problems (13). When analytical problems occur once the toxicology studies are underway, it can be difficult to interpret the data from the studies and, if serious enough, the analytical problems can invalidate the study results. One way to prevent this problem is to develop an analytical plan with a chemist for both the test material and the test material in vehicle. The applicable Good Laboratory Practice regulations should be followed if the study is to be submitted for regulatory submission.

1. Analysis of Test Material

Chemical characterization of the test material before the initiation of toxicology studies involves a review of the manufacturing process to identify and characterize impurities and residual solvents in the final product. The sample of test material used for toxicology-testing must be representative of the commercial product and must meet the relevant compendial standards for that particular excipient. A sufficient quantity of one production lot of test material should be used for the entire toxicology testing program. If additional lots of test material are used, then they should be analyzed in the same manner.

Further chemical analysis is necessary once the sample of test material is obtained. The purity or composition of the sample of test material should be determined in a stability study whereby the sample is stored at a specified temperature (e.g., room temperature, frozen, or other) and purity–composition analysis is conducted at regular intervals. It is necessary to show that the sample of test material is stable under specific storage conditions for the length of the toxicology-testing program. The stability study can be conducted concurrently with the toxicology-testing program.

2. Analysis of Test Material in Vehicle

It is necessary to develop an analytical method or to refine an existing method with adequate precision and accuracy to measure the concentration of test substance in a vehicle (i.e., diet, corn oil, water) for homogeneity, stability, and concentration studies. The limit of detection of the analytical method may preclude the use of certain vehicles because low concentrations of some test materials may not be detected in a particular vehicle.

The homogeneity and stability of the test material in vehicle should be determined before the start of a toxicology study. The homogeneity is determined by preparing a diet or dosing solution mixture from which samples are analyzed in the top, middle, and bottom of the mixture at the highest and lowest concentrations to be tested in a toxicology study. The middle sample is usually retained for stability analysis; this sample is reanalyzed at regular intervals for the length of time needed to demonstrate stability. The results of the homogeneity and stability studies can be used to determine if a particular vehicle is appropriate, or if the mixing procedure will homogeneously distribute the test material in the mixture. The results of the stability study can also indicate if the mixture of test material and vehicle forms degradation products over time. If one vehicle is not stable or homogeneous when mixed with the test material, then another vehicle or another mixture procedure can be tested before the toxicology studies begin.

The concentration of test material in the vehicle for all concentrations tested should be verified on a regular basis during the course of a toxicology study. The number of times that analytical verification is repeated during the study is at the discretion of the study sponsor.

IV. DETERMINATION OF NO-OBSERVED ADVERSE EFFECT LEVEL

Once the toxicology study has been completed, it is necessary for the scientist to evaluate all of the data and determine the dosage at which no-adverse effects occur (see Chapter 11). However, for new pharmaceutical excipients or food additives administered by the oral route of exposure, nutritional or physiological effects can occur during repeated-dose toxicity studies. High levels of test material in the animals' diet can disrupt normal homeostatic processes and produce functional or morphological changes that may or may not be toxicologically significant. Careful evaluation of the data are required to determine if these effects are to be considered when determining the NOAEL for a particular study. The conduct of additional studies may be necessary to ascertain if an effect is toxicological, nutritional, or physiological.

A. Functional Changes

Decreases in body weight are one of the most sensitive indicators of an animal's health (13). If the test material is incorporated into the food or water, then it may affect palatability. This would produce a decrease in the animal's food or water intake and produce a corresponding decrease in body weight. Alternatively, in the absence of a change in food consumption, decreased body weight may be due to an indirect nutritional effect (i.e., dilution of the number of calories per gram of diet) or a direct nutritional effect (i.e., the test material directly affects the absorption, metabolism, or excretion of nutrients). If a change in body weight is noted in the absence of other toxicological effects and is not due to decreased food intake, then the effect may be nutritional. Special statistical analysis of existing data or a paired feeding study may be conducted to determine if a decrease in body weight occurred from reduced caloric intake.

Laxative effects commonly occur in toxicology studies conducted with food additives (17). Therefore, it is likely that this effect may also occur in studies conducted with new pharmaceutical excipients administered by the oral route of exposure. Diarrhea occurs owing to osmotic or physiological changes from the presence of a test substance within the gastrointestinal tract. When laxative effects occur in animals without other adverse effects and at levels that greatly exceed anticipated human exposure, then they usually are not of toxicological concern. For example, laxative effects have occurred in rodent-feeding studies conducted with sodium carboxymethylcellulose; however, a threshold for this effect in humans was identified (18).

B. Morphologic Changes

Cecal enlargement is a common finding in rodents fed high levels of noncaloric substances such as dietary fibers and other poorly digestible carbohydrates such as lactose, mannitol, and chemically modified starches (19). The cecal enlargement may occur alone or in conjunction with other effects, such as diarrhea, increased kidney weight, and pelvic nephrocalcinosis (17,19). The mechanism for this effect is poorly understood, but may involve osmotic changes or increased calcium absorption (17,19).

Increased liver weight may be a physiological response if it is due to microsomal enzyme induction following oral administration of substances that are metabolized extensively by the liver (17). Further studies may be necessary to determine if the microsomal enzyme induction caused by the test substance would affect metabolism of an active pharmacological ingredient in a drug product.

V. OPTIONAL ADDITIONAL TESTS OR ENDPOINTS

The decision whether to incorporate additional toxicological endpoints into a standard toxicology protocol or to conduct additional tests should be decided on a case-by-case basis. In general, this decision is based on review of the background information or, once the testing program is underway, it is based on the results of current toxicology studies. It is not always possible to predict the target organs of toxicity from review of the background information; hence, the toxicology-testing program for a new pharmaceutical excipient should be flexible enough to allow conducting additional endpoints or tests as necessary. Additional studies may also be necessary to assess whether a particular effect is nutritional or toxicological. The information discussed in this section is not meant to be a comprehensive list and will presumably change with new developments in this area of science.

A. Additional Toxicological Endpoints

The use of additional endpoints in a standard toxicology protocol can add value to a study by minimizing the cost, time, and number of animals needed to investigate the potential toxicity of a new pharmaceutical excipient. It is important that adequate procedures and statistical analyses are available to analyze the additional data collected. The extra work done by the technical staff in order to obtain the additional data should not jeopardize the collection of routine data for the toxicology study.

Additional endpoints can be incorporated into many of the toxicology tests recommended for new pharmaceutical excipients administered by the oral route of exposure. For example, a one-generation rat reproduction study can be conducted in combination with a 90-day rat toxicity study. Behavioral toxicology could be another endpoint evaluated in a one-generation rat reproduction study. Additional endpoints for the in-life phase of 28-day, 90-day, or chronic toxicity studies in rodents include: urinalysis, neurotoxicity measurement, or additional blood collections for measurement of hematology or clinical chemistry parameters. At study termination, additional endpoints may include clinical chemistry (beyond the routine measurements), urinalysis, genotoxicity, neurotoxicity, and immunotoxicity. Several references are available that discuss the choice of specific endpoints (13,20).

Additional animals can be added to one or more dose groups if necessary for interim sacrifice. Alternatively, recovery animals can be added to show if toxicological effects are delayed or irreversible.

B. Additional Toxicology Tests

Additional toxicology tests to evaluate the safety of a new pharmaceutical excipient administered by the oral route of exposure are currently listed in Appendix

2 of the toxicology testing program (see Table 1), although this program is flexible enough to allow the conduct of additional tests at any stage in the program. For example, if the results of the Ames test in Appendix 1 indicate that the new pharmaceutical excipient is mutagenic, then before conducting other mammalian toxicology tests, it may be appropriate to conduct the genotoxicity tests in Appendix 2 as well as others to determine the potential genotoxicity of the new compound. Additional genotoxicity tests should be decided on a case-by-case basis and may include the conduct of in vivo and in vitro tests such as unscheduled DNA synthesis, sister chromatid exchange, and the dominant lethal study. Assessment of reproductive function can be further defined in a segment 3 (perinatal/postnatal) study, fertility assessment by continuous breeding (FACB), hormone assays, or endocrine disruptor screening batteries. The purpose of conducting these additional assays is twofold: (a) to verify the existence of a particular toxicological effect, and (b) to identify a dose–response curve and NOAEL for a toxicological effect.

C. Other Studies

Other types of studies may need to be conducted either during or following performance of the toxicology-testing program. The performance of additional studies, such as pair-feeding studies, may be helpful in determining whether an effect is nutritional or toxicological. In addition to animal studies, human clinical studies may be conducted to study potential nutritional effects of a new pharmaceutical excipient.

The interaction of food additives and the bacterial flora of the gastrointestinal tract may influence the results of toxicology tests (17). Presumably, this effect may also occur for new pharmaceutical excipients administered by the oral route of exposure. Many food additives can be metabolized to short-chain fatty acids (acetate, propionate, or butyrate) by bacteria in the large intestine (21). Several methods are available to determine the pH, residual nonstarch polysaccharides, and short-chain fatty acid production from human fecal fermentation of dietary fibers (22,23). The results of these studies can provide information on the energy content of dietary fiber. Other experimental designs are available to study the effect of a new pharmaceutical excipient on the gut microflora (17).

VI. CONCLUSION

The information obtained from the steps discussed in this chapter—(a) assessment of background information, (b) conduct of a toxicology-testing program, and (c) determination of the NOAEL—should be sufficient to assess the safety of a new pharmaceutical excipient by the oral route of exposure in humans.

REFERENCES

1. M Steinberg, JF Borzelleca, EK Enters, FK Kinoshita, A Loper, DB Mitchell, CB Tamulinas, ML Weiner. A new approach to the safety assessment of pharmaceutical excipients. Regul Toxicol Pharmacol 24:149–154, 1996.
2. IPEC Europe Safety Committee. The proposed guidelines for the safety evaluation of new excipients. Eur. Pharm Rev Nov:13–20, 1997.
3. Official Journal of the European Communities, 91/155/EEC, March 5, 1991
4. Official Journal of the European Communities, 93/21/EEC, May 4, 1993.
5. Code of Federal Regulations, OSHA Hazard Communication Standard, 29CFR 1910.1200.
6. JC Connelly, R Hasegawa, JV McArdle, ML Tucker. ICH guideline. Residual solvents. Pharmeuropa 9(suppl): S1-S68, 1997.
7. CA Lawrie. Different dietary patterns in relation to age and the consequences for intake of food chemicals. Food Addit Contam 15(suppl):75–81, 1998.
8. U.S. Food and Drug Administration (FDA) Bureau of Foods. Toxicological Principles for the Safety Assessment of Direct Food Additives and Color Additives Used in Food. U.S. Dept. Commerce, NTIS PB83-170696, 1982.
9. FDA. Toxicological Principles for the Safety Assessment of Direct Food Additives and Color Additives Used in Food. Redbook II (draft). Rockville, MD: FDA, 1993.
10. Organization for Economic Co-Operation and Development (OECD). OECD Guidelines for the Testing of Chemicals. Paris: OECD Publications, 1993.
11. JB Hallagan, RL Hall. FEMA GRAS—a GRAS assessment program for flavor ingredients. Regul Toxicol Pharmacol 21:422–430, 1995.
12. I Munro. A procedure for the safety evaluation of flavoring substances. WHO Food Addit Ser 35:423–465, 1996.
13. NH Wilson, JR Hayes. Short-term repeated dosing and subchronic toxicity studies. In: AW Hayes, ed. Principles and Methods of Toxicology. 3rd ed. New York: Raven Press, 1994, pp 649–672.
14. LA Kotkoskie, C Freeman. Subchronic oral toxicity study of Aquacoat ECD ethylcellulose aqueous dispersion in the rat. Food Chem Toxicol 36:705–709, 1998.
15. MJ Groves. Butylated hydroxyanisole and butylated hydroxytoluene. In: A Wade, PJ Weller, ed. Handbook of Pharmaceutical Excipients. 2nd ed. Washington, DC: American Pharmaceutical Association, 1994, pp 45–48.
16. JA Foran, ed. ILSI Risk Science Institute (ILSI). Principles for the Selection of Doses in Chronic Rodent Bioassays. Washington DC:ILSI Press, 1997, pp 9–55, 69–78.
17. International Programme on Chemical Safety. Environmental Health Criteria 70. Principles for the Safety Assessment of Food Additives and Contaminants in Food. Geneva: World Health Organization, 1987.
18. JECFA. Toxicological evaluation of certain food additives and contaminants. WHO Food Addit Ser 26:81–123, 1990.
19. FJ Roe. Relevance for man of the effects of lactose, polyols and other carbohydrates on calcium metabolism seen in rats: a review. Hum Toxicol 8:87–98, 1989.
20. RL Suber. Clinical pathology methods for toxicology studies. In: AW Hayes, ed.

Principles and Methods of Toxicology. 3rd ed. New York: Raven Press, 1994, pp 729–766.

21. C Remesy, C Demigne, C Morand. Metabolism and utilisation of short chain fatty acids produced by colonic fermentation. In: TF Schweizer, CA Edwards, eds. Dietary Fibre—A Component of Food. Nutritional Function in Health and Disease. London: Springer–Verlag, 1992, pp 137–150.

22. JL Barry, C Hoebler, GT Macfarlane, S Macfarlane, JC Mathers, KA Reed, PB Mortensen, I Nordgaard, IR Rowland, CJ Rumney. Estimation of the fermentability of dietary fibre in vitro: a European interlaboratory study. Br J Nutr 74:303–322, 1995

23. CA Edwards, IR Rowland. Bacterial fermentation in the colon and its measurement. In: TF Schweizer, CA Edwards, eds. Dietary Fibre—A Component of Food. Nutritional Function in Health and Disease. London: Springer–Verlag, 1992, pp 119–136.

7
Routes of Exposure: Topical and Transdermal

Matthew J. Cukierski*
ALZA Corporation, Mountain View, California

Alice E. Loper
Merck Research Laboratories, West Point, Pennsylvania

I. INTRODUCTION

The dermal route of administration poses a number of scientific, technical, and regulatory challenges to those charged with characterizing the safety of new excipients for topical or transdermal use. Many of these challenges will be discussed in this chapter. The goal of safety testing is the protection of patients from potential harm caused by a new excipient that will be applied to the skin. This chapter highlights scientific, technical, and some regulatory nuances specific to the dermal route of administration for the safety assessment of new pharmaceutical excipients. Reliance on dermal drug-testing strategies, which are not always straightforward, may result in a scope of testing that exceeds that necessary for an excipient.

The most difficult aspects of developing a preclinical (nonclinical) toxicology strategy for a new pharmaceutical excipient are (a) to determine if certain testing is necessary or relevant, and (b) to successfully use the results of the toxicology studies to help direct clinical study designs. Recently, the Safety Committee of the International Pharmaceutical Excipients Council (IPEC) published proposed guidelines for the safety testing of new pharmaceutical excipients (1).

---

* *Current affiliation*: Coulter Pharmaceutical, Inc., South San Francisco, California

Table 1 Summary of Excipient Guidelines Excerpted from
Reference 1 for Human Clinical Transdermal and Topical Exposure

Tests	Transdermal or topical
Base set (Clinical exposures of less than 2 wk)	
Acute oral toxicity	R
Acute dermal toxicity	R
Acute inhalation toxicity	C
Eye irritation	R
Skin irritation	R
Skin sensitization	R
Acute parenteral toxicity	—
Application site evaluation	R
Pulmonary sensitization	—
Phototoxicity/photoallergy	R
Bacterial gene mutation	R
Chromosomal damage	R
ADME-intended route	R
28-day toxicity—intended route (2 species)	R
Clinical exposures for 2–6 wk	
90-day toxicity—intended route (most appropriate species)	R
Teratology (rat or rabbit)	R
Additional assays	C
Genotoxicity assays	R
Clinical exposures of longer than 6 wk	
Chronic toxicity (rodent, nonrodent)	C
Single generation reproduction	R
Photocarcinogenicity	C
Carcinogenicity	C

R, a required test; C, conditional based upon anticipated length of therapy and
potential uses of the excipient.

The IPEC recommendations (Table 1), used in conjunction with other guidelines
[e.g., International Committee on Harmonization, ICH ; Table 2; American Society for Testing and Materials, ASTM (8); International Organization for Standardization, ISO (9); Organization for Economic Cooperation and Development, OECD (10)] and regulations (11–13) give the toxicologist a sense of the time and resources needed to evaluate a new excipient. Throughout this chapter, it is assumed that the final dosage form (topical or transdermal formulation) will be subjected to a full battery of nonclinical safety studies that will support product

Table 2 Nonclinical Safety Studies for the Conduct of Human Clinical Trials for Pharmaceuticals Recommended for Adoption at Step 4 of the ICH Process on 16 July 1997 (Adult Patient Populations).

Clinical trial duration (wk)	Duration of repeated dose toxicity studies (wk) to support phase I and II clinical trials in the EU; Phase I, II, and III Clinical Trials in the United States and Japan[a]	
	Rodent	Nonrodent
Single dose	2–4[b]	2
≤2	2–4	2
≤4	4	4
≤12	12	12
≤26	26	26[c,d]
>26	26	36[c,d]
	Duration of repeated dose toxicity studies (wk) to support phase III clinical trials in the EU and marketing in all ICH regions	
≤2	4	4
≤4	12	12
≤12	26	26
>12	26	26 or 36[d,e]

[a] In Japan, if phase II trials are not as long as phase III trials, use phase III trial recommendations to support phase II.

[b] 2 wks are minimum duration in EU and United States, 4-week rodent required in Japan.

[c] 36-wk duration not yet adopted by ICH.

[d] Data from 26-wk rodent studies should be available before initiation of clinical trials >3 months; data from 36-wk nonrodent study should be available before clinical trial duration exceeds that which is supported by available toxicity data.

[e] The following are needed before initiation of pediatric trials: all adult data; all repeated dose toxicity, reproductive, and genotoxicity; and assessment of the need for carcinogenicity.

Table slightly modified from ICH Harmonized Tripartite Guideline.

Source: Refs 2–7.

registration. This is a fundamental assumption on which all other recommendations in this chapter are based.

II. GENERAL BACKGROUND INFORMATION

A. Definition of Terms

Topical delivery, as used in this chapter, is narrowly defined as application of a topical formulation to the surface of the skin for therapy of localized conditions, rather than the broader definition of topical application to skin and mucosa (14). *Topical formulations* are defined as fixed compositions of xenobiotic(s) and excipients, usually solutions, creams, ointments, gels, aerosols, liposomes, porous polymeric beads (''microsponges''), plasters and tapes, and cyclodextrin complexes (15). Topical excipients are likely to be applied to areas of skin with some underlying pathology that may affect the barrier function of the skin, and thus local or systemic exposure to the excipient. Because skin pathology may be widespread, the area of skin exposed to topical formulations in clinical use is unlikely to be well-defined, adding to the complexity of appropriate dosage selection for safety testing.

Transdermal drug delivery encompasses application of drug to (usually) normal skin, either for systemic therapy or therapy of underlying tissue (e.g., muscle). The *transdermal formulation* is defined as a fixed composition of xenobiotic and excipients, but is often structured as a device for controlling rate of delivery of xenobiotics and sometimes excipients to a defined area of the skin surface. Alternatively, these may be referred to as *transdermal delivery systems* or *transdermal therapeutic systems* (TTS). Transdermal therapeutic systems range in design complexity from relatively simple matrix systems to multilaminant patches (16). A multilaminant transdermal patch, such as the Nicoderm CQ nicotine transdermal system (Marion Merrell Dow) contains a release liner, a contact adhesive layer, a (drug formulation) rate-controlling membrane, a drug reservoir (containing drug formulation), and an occlusive backing. In contrast to the multicomponent TTS, simple semisolid delivery systems (e.g., nitroglycerin ointment) applied to a defined surface area have been used for transdermal drug delivery.

The transdermal patch components may come in direct contact with the drug or skin. In addition, the drug formulation or adhesive may be composed of numerous materials in solution or suspension. With the exception of the active compound, all these components, both solids and liquids fall under the definition of an excipient. As such, the safety of each component—in context to its role in the transdermal delivery system—needs to be considered to ensure that the safety studies conducted with the final product can be properly interpreted.

Films or other structural components of a transdermal system are not simply ''packaging'' for the drug, but are integral components in the dosage form and influence both the performance and safety of the product. For example, if a drug rate-controlling polymeric film malfunctions and permits too rapid drug delivery, a toxic dose of the drug could be delivered or, for low-toxicity drugs, the patient's duration of therapy would be insufficient resulting in a "drug holiday."

As defined here, transdermal delivery of either active drug or diffusable excipients is driven solely by passive diffusion of drug down a gradient of concentration (or more correctly, chemical activity), from the device, across the skin, and into the systemic circulation through the dermal capillary beds.

In many currently marketed topical and transdermal formulations, excipients are included that serve as more than formulation aids to ensure drug stability, drug delivery out of the formulation, or patient compliance, and formulation esthetics. These excipients have the additional function of altering stratum corneum permeability to a xenobiotic and are referred to as "enhancers" for both topical and transdermal delivery (17–26). Although the measurable, net effect of an enhancer is an increase in the mass of drug delivered through a unit area of the stratum corneum in a given time, the mechanism(s) by which enhancers produce this effect on stratum corneum are reported to be varied.

In contrast to a passive transdermal system, an "iontophoretic transdermal formulation" or "electrotransport system" contains ionically charged species in a three-part system composed of anode and cathode reservoirs and a battery–microprocessor unit. The ionized xenobiotic and other ions of the same charge, such as ionizable excipients, particularly those of molecular weights less than about 4 kDa, move through the skin when an electrical potential gradient is applied (27–29). Even excipients that normally penetrate skin poorly may exhibit increased skin penetration under the influence of an external electrical field (30–32).

Electroosmosis during iontophoresis is a phenomenon that can produce movement of neutral (uncharged) excipient molecules into the skin during application of iontophoretic current (33–36). Because normal stratum corneum is poorly permeable to ionized species or to large, hydrophilic molecules (37–38), iontophoretic devices profoundly affect the barrier properties of the skin. Thus, testing of a new excipient in the final iontophoretic formulation assumes added importance, for epidermal and dermal concentrations during electrically assisted delivery may substantially exceed those that can be achieved by topical or systemic delivery. Other specialized delivery methods, such as ultrasound disruption of the skin barrier, sonophoresis (39–40), or electroporation by a high-voltage external electric field (41) may also result in unanticipated exposure to excipients.

B. Background Information

Of primary interest to the toxicologist when designing a comprehensive safety program for a new dermal excipient are (a) the new excipient's intended use and the potential length of therapy, (b) approximate concentration in the end product (if known), (c) its history of clinical, veterinary, or research use, (d) its chemistry (including impurities, residual solvents, shelf-life stability, and degradation products), (e) any known toxicity (related to the class of chemical), (f) its ability to flux across skin, (g) its regulatory status, and (h) its potential for misuse or abuse. The safety evaluation of a new topical or transdermal excipient begins with comprehensive literature searches, followed by skin flux (and possibly, absorption, distribution, metabolism, and excretion [ADME]) studies, and in vitro cytotoxicity screening studies. If sufficient history of use and safety cannot be supported by these searches and short studies, then definitive in vitro and in vivo safety studies must be considered.

C. Computerized Literature Search

A review of the available literature is the basis for determining the extent to which relevant data exist for a new excipient. Structure-based searches can determine if structural similarities or structure–activity relationships exist between the new material and materials with known local or systemic toxicity. For topical and transdermal products, a material that is ''new'' as a pharmaceutical excipient may have a history of use in either medical tapes, bandages, or devices, or in cosmetic products. Not only the presence of the excipient in the product, but if possible, the level of excipient, the rate of application of excipient, and area of application of the medical device or cosmetic product should be documented. Any intended differences in the method of administration of the excipient in the device or cosmetic product, and the excipient in a new pharmaceutical product, must be considered in evaluating the extent of safety testing that will be required.

Use in foodstuffs or in food contact applications may provide additional data on the potential toxicity of the material. In particular, selection of polymeric materials for the construction of transdermal devices, especially backing or release liner materials that will not directly contact the skin, may be guided by food contact applications. The local skin tolerability of food–contact-approved excipients with low molecular weight, diffusable components, such as plasticizers, must be determined if diffusion and equilibration of these components during product shelf-life will produce contact with the skin. However, oral tolerability does not guarantee topical tolerability, which depends on potential skin contact time and intended use.

When new or novel materials are under evaluation for use in a transdermal therapeutic system, safety information may be limited and available only through the manufacturer. The following steps outline a strategy for collecting background information or baseline data on potential transdermal delivery system components:

1. Clearly identify and understand the function of the component.
2. Search the literature for safety information for each component (include open literature and freedom of information [FOI] documents on approved products containing the material). Vendor or manufacturer supplied safety data should also be requested.
3. Characterize the component, chemically and physically, either through available literature or testing.
4. Determine if "leachables" can be extracted from the material and possibly be delivered across the skin or gastrointestinal (GI) tract.
5. Identify which other components (including the active compound) will come in contact with the material in a finished transdermal system.

If adequate, reliable, safety information is available, additional testing on the material itself may be unnecessary.

D. Species Selection

The selection of animal species for toxicity testing is generally dictated by regulatory bodies and their guidelines. Many regulations or guidelines require both rodent and nonrodent species; however, the choice and number of species for dermal toxicity testing does not always fit regulatory paradigms. For example, it might be scientifically justified to use two nonrodent species (rather than the usual rodent, nonrodent), or just a single nonrodent species. Ultimately, the choice of species is based on sound scientific judgment (see Chapter 5). Table 3 summarizes species commonly used in dermatological testing.

Table 3 Selection of Species[a] for Toxicology Testing by the Dermal Route of Exposure

Study	Species
Skin irritation	Mouse, guinea pig, rabbit
Phototoxicity	Mouse, guinea pig
Toxicokinetics	Guinea pig, rabbit, dog, swine
Wound healing	Swine
Sensitization	Mouse, guinea pig, rabbit, swine
Subchronic dermal toxicity	Rabbit, guinea pig, dog
Chronic dermal toxicity	Rabbit, dog
Subchronic systemic toxicity	Rat, rabbit, dog
Chronic systemic toxicity	Rat, rabbit, dog
Carcinogenicity	mouse
Photocarcinogenicity	mouse

[a] Substitutions based on scientific criteria are allowed.

The vast majority of dermal toxicity testing is carried out in mice, guinea pigs, and rabbits. Rats and dogs are more frequently used for systemic toxicity studies, because large toxicology databases are available on these species. Similarities in swine and human skin have led to the use of swine in specialized safety testing, such as wound healing or toxicokinetics. Although the rhesus monkey may be a useful model for subchronic or chronic study of xenobiotics with similar enzyme or receptor affinities in monkey and human, it is not a recommended species for the study of excipients without specific pharmacological effects.

Hairless strains of mice, rats, and guinea pigs are sometimes used for dermal pharmacology and toxicology studies because shaving or depilation is not necessary. Woodard et al. investigated the skin morphology and pharmacology of Crl:IAF(HA)-hrBR-IAF (Charles River) hairless guinea pigs (42). So-called hairless guinea pigs, such as the IAF outbred strain, are useful for TTS studies because adhesives generally adhere less aggressively to hairless guinea pig skin than to rabbit skin (partially owing to regrowth of rabbit fur), and there is no need for shaving or depilation before applying topical formulations. The IAF guinea pig is a useful animal model for both irritation and sensitization studies (43,44). Experience with the IAF strain over the past decade has proved its value in topical drug delivery in general and TTS testing in particular (MJ Cukierski, personal communication, 1997).

As discussed (45,46), clinical studies are needed to complete a dermal safety assessment program. Inherent biological differences between the test systems used and humans may result in experimental outcomes that are not indicative of human responses. For example, the Draize method of skin irritation testing in rabbits can predict severe irritants to humans, but tends to overpredict the potential of mild or moderate human irritants (47). Even well-controlled clinical trials do not always identify potential adverse outcomes. In some instances, toxicity (such as sensitization) may not be identified until after a product is marketed and used by many patients.

III. PHARMACOKINETIC AND TOXICOKINETIC STUDY DESIGN

The dermal route of exposure, similar to the nasal, buccal–sublingual, inhalation, or intravenous routes, bypasses the portal venous circulation and initial hepatic metabolism. Dermal delivery of an excipient to the systemic circulation without the first-pass effect resulting from oral administration can produce different metabolic and toxicological profiles when compared with oral, intraperitoneal, or rectal administration. Also, the rate and extent of absorption by the dermal route may produce dramatically different maximum or sustained plasma concentrations than other routes.

Carefully designed pharmacokinetic studies are required to characterize the absorption, distribution, metabolism, and excretion (ADME) of topical or transdermal excipients for toxicological evaluations (48). Additionally, verification of localized skin exposure to an excipient, its impurities, degradates, and metabolites supports interpretation of dermal irritancy studies, whether these materials reach the skin by topical application or an alternative route. Ultimately, the pharmacokinetic and ADME studies support the selection of doses and exposure levels for safety testing that will provide an adequate margin of safety for initiation of clinical trials with the excipient. Proof of exposure validates the safety-testing strategies.

Pharmacokinetic and ADME studies initiate the chain of decisions about appropriate safety testing of excipients, as shown in the generalized overview of key decision points in a safety assessment program which are outlined in Figure 1.

A. ADME by the Topical or Transdermal Route

Topical and transdermal excipients may either produce substantial skin or systemic exposure, or may never cause toxicity owing to the failure to penetrate the stratum corneum (38). The distinction between applied dose and dose delivered (or absorbed) is critical when evaluating published toxicity data or interpreting new experimental results.

The specific pathway for permeation (transcellular, transfollicular, paracellular, or other) also influences the skin flux (49–54). Even compounds that do not readily flux across normal, intact skin on their own may penetrate abraded, burned, or diseased skin. Topical formulations are often applied to compromised skin, whereas transdermal delivery systems are not. Permeation enhancers (those added intentionally or unintentionally), or disruption of the skin barrier by iontophoresis, sonophoresis, or electroporation may increase skin flux of excipients.

Dose selection for pharmacokinetic or toxicokinetic studies in support of safety studies is aided immensely when there is some estimate of expected clinical exposure. Isolated human skin mounted in diffusion cells with warmed (usually), well-mixed or perfused receptor phase at the dermal surface of the skin has been widely used to predict potential human skin permeability (55–60). The methods of skin preparation and storage, diffusion cell design, and receptor phase composition vary among laboratories. Selection of a proper test system is key to obtaining valid data (58,61,62). Some specific systems are discussed in the following; the advantages and disadvantages of common in vitro diffusion techniques

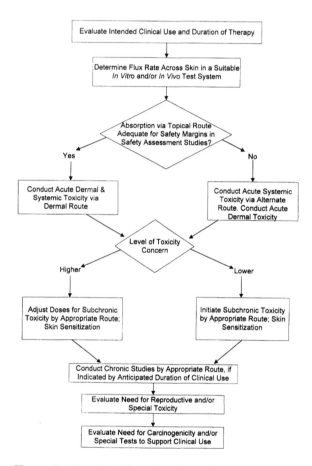

Figure 1   Procedural flow chart for testing excipients.

are summarized in Table 4. The choice of the most appropriate system should be based on properties and uses of the new excipient.

1.   In Vitro *Human Skin Transport*

When determination of the barrier function of the stratum corneum is the primary goal, partial-thickness skin (stratum corneum, epidermis, and partial-thickness dermis) may be obtained with a dermatome from an organ donor. It may be possible to specify donor skin collection from anatomical sites likely to be exposed to the test material in clinical use, as regional variations in skin permeability have been reported (63,64). Barrier properties of fresh versus frozen skin have

Table 4 Nonclinical In Vitro and In Vivo Skin Flux Methodology

Test System	Objective
In vitro	
Split-thickness fresh donor skin	Assess skin penetration, skin metabolism
	Strip or abrade to assess stratum corneum barrier
Split-thickness frozen donor skin	Assess skin penetration
	Strip or abrade to assess stratum corneum barrier
Heat-separated stratum corneum/ epidermis	Assess skin penetration without dermal barrier to diffusion
Isolated stratum corneum	Assess skin penetration through membrane-supported stratum corneum without epidermal or dermal barrier to diffusion
Cultured human skin cell preparations	Assess skin metabolism/penetration. More convenient and available alternative than fresh skin
Full-thickness Fuzzy rat skin	Substitute for skin penetration experiments. Metabolism likely to differ from human
In vivo	
Fuzzy rat or swine topical vs. IV pharmacokinetics	Estimate skin penetration rate in the intact animal to correlate with *in vitro* measurements. Examine skin distribution
Isolated porcine skin flaps	Isolate skin flap for total collection of absorbed compound and metabolites
Human skin transplant on immunodeficient mice	Test skin penetration and disposition in a perfused model of human skin.

been widely debated in the literature; the effect of freezing on transport appears to be closely related to the test material (65). If the stratum corneum is the rate-limiting step in diffusion in vitro, then the partial-thickness skin preparation may be adequate. When the nonperfused, partial-thickness dermis contributes substantially to the resistance to flux, then either heat-separated stratum corneum–epidermis (55,56), or isolated stratum corneum supported on a dialysis membrane (23), may be used. Obviously, the more the skin specimen is manipulated, the greater the risk of damaging the barrier function of the skin. An appropriate control before initiation of an excipient transport study is the application of a small amount of tritiated water to skin mounted in diffusion cells (66). Measures of the electrical resistivity of skin can also be used to determine the integrity of the stratum corneum before application of the test material (67).

If testing requires that the metabolic function of skin is maintained, skin must be freshly obtained from surgical procedures or donors, under conditions that maximize viability (68). Prolonged storage under refrigeration, freezing, and heat-separation of epidermis and dermis have detrimental effects on skin viability (69). Cultured cell preparations are also available for metabolic or pharmacology studies, but the expression of enzymes or organization of the epidermis in cultured skin systems is highly dependent on the culture system (70,71).

The application of test material to the stratum corneum surface of the skin should also be consistent with the intended clinical use and proposed methods of application in toxicology studies. Many diffusion cell designs support application either with or without occlusion. Formulation of test materials should result in a presentation to the skin similar to what is anticipated for safety and clinical studies. For example, for very highly permeable substances, the vehicle in which the test material is applied may begin to contribute to resistance to diffusion and decreased flux with time, as was demonstrated for timolol base in a gelled mineral oil vehicle (72).

In vitro studies with disruption of the stratum corneum may provide information about the importance of this barrier. Tape-stripping is quite effective in comparison with lines abraded with a needle, or by thermal or irritant damage, even though only about two-thirds of the stratum corneum is removed by tape-stripping (73,74).

Many diffusion cell designs have been published, including the widely used Franz cell and flow-through cell designs. Electrodes can be incorporated for estimation of iontophoretically enhanced flux (28,33,34,75). Adequate mixing of the receptor phase is required to prevent an unstirred boundary layer at the dermal surface of the skin, which contributes resistance to diffusion and underestimates skin flux. Temperature control of the diffusion cell is required to maintain skin surface temperature at approximately 32°C and, if possible, receptor phase temperature at 37°C.

The liquid receptor phase for the dermal reservoir compartment of the diffusion cell must maintain "sink" conditions (excipient concentrations no greater than 10% of saturation solubility in the receptor phase). Inadequate solubility in the receptor phase will result in additional resistance to diffusion, resulting in underestimation of flux across the stratum corneum. Isotonic solutions at approximately neutral pH with physiologically compatible buffer systems and antimicrobial agents are preferred. Serum albumin may be added to increase solubility if the compound is highly protein bound. Aqueous cosolvent systems can be used for extremely water-insoluble materials, if the cosolvent chosen is a material such as polyethylene glycol which will not significantly partition into the skin and affect skin permeability. Prewarming and degasing the receptor phase with sonication, vacuum, or both is advisable to reduce the formation of air bubbles beneath the dermal side of the skin. The receptor phase should be sampled for assay at appropriate times to define the lag phase and steady-state skin flux (76).

2. Alternative In Vitro or In Vivo Studies

The results of the isolated human skin-transport studies provide a first estimate of likely clinical exposure, which can be correlated with in vitro and in vivo skin permeation rates from potential animal species for safety evaluation. Also, a variety of test systems, such as the sparsely haired Fuzzy rat, a mutant strain of the Wistar Furth rat, may give penetration rates similar to those of isolated human skin (77–79). Other models of isolated human skin transplanted to immunodeficient mice (80) or isolated porcine skin flaps (81) have been used to refine estimates of skin penetration or disposition of compounds in the skin. The skin flap models are difficult to establish and maintain, limiting their use to the determination of the absolute fraction of dose absorbed and extent of metabolism in the skin. With the possible exception of the isolated human skin flap, potential strain differences in metabolism, for instance, between Fuzzy rats and rodent safety assessment species, must be considered along with differences in metabolism between humans and animal species (82).

In summary, there are numerous in vitro and in vivo methods for estimation of skin flux of excipients (see Table 4). Selection of the most appropriate method(s) should be based on the physical–chemical properties of the new excipient and its proposed clinical applications.

3. Single and Multiple Dose Dermal Pharmacokinetics or Toxicokinetics

Dermal pharmacokinetics studies assess the expected systemic exposure from single- or multiple-dose applications of the appropriately formulated excipient in safety assessment studies. These data determine if acute systemic toxicity or repeated-dose toxicity studies may be accomplished by the dermal route (see Fig. 1). For any pharmacokinetic study the excipient must be quantified in the systemic circulation or in tissues. Although dramatic advances in analytical methodology have been made in the past decade (83,84), the isolation and quantification of many excipients is not routine. Besides chromatographic and spectroscopic methods, specific immunoassays may provide good sensitivity, if they do not cross-react with excipient metabolites or related substances. Custom radiochemical synthesis of isotopically-labeled excipients is an alternative.

a. Single Topical Dose Pharmacokinetics and Toxicokinetics The in vitro skin diffusion studies described in the foregoing for both isolated human and animal skin can provide estimates for test article concentrations, areas of applications, and selection of test species for pharmacokinetic or toxicokinetic evaluation. Maintaining the dose within a defined surface area and prevention of ingestion or contact with mucosal surfaces is key to reliable estimation of systemic exposure by the dermal route in animal studies. The length of application of a single dose and times of serial sampling for blood, urine, feces, or bile must be adequate to determine the time to reach steady-state skin transport, and after removal of test article, the apparent elimination rate of the excipient from the body.

Groups of animals may be euthanized and necropsied at selected time points to determine potential target-organ concentrations of the excipient, impurities, and metabolites, because skin is a metabolically active organ (68,85,87).

Calculation of the absolute rate of percutaneous absorption and dermal bioavailability requires comparison of the plasma concentration–time profile (AUC) of the excipient after dermal and intravenous doses of the excipient (88–91). The rate of excipient accumulation in skin and elimination from the skin after removal of the test article can be inferred from pharmacokinetic analysis. This information, as well as the basic information about metabolism and clearance after an IV dose, contribute substantially to understanding the pharmacokinetic properties of the excipient.

b. Multiple Topical Dose Pharmacokinetics and Toxicokinetics   If systemic exposure can be adequately measured from single-dose application, it is sometimes possible to estimate the expected accumulation on multiple applications in repeated-dose safety studies. However, the duration of single-dose application may be insufficient to produce steady-state plasma concentrations, indicating either a very long half-life for elimination, or a very long accumulation time for the excipient to reach steady-state concentrations in the skin. Repeated topical applications of excipient to the same site may be conducted with sampling on the first day and last day of dosing, with an adequate sampling schedule to define the terminal elimination phase of the excipient from the systemic circulation. The length of dosing is variable, depending on the length of the toxicology study it is intended to support, or on the expected time to reach steady state. Regulatory guidelines that address xenobiotic ADME studies may be consulted for the need for repeated-dose tissue distribution studies in cases in which tissue half-life exceeds the systemic elimination rate of excipient and is more than twice the dosing interval in the toxicity studies, or significant systemic accumulation occurs (92). The decision to conduct these multiple-dose studies prospectively or concurrently with formal safety studies must be made on an individual basis, depending on the potential effect of accumulation on selection of doses for the safety study.

The systemic absorption of drug from abraded or tape-stripped skin may be measured in either single- or multiple-dose studies to provide information for the excipient under potential conditions of use or misuse of the excipient in a topical or transdermal product. If the excipient is likely to be used in formulations for treatment of specific dermal pathologies, skin penetration in animal models of dermal pathology can be considered to determine potential differences in exposure between the target patient population and normal volunteers. The difficulties inherent in developing animal models of human skin pathology are discussed in other texts (93,94).

B. Application Site Exposure

Information about skin penetration or distribution of the excipient from pharmacokinetic studies can be used to justify dose selection (concentration or area of application) in acute or repeated-dose dermal studies. Determining drug concentrations in stratum corneum tape-strips from application sites at selected times after dosing has been advocated as a method for determining skin penetration in both toxicology and clinical studies (63,95). This technique may be more useful in comparative measurement of skin penetration for the same excipient under various conditions than for absolute measures of skin penetration. For poorly penetrating excipients, removal of the excess drug from the surface of the stratum corneum, without removing the stratum corneum reservoir of drug (96), may be problematic.

Serial cryosectioning from dermal to epidermal surface of skin biopsies combined with histological analysis can provide samples for quantitative analysis of excipient concentrations at various skin depths (97). Autoradiography has been used to provide at least a qualitative measure of the exposure of epidermis, dermis, and skin appendages, such as hair follicles and sebaceous glands (53,98). Cutaneous microdialysis has been investigated, but substantial technical hurdles remain for use in quantitation of skin penetration (99).

C. Oral or Parenteral ADME Studies

If it is established from in vitro, in situ, or in vivo studies that systemic exposure is poor with topical administration, then oral or intravenous administration of the excipient must be considered for adequate safety evaluation. Oral disposition studies of any excipient, poorly penetrating or otherwise, can relate exposure in toxicology studies to exposure after accidental ingestion. Designs for oral or parenteral pharmacokinetic or toxicokinetic studies are described in other chapters. In addition to the usual blood or plasma concentration measurements, it is helpful to quantitate the skin concentrations of excipient and any impurities, degradates, or metabolites, following single- and repeated-dosing. These studies may support safety for application to abraded skin or skin with otherwise defective barrier properties.

IV. FACTORS TO CONSIDER IN DESIGN OF SAFETY STUDIES

For each of the major types of safety studies described in this section, references to both ICH and IPEC guidelines are included. It must be stressed that ICH guidelines apply to active pharmaceuticals and not ''inactive'' excipients. However, the ICH guidelines are very useful in determining what should or should not be considered for a comprehensive safety program for both single excipients and final products. The IPEC guidelines apply specifically to excipients and should be consulted closely during the qualification of a new excipient. The ICH and IPEC guidelines differ in clinical exposure duration criteria. IPEC uses clinical exposures of less than 2 weeks, 2–6 weeks, and longer than 6 weeks to determine testing durations in animal studies, whereas the ICH guidelines are more complex (see Tables 1 and 2). In general, the IPEC guidelines are more flexible because excipients are not active drug products and typically pose less risk than do drugs. ICH guidelines clearly state that single-dose toxicity should be evaluated in two mammalian species before human exposure (4). Dose escalation studies are acceptable alternatives to a single-dose design.

A. Acute Systemic Toxicity by the Dermal Route

For new topical or transdermal excipients, the first in vivo safety study—acute systemic toxicity—defines an excipient's potential to induce systemic toxicity after single-dose or short-term exposure. An excipient's toxicity is delineated by the choice of test system, strain, sex, age, dose, duration of dosing, rate of dosing, coformulation with other excipients or vehicles, and route of administration. Skin temperature, degree of skin hydration, and open air or occlusive application can also influence the skin flux of compounds and, thus, their toxicity.

For topical or transdermal excipients to cause systemic toxicity, the excipient must transit the skin, be distributed within the body by the bloodstream or lymphatic circulation, and finally, be present at, or accumulate to concentrations high enough to evoke toxicity. By using the intended clinical route of administration in safety studies, the ADME and, ultimately, the toxicity of the test article will be a closer approximation of the clinical situation than if an alternative route is used. In all testing protocols, dose should be clearly defined in terms of dose applied and dose delivered. The need for occlusive or nonocclusive application and formulation requirements for presentation of the excipient must also be defined. In the simplest case, the flux (e.g., $(\mu g/cm^2h^{-1})$, surface area of application (cm^2), and duration of application (h) are used to calculate the total dose delivered across the skin.

For dermal applications, general acute toxicity protocols for rodents and nonrodents are the same as those outlined by Auletta (100) for oral testing, with the exception of the route and limit dose. For nonrodents (dogs and nonhuman primates) a limit test with two to five animals per sex per group is conducted if the test material is expected to have low or no toxicity. If toxicity is anticipated, an "up-and-down" study design can be conducted. Because most topical and transdermal excipients are expected to have little or no systemic toxicity, the limit test will be used most frequently.

Dermal safety evaluations are unique in the methodology for test article application. Skin contact must be maintained while minimizing the risk of ingestion or premature loss of dose. Laboratory techniques to apply test articles to the skin have been described (100,101). Hairless animals can be used, but usually, species requiring removal of fur from application sites are used. Shaving or depilation can be done 1 or 2 days before test article application. However, chemical depilitants can induce irritation of their own, and their use should be monitored closely. Shaving should be done only with a high-quality small animal clipper, preferably with an integrated hair vacuum. Clipper blades should be very sharp, and excess clipper blade oil should be removed from the animal's skin before any test article is applied. Test article application and wrapping techniques vary between laboratories. The goal is to avoid causing skin irritation with the shaving, test article application, wrapping, and unwrapping procedures. A typical application procedure for guinea pigs or rabbits follows (MJ Cukierski, personal communication, 1998). Skin sites are shaved and cleaned (usually with an alcohol wipe). The skin must be dry, clean, and free from any particulates (e.g., bedding) before test articles can be applied. If nonadhesive Finn Chambers are used for application of viscous test article formulations, they are usually applied to a small piece of Micropore surgical tape (3M) before being filled and applied to the skin. The liquid test article is placed in the Finn Chamber(s) and then applied to the dorsal surface of the animal in a sequence consistent with the protocol. If Hill Top Chambers are used, they have their own adhesive. After application of the chambers, the animal is wrapped with Conform stretch bandage (Kendall) that is secured to the skin with Micropore surgical tape. Vetrap bandaging tape (3M) is then applied over the Conform stretch bandage. Usually, collars are not necessary if the application sites are sufficiently close to the head. Occasionally, individual animals (usually rabbits) become adept at removing bandaging by kicking or chewing. The addition of a stretch stockinet or a collar usually will prevent test article removal in these animals. All bandaging must be snug, but not so tight that it impedes normal breathing motion, or lung infections or death may occur. Frequent observations of wrapped animals should be outlined in the study protocol to assure the health of the animals and validity of the data. Rather than unwrapping, bandaging materials are usually cut from the animals using bandaging shears. At the time of system removal, the location of the test article(s) can be indicated by the use of a gentian violet skin-marking pen (such as the Twin Tip surgeon's pen).

1. Acute Dermal Toxicity Assay (Irritation)

The purpose of an acute dermal toxicity assay is to assess if a test article will cause direct injury to the skin when applied for brief durations at doses or concentrations that may exceed the planned clinical use. The assay is conducted in a manner similar to that just described for the acute dermal systemic toxicity assay; however, in this study design, skin response is the prime experimental endpoint. Lawrence (102) has recently outlined some applications of in vitro human skin models for dermal irritancy.

a. Protocol Designs and Practical Considerations For topical excipients, the duration of exposure can be highly variable; for transdermal delivery system excipients, the exposure usually mimics the clinical wear duration. For liquid test articles, the procedures for test article application are the same as described in the acute systemic toxicity section.

Polymeric materials, films, foils, gels, and other solid materials can be assessed for local skin irritation potential by direct application to the skin of experimental animals. For bulk polymeric materials, such as pellets or powders, thin films can be pressed to facilitate testing, assuming the fabrication steps (i.e., heat and pressure) will not alter the chemical properties of the test material. The probability of direct skin exposure and exposure time in clinical use must be assessed. For example, an occlusive backing material of a transdermal therapeutic system may not come in direct contact with the patient's skin during normal use. However, it is possible that the backing could come in direct contact with the patient's (or partner's) skin for extended periods while sleeping.

The skin area available for testing is generally limited by the choice of experimental animal and the expected clinical exposure. Hartley guinea pigs and New Zealand white rabbits are usually the species of choice for direct skin irritation studies, but other species, such as swine, can be used for special circumstances. In practice, samples of the test material are applied to intact and slightly abraded skin sites for durations ranging from hours to days. There are various methods of adhering thin films to the skin using tapes, bandages, and a host of wrapping materials. Direct skin contact must be maintained to prevent false-negative results. As long as the test article remains in contact with the skin for the desired duration, and the test animal is not unduly stressed by the wrapping materials, no one technique stands out as superior.

Trauma must be minimized when removing test articles or adhesive bandaging to prevent false-positive results. After the required exposure, test articles are removed from the skin and the sites are graded for local irritation.

b. Skin Scoring and Histopathology Skin sites are usually scored using the Draize scoring scale (Table 5). Animals are examined under simulated daylight illumination, and skin sites are scored at 0.5, [2], 24, and 48 h after test article removal. Erythema–eschar and edema are scored and then an irritation index is calculated.

Table 5 Draize Scale Lesion Scoring

Evaluation of Skin Reactions			
Erythema and eschar formation		Edema formation	
Score	Reaction	Score	Reaction
0	No erythema	0	No edema
1	Very slight erythema	1	Very slight edema
2	Well-defined erythema	2	Well-defined edema
3	Moderate to severe erythema	3	Moderate edema (raised $\sim$1 mm)
4	Severe erythema to eschar formation	4	Severe edema (raised >1 mm and extending beyond area of exposure)

Total possible score for erythema/eschar = 4; Total possible score for edema = 4.

Skin Irritation Index	
Index	Category
0.0–0.5	None or negligible
0.6–2.0	Mild
2.1–5.0	Moderate
5.1–8.0	Severe

Source: Ref. 103.

Skin scoring should be blinded, thus an experienced person to score the sites and a technician to handle the animals and record results are needed to assure blind scoring. Confirmation of animal numbers is imperative, as any non systematic errors during the actual blind-scoring procedure can easily invalidate a study.

In addition to macroscopic skin scoring (Draize's scale), cutaneous toxicity can be examined histologically. Table 6 lists the main histopathological features corresponding to the Draize (macroscopic)-scoring scale. Skin histopathology techniques, including collection, processing, and evaluating skin samples, have been reviewed (104).

2. Skin Sensitization

In ICH guidelines, skin sensitization tests are broadly covered by the "local tolerance studies" recommendations (4), although it is clear that the mechanisms that lead to skin sensitization are actually "systemic." OECD Guideline No. 406

Table 6 Main Histopathological Characteristics of the Draize Scale Lesions

Grade or value	Histopathological features	
	Epidermis	Dermis
0	Normal	Normal
1	Slight spongiosis	Mild congestion
2	Spongiosis	Moderate congestion
	Acanthosis	Diffuse edema
	Hypergranulosis	Inflammation (mild)
3	Marked spongiosis	Severe congestion
	Marked acanthasis	Hemorrhage
	Marked hypergranulosis and hyperkeratinization	Marked diffuse edema
		Acute inflammatory infiltrates
4	Erosions (localized)	Marked acute inflammatory infiltrates
	Ulcer (localized)	Hemorrhage
	Superficial exudates	Hemosiderin deposits
	Hyperplasia or atrophy	Capillary proliferation
		Vascular damage

addresses skin sensitization (10). Recommendations set forth by IPEC include skin sensitization in their "Base Toxicity Set" of studies that should be conducted before human exposure (1). Alternative methods to skin sensitization testing have been reviewed (105).

For a new pharmaceutical excipient, testing the sensitization potential of the excipient by itself is highly recommended. It is always prudent, and to satisfy regulatory requirements, often necessary, to conduct skin sensitization studies before any multiple exposure clinical protocol.

Dermal irritation and sensitization reactions are the two most readily apparent forms of toxicity when assessing new (dermal) pharmaceutical excipients or drugs. For convenience, these topics are often discussed separately; however, local irritation and immune-mediated sensitization reactions frequently overlap. Their respective contributions to observed cutaneous reactions are often difficult to distinguish. Sensitizers may or may not be irritating, but irritation, either chemical or physical, can help potentiate a weak sensitizer. Compounds that induce local pharmacological skin effects, such as histamine release, can cause erythema and interfere with skin scoring (particularly during challenge applications or injections), thus the need for adequate controls during such evaluations.

The T-cell–mediated mechanisms (type IV hypersensitivity) will predominately be responsible for skin reactions seen in the laboratory while following a standard multiple induction and challenge guinea pig sensitization protocol. General references should be consulted for a thorough description of the role of T-cell and antibody-mediated skin reactions (106–110).

Definitively identifying weak sensitizers using classic experimental endpoints may be insufficient. For example, colored test articles that stain the skin can obscure erythema. Complementary in vivo screening studies or in vitro assays may be needed to confirm the conclusions based solely on skin scores.

a. In Vivo Sensitization Protocol Designs and Practical Considerations
In practice, three general types of in vivo skin sensitization protocols can be followed: (a) intradermal injections, (b) direct skin contact (topical), and (c) a mixture of intradermal injections and topical applications. The guinea pig and mouse are most often the animals of choice for in vivo sensitization assays, although other species; such as swine, have been used. There are roughly 15 named guinea pig sensitization protocols (or variations of protocols) from which to choose (111,112). Differences in the protocol designs include the method of exposure (topical, intradermal or a mixture of both routes), duration, frequency or timing of induction, and challenge applications; the use and type of adjuvants; choice of vehicles; and test article dosages (100,111,112). The Buehler test and guinea pig maximization test (GPMT) are most frequently conducted to identify potential human sensitizers, although other designs are acceptable under certain circumstances (113).

New pharmaceutical excipients under investigation for use in transdermal formulations are often evaluated as follows: After completing the recommendations listed in Sec. II, the excipient is first screened in a mouse sensitization assay, such as the local lymph node assay or mouse ear swelling test (MJ Cukierski, personal communication, 1998). If negative, the results of the mouse assays will typically lead to additional testing in guinea pigs. Positive results in a mouse assay are sufficient to identify a compound as a potential human sensitizer. The decision to conduct additional guinea pigs assays after a positive mouse assay depends on its role in a particular drug formulation.

Following the mouse screening studies, one or two guinea pig sensitization assays are conducted. Before the sensitization study is started, an irritation assay must be conducted to determine the highest nonirritating concentration of the test article (see Sec. IV.A.1.b). A protocol using intradermal injections of the test article, with and without adjuvant, is conducted first, and may be followed by a second study using topical applications of the test article. Intradermal injection minimizes uncertainty about absorption and absolute dose delivered but obviously does not mimic the clinical situation.

A confounding factor for both topical and intradermal injection studies is that the test article may have to be formulated with other excipients for application or injection. The immunogenicity of the additional excipients (vehicles) should be known to prevent spurious results.

From a practical standpoint, the success of an intradermal sensitization study will depend on the quality of intradermal injections. Intradermal injections require practice, patience, and carefully restrained guinea pigs. The needle (typically 25 or 26 gauge) is inserted into the dermis, and the steady force of the injection should cause a small bleb of test formulation to form in the skin. Approximately 0.1 mL or less should be injected per site. If the skin does not form this bleb, the injection may be subcutaneous and should be repeated. If multiple injections are necessary (for example, during challenge with multiple concentrations of test article), the injections should be evenly spaced on the back, near the neck. When removing the needle, some test material may leak out of the injection site, reducing the total dose administered. This can be minimized by using the smallest needle possible and careful technique. Care should be taken not to compress the injection sites (particularly during transport of the guinea pigs back to their cages) or to bandage them immediately after injection.

For intradermal injection studies, the test article is formulated in solution or suspension, with or without adjuvant, and injected intradermally in the dorsal (shoulder) region of the guinea pig to ''induce'' sensitization. A typical induction regimen is nine injections over 21 days (Monday, Wednesday, Friday, for 3 weeks). A positive control (e.g., dinitrochlorobenzene) and negative control (injection vehicle) are always included to ensure that test systems are responsive and that the technical procedures are correctly performed. After the final induction, the animals are not treated for 10–14 days. After the 2-week ''rest'' period, the animals are challenged with various nonirritating concentrations of the test article. If an adjuvant was used during induction, it is not used during the challenge phase. Challenge injections (or topical challenge applications) should be at naive sites, on the back of the animal. Multiple challenge concentrations (usually three to five serial dilutions) can be tested in the same animal. Both skin and any systemic reactions are observed at approximately 2, 24, 48, and 72 h postchallenge. If systemic toxicity occurs (i.e., anaphylactic or anaphylactoid reactions), it will typically happen immediately after (or during) injection or soon after topical application. Injection sites are scored using standardized criteria for formation of erythema and eschar (0–4) and edema (0–4) over the course of 3 days. Positive responses are usually defined as scores of 2 or higher out of a maximum score or 8 at 48 or 72 h that (a) clearly do not represent irritation, (b) generally increase 24 h after injection, and, (c) may persist for up to 72 h. These are not absolute criteria and sound scientific judgment and experience are required for interpretation of sensitization results. If the results of the first challenge are equivocal, additional challenges can be conducted. It must be kept in mind that each challenge injection can be considered another induction, and interpretations of subsequent challenges must account for this fact. Additional experimental endpoints such as lesion measurements or descriptions, site histology, blood samples for in vitro lymphocyte assays, antibody assays, and others can be added to protocols on a case-by-case basis.

Following the intradermal sensitization study, the test article can also be tested in a topical application protocol or, if needed, as part of a functional transdermal delivery system. Scientific considerations and practical procedures for conducting topical sensitization studies have recently been reviewed (45,47). Conducting a topical sensitization study is not unlike the intradermal study just described; however, there are several practical considerations that require attention. Among the most important are the following: Skin scoring should be blinded; therefore, the protocol should have a well-defined site randomization schema so that the person conducting the scoring cannot determine the test article application pattern. Application sites must be clean and dry before application of topical or transdermal formulations. Alcohol wipes can be used for cleaning, but the site must be dry before test article application. If chambers are used to contain liquid or paste formulations, the wrapping procedures should be consistent, as described in Sec. IVA.1. Guinea pigs may rub their backs against feed hoppers inside their cages while attempting to hide or burrow, causing irritation or minor skin injuries that sometimes confound skin scoring.

b. Interpretation of Results and Strategy for Positive Assay Results   Positive sensitization reactions to particular compounds may range from weak to strong and can differ among species owing to inherent differences in the immune system or in methods of exposure. Although common test systems, such as guinea pigs, are capable of distinguishing weak from strong sensitizers, and there is an established correlation between sensitization reactions in guinea pigs and humans (114–116), another factor to consider is the notion of a threshold concentration needed to induce or trigger a sensitization response (117). A hapten needs to be present in the skin above a certain threshold concentration before it can be processed by the immune system and result in sensitization (108). If the commercial use of a potential sensitizer is significantly below this threshold, it may be acceptable to use the compound. A weak sensitizer in routine nonclinical studies should not automatically be eliminated from use in topical or transdermal products. Only well-controlled human sensitization studies can predict if a weak sensitizer in guinea pigs will pose a clinical concern. Conversely, negative results in typical mouse and guinea pig assays do not ensure that the compound will not be a human sensitizer.

3. Acute Systemic Toxicity by the Oral Route

Whereas the intended clinical route of delivery is the preferred route of exposure to assess the toxicity of an excipient, dermal administration may not be sufficient to fully characterize the systemic safety of an excipient. Oral administration can be considered to establish adequate margins of safety for systemic exposure to poor skin penetrants. Chapter 6 discusses the details of oral toxicity testing of excipients. When comparing oral with dermal administration of the same material, excipients will be subjected to very different conditions that may increase or decrease the toxicity of the excipient, including (a) pH; for example, gastrointestinal (GI) pH ranges from approximately 1.8 to 7.8 (118), whereas the skin pH of healthy adult white males ranges from 5.4 to 5.9 (119); (b) exposure to enzymes; (c) bacterial flora variations; (d) amount of water available; and (e) extensive mixing in the GI tract.

Even if acute systemic toxicity can be conducted by the dermal route, oral administration provides additional information about excipients in formulations intended for use in children or adult patient populations who may have young children in the household. Oral toxicity information could be useful in the case of a child ingesting a topical formulation or swallowing a single or multiple transdermal patches (120).

Depending on the physical form of the excipient and its chemical properties, formulation with other excipients may be required to render it orally absorbed or tolerable. A limit dose of 5000 mg/kg by the acute oral route is employed when exposure by dermal application cannot be determined. For oral administration, general acute toxicity protocols for rodents and nonrodents are outlined by Auletta (100) and differ slightly from each other (see Chapters 5 and 6).

4. Acute Systemic Toxicity by Parenteral Routes

If oral absorption is poor, risk of oral exposure is negligible, or metabolic disposition differs greatly between oral and parenteral routes, then parenteral administration may be a more valid alternative for evaluation of acute systemic toxicity. If an excipient is administered intravenously, the dose can be accurately controlled, but may be delivered at a much faster rate than would be possible with transdermal administration. The test article will likely require formulation with other excipients for injection. Compounds that might be relatively innocuous on the skin could be damaging to vascular endothelium. The intravenous route avoids skin metabolism and provides no immediate exposure to the dermal immune system.

Additionally, nonabsorbable polymeric materials can be tested according to *USP*, ISO, ASTM, or other recognized biocompatibility guidelines. Testing may be performed using either injected extracts or by implantation of solid samples of the material. In vitro cytotoxicity of extracts and solid samples can be evaluated as an adjunct to acute in vivo testing.

5. Paradigm for Safety Assessment of TTS Excipients

The FDA regulates transdermal therapeutic systems as new chemical entities (12). In the United States, even well-characterized, marketed drugs and excipients that are incorporated into transdermal dosage forms must be submitted in a New Drug Application (NDA) and be approved for marketing by the FDA. The scope of toxicology testing on such a product is usually narrowed to investigate only delivery system-specific toxicity. Delivery-specific toxicity might, for example, include studies such as skin metabolism, skin irritation, and delayed-type hypersensitivity (allergic contact dermatitis). The obvious rationale for this approach is that in addition to any potential toxicity caused by the active compound, each excipient may interact with the active compound, other excipients, or the body itself to cause unanticipated toxicity. This is a case where the sum of the parts (safety of individual components) may not equal the whole (safety of the transdermal system). Comprehensive safety assessment plans for transdermal delivery systems have been described (45,47,59,113).

Despite requirements for testing the finished delivery system, it is always useful and sometimes necessary to test individual components. Without a priori knowledge of the safety profile of each individual component, determining the cause of adverse reactions in a complete transdermal delivery system can quickly lead to a series of time-consuming, expensive studies to identify the source of toxicity. In the early stages of product development, unexpected adverse reactions (usually skin related) in toxicology studies can delay the phase I–II clinical trials. A potentially successful product development program could be unnecessarily terminated owing to a single transdermal delivery system component.

Transdermal excipients, such as stabilizers, emulsifiers, permeation enhancers, antioxidants, buffers, and adhesives, can usually be tested using standard protocols of oral or dermal toxicity as previously described. The structural components, such as solid membranes or foils, and packaging materials should be tested using protocols adapted from the biomaterials or device industry. Although the vast majority of solids used in drug reservoirs, backing layers, release liners, rate-controlling membranes, and such, are considered biologically "inert," it should be shown that the materials do not directly or indirectly contribute to systemic toxicity or to skin reactions. Extractable chemicals, including residual solvents, that could leach from these materials, and interactions between excipients, are the two primary safety concerns. A suggested safety-testing schema for solid polymeric materials follows.

A systematic approach to safety assessment, with a full assessment battery on each of component of the transdermal system is not always practical, economically feasible, necessary from a regulatory standpoint, or scientifically justified. By ignoring how the material will be used in the final product, considerable time, energy, resources, and experimental animals could potentially be wasted. The information as described in Sec II of this chapter may be sufficient.

If few or no safety data are located, it may be necessary to conduct testing after careful consideration of the potential risk the material poses under normal use and misuse or abuse conditions. To guide material selection for transdermal delivery systems, the following strategy may be used:

1. Conduct cytotoxicity screening assays
2. Conduct *USP* class V or VI biological reactivity tests (121)
3. Based on the results of cytotoxicity and *USP* tests, rank materials against each other and determine if they are appropriate for their intended use
4. Conduct finished product testing

Cytotoxicity assays, although limited in their ability to predict hazardous materials in humans, do have value in the overall safety assessment process for polymers and elastomers. A wide variety of assays can be conducted (9,122, 123), primarily as a tool to rank materials against each other in terms of their potential to cause cell damage. Gad (124) discussed the merits and details of cytotoxicity testing and accurately conveys the notion that these assays are useful screening methods, but are not pass or fail assays. The *USP* elution assay (125) and MTT cell viability assay (123) are routinely conducted before or concurrent with *USP* class VI biological reactivity tests (Cukierski, personal communication, 1998).

Most polymeric materials used in transdermal delivery systems can be screened for biological reactivity using procedures outlined in the *USP 23*, ⟨88⟩ Biological Reactivity Tests, In Vivo (121). These tests are used to assess the biological response in animals to polymeric or elastomeric materials (or extracts of those materials) that will have direct or indirect patient contact. The *USP 23*, ⟨88⟩ Biological Reactivity Tests, In Vivo are useful assays to rank materials for potential toxicity or to verify known safety of materials from different lots or manufacturers. After testing, polymeric materials are designated class I to VI depending on the criteria defined in the guidelines. However, *USP 23* ⟨88⟩ guidelines specifically state that the classification schema does not apply to plastics that, "may be used as an integral part of a drug formulation" (121). Nevertheless, the *USP* biological reactivity tests are useful screening assays and should be considered as first-line tests for potential toxicity.

The *USP* class VI designation requires the most comprehensive testing. It consists of three separate assays: (a) a systemic injection test in mice, (b) an intracutaneous injection test in rabbits, and (c) a muscle implantation test in rabbits. As with the cytotoxicity assays, results of these screening studies should not be used as the sole criteria to accept or reject prospective materials from use in a TTS. However, if a candidate material is highly cytotoxic and fails all three

USP class VI tests, its use in direct skin or drug contact applications should be reconsidered.

B. Repeated Dose Toxicity

1. Subchronic (28-Day) Toxicity

Twenty-eight–day toxicology studies conducted in rodents and nonrodents are included in IPEC guidelines for nonclinical studies in support of clinical trials less than 2 weeks in duration (1). The ICH nonclinical duration requirements (for drugs) in support of phases I, II, and III are different for phase I–II trials compared with Phase III trials of the same clinical duration. There are also minor differences between regulatory bodies for the required duration of nonclinical studies in support of the same clinical trial design. Table 2 lists the nonclinical study duration guidelines as proposed for adoption at step 4 of the ICH process.

   a. Species Selection for Subchronic Toxicity by the Dermal Route   As previously described, the selection of species for repeated dose dermal toxicity testing is governed by both regulations and the nature of the test article. Although mice and rats can be used for subchronic and chronic dermal toxicity testing (e.g., skin-painting studies), they can pose technical problems. Guinea pigs, rabbits and dogs are often more appropriate experimental models than rodents for long-term dermal toxicity studies owing to their size and ease of handling. The anatomy and physiology of these species are well known to toxicologists, and they are common dermal toxicity models. Substituting a nonrodent for a rodent species is usually acceptable to regulatory agencies if there are well-defined scientific or technical reasons for the selection.

   Repeated-dose dermal toxicity studies serve a dual purpose. If the test material is absorbed through the skin, assessing systemic toxicity by standard toxicity endpoints is possible. In addition, assessment of local skin reactions, such as cumulative irritation, can be made following multiple skin applications. Study designs can be modified to deliver test article to the same site(s), or to naive sites at each application. For a standard 4-week toxicology protocol outline, see Auletta (100). For solid materials (e.g., polymeric films), other assays may be more appropriate (refer to Sec. IV.A.5 of this chapter).

   b. Dose Selection   In dermal toxicity studies using test article formulations known to flux through skin, dose can be adjusted in two ways: (a) the test article concentration can be kept constant and the total surface area covered is varied, or (b) the application surface area is held constant and the test article concentration is varied. In most instances, the first scenario is preferable. Keeping the test article concentration constant among different dose groups allows direct comparisons of both systemic toxicity and local skin reactions between groups. However, if a solution of excipient is being tested at a concentration below its saturation solubility in the vehicle used for administration, skin flux and thus exposure will increase with increasing concentration of excipient in the vehicle. In this case, a strategy based on option (b) may be preferable. A dose of 1000 mg/kg per day is often considered an upper limit for a 4-week duration, but larger doses can be applied if needed.

It is difficult to use transdermal delivery systems to test single components, because all systems yet fabricated are multi-component delivery systems. However, prototypes have been used to examine essentially a single component or the interaction of two components. If a prototype transdermal delivery system (without active compound) is used to test an excipient, such as an adhesive, then various quantities of patch prototypes can be applied to create different dose groups. For example, group 1 might have two patches applied, group 2, four patches, and so on.

Testing for an excipient intended for an iontophoretic transdermal delivery system may require tests of the entire system to obtain adequate dermal exposure to the excipient. Also, repeated administration to the same site, or on a rotating-site basis, will determine if study outcome will be altered by the dermal effects of repeated exposure to excipient in the presence of electrical current. In clinical trials using iontophoretic systems, test articles are typically placed on previously unexposed skin sites (as much as possible) for each subsequent test article application.

c. Test Article Application   A dosing plan (often included as a diagram) should be included in any repeated-dose dermal toxicity protocol. This plan shows the specific location that test and control articles will be placed on each animal. It should also show the pattern of any intentional changes to application sites at defined intervals during the course of the study, or ''site rotation.'' There are three general patterns of test and control article application that can be described in a protocol: (a) apply the test and control articles to the same location for the entire duration of the study (if possible), (b) apply to the same location for as long as possible (until irritation becomes problematic) and then move to new location for as long as possible, and (c) systematically rotate application sites during the study.

Each of these alternatives has obvious advantages and disadvantages. For example, applying the test article to the same location for 4 weeks will clearly show if cumulative irritation is a potential problem. If local irritation does occur early in the study and the application site is moved, then the study can continue without compromising systemic endpoints. Systematically rotating application sites will minimize the chance of identifying cumulative local irritation, but may be more indicative of clinical use patterns, which are typically 72 h or less on the same skin site. Rotation may be a preferable study design for occlusive applications, as occlusion alone may be poorly tolerated after many successive applications to the same site. Continuing to apply test material to severely irritated skin is strongly discouraged for both scientific and ethical reasons. If the skin becomes very irritated, the absorption kinetics of the test article may be increased or decreased based on the type and degree of injury. This outcome would also necessitate medical intervention (i.e., antibiotics and analgesics) that is required by animal welfare regulations and Institutional Animal Care and Use Committees.

For solid materials (e.g., polymeric films), other assays such as the *USP* class VI tests in conjunction with cytotoxicity assays, genotoxicity assays and final product testing are more relevant than long-term, direct dermal contact studies.

d. Skin Scoring and Histology   Skin sites are usually scored using the Draize scoring scale, with evaluations up to 48 or even 72 h after test article removal. For 4-week toxicity studies, skin site scoring can be accomplished in various ways. A satellite group of animals can be added to the protocol design such that their skin sites are scored up to 48 h following removal of the last test article. These animals can then be euthanized as per the protocol. Another alternative is to score skin sites on or about day 25 of the 28-day study. Skin sites, including test article sites and untreated (control) sites, should be collected and processed for histopathological examination (see Sec. IV.A.1.b). Full histopathological examination of all target organs may be required if systemic toxicity is being conducted primarily using the dermal route. Alternatively, systemic target-organ toxicity may come from oral or parenteral safety assessment studies, as described in subsequent sections.

2. Chronic Toxicity (90 day) to Support Limited Repeat Human Exposure

Ninety-day toxicity studies in rodents and nonrodents are recommended by ICH guidelines for phase I–II clinical trial durations between 2 and 12 weeks. Ninety-day nonclinical studies will only support a phase III clinical trial of up to 4-weeks duration in the European Union, and 4 weeks or less of use of a marketed product in all ICH regions (see Table 2). Recommendations set forth by IPEC include 90-day toxicity (in the most appropriate species) in support of clinical trials for limited and repeated exposure in humans over 2–6 weeks.

There are few practical differences in conducting a 13-week dermal toxicity study versus a 4-week dermal toxicity study. For rodents, the number of animals per sex per group is usually increased in the 13-week study (e.g., from five to ten), and additional samples such as blood or urine can be collected to monitor general health or test article concentrations. For example, the ASTM publishes standard test methods for conducting 90-day dermal toxicity that outline most practical considerations for constructing a protocol (8). Nonrodent study designs are essentially the same as those for 4-week studies.

3.   Chronic Toxicity to Support Long-Term Human Exposures

a. Six-Month Toxicity   As recommended by ICH guidelines, 26- week toxicity studies in rodents are needed to support phase I–II clinical trials equal to or greater than 26 weeks in duration. Twenty-six week rodent studies will also support phase III clinical trials equal to or greater than 3 months. Twenty-six week nonrodent studies will support phase I–II clinical trials less than or equal to 26 weeks duration. Nonrodent studies of 36 weeks should be available before phase III clinical trials exceed 12 weeks in duration.

In contrast with the ICH guidelines for active drug substances, IPEC recommendations list chronic toxicity studies as conditional based on the excipient and its intended pattern of use. Under the IPEC recommendations, chronic toxicity is still conditional, based on professional judgment, when anticipated human exposures would exceed 6 weeks.

In practice, the assay would be conducted in the same manner as a 4- or 13-week dermal toxicity study. In addition to the standard experimental endpoints contained in studies of 26 weeks or less, the collection of additional samples should be considered. Addition of blood sampling for pharmacokinetic or immunological testing are useful endpoints. At necropsy, additional skin samples could be collected and used for in vitro flux studies and special histological- or immuno-histochemical-staining procedures. Skin site evaluations are the same as for the previously described 4-week study design.

b. Twelve-Month Toxicity   Twelve-month dermal toxicity studies are not currently recommended by ICH guidelines and are not routine studies. The decision to conduct a 1-year study for an excipient by the dermal route would have to be based on special circumstances (for example, chronic dermal use in children or possible lifetime exposure). It is likely that if such a study were conducted, it would be as a satellite group in a carcinogenicity or photocarcinogenicity study. Before undertaking such a laborious and expensive study, the appropriate regulatory agency should be consulted for concurrence on the need to conduct this study. Appropriate species for such a study would include guinea pig, rabbit, and dog. As mentioned in the previous section, IPEC recommends that chronic toxicity studies (>12 weeks) should be conditional based on the nature of the excipient and its intended pattern of use and professional judgment of the toxicologist.

In practice, the assay would be conducted in the same manner as a 4-, 13-, 26-, or 36-week dermal study. As mentioned for the 26-week situation, additional experimental endpoints (pharmacokinetic, immunological, histological, physical testing [e.g., skin tensile strength]) should be considered.

C. Reproductive and Developmental Toxicity

Three ICH guidelines address reproductive and developmental toxicity testing for medicinal products. Two of these guidelines (2,3) deal with nonclinical study design, while another (4) describes the timing of nonclinical studies in relation to clinical trials. Noteworthy differences currently exist in clinical trial inclusion criteria between the EU, MHW, and FDA with respect to the availability of non-clinical reproductive and developmental toxicity data. These differences could impact the timing of nonclinical reproductive and developmental toxicity studies in support of clinical trials depending on the governing regulatory body. As stressed throughout this chapter, ICH guidelines are directed specifically at active compounds, and not excipients. However, it is important to be aware of these subtleties for excipients used in chronic therapy or in special patient populations.

Guidelines recommended by IPEC do not require reproductive toxicity testing if the intended clinical exposure is less than two weeks. For clinical exposures greater than two but less than six weeks, rodent and non-rodent teratology studies are recommended. For clinical exposures greater than six weeks, the study of fertility and early embryonic development (to implantation) is recommended in addition to the teratology studies.

1. Dermal, Reproductive, and Developmental Toxicity

If reproductive and developmental toxicity studies are warranted based on the intended use of the excipient and the available data, the next major decision is to determine the route of exposure for the excipient. General regulatory guidance is available from ICH for systemic exposure and procedures to assess the reproductive toxicity profile of a compound (2–3). The clinical route of administration of an excipient should be used in reproductive toxicity studies. Alternative routes which result in a greater body burden and the same metabolic profile as the clinical route can be considered. If an excipient is metabolized by the skin, GI tract, liver, or other metabolic site, such that the metabolites are the same qualitatively, and also quantitatively relative to the anticipated clinical exposure from dermal dosing, then the oral or IV route would be preferable from both logistical and safety standpoints.

The details and merits of specific in vivo or in vitro study designs extend beyond the scope of excipient testing and will not be addressed in this chapter. General overviews of the principles of reproductive toxicology testing are available, including in vitro testing alternatives (126–132).

In practice, conducting a reproductive and developmental toxicity battery by the dermal route may be more technically challenging than by other routes; however, there no special scientific or technical procedures inherent to dermal excipient testing when compared with dermal drug testing. However, there are logistical or technical differences between dermal and oral or IV studies.

The primary scientific concerns for dermal reproductive and developmental toxicity testing designs have been discussed (133). If skin flux or pharmacokinetic data are not available, or if the excipient does not transit skin, then the dermal route is inadequate for risk assessment, and a different route of exposure should be chosen.

The next major concern is the effect of local skin reactions on the outcome of the study. Local skin irritation is not considered maternal toxicity for purposes of developmental toxicity studies. However, severe, local irritation can lead to a variety of secondary systemic effects (e.g., weight loss or infection) that can lead to developmental toxicities. Rotation of application sites may help reduce cumulative skin irritation.

Stress from handling or restraint is known to cause both developmental toxicity and increased fetal death in rats (134). Dermal dosing involves the placement of chambers (e.g., Finn or Hill Top Chambers), occlusive dressings, or solid dosage forms directly on the skin. Most of these applications require bandages or adhesives to keep them in place and protect them from removal by the animals. Shaving application sites and changing these dressings daily can be stressful on animals. Inclusion of a sham-treated control group in study designs is strongly recommended. Depending on the complexity of the application process, a second (untreated) control might be required to fully characterize stress-induced outcomes.

Rodents, usually rats, and rabbits are the two species recommended for reproductive and developmental toxicity testing. Conducting dermal reproductive and developmental toxicity studies with rats is possible (135–142), but poses certain logistical challenges. Using occlusive wrappings on gang-caged rats or rabbits during fertility studies is not advisable, for they tend to chew on and remove bandaging. Rabbits can be dosed quite easily for reproductive and developmental toxicity studies, but they may require collars if the test article is not under an occlusive wrap or if two rabbits are housed together (see Sec. IV.A.1 for wrapping procedures).

The dermal route of administration virtually eliminates nonhuman primates from consideration for reproductive toxicology studies, as these animals will typically remove topically applied test article and any and all bandaging as soon as they are returned to their home cage. Nonhuman primate restraint chairs should only be used if absolutely necessary and if approved by an Institutional Animal Care and Use Committee (IACUC). Because these chairs can be used for only relatively brief periods, this procedure should be attempted only for compounds that rapidly flux through the skin. Even though nonhuman primates can be acclimated to restraint chairs with great success, the use of monkeys in dermal reproductive toxicology testing using this method is labor-intensive, expensive, and potentially dangerous to the animal care staff.

2. Developmental Toxicity Using Alternative Dosing Strategies

Because all reproductive toxicology assays involve daily dosing over specific time intervals, animals are usually subjected to daily handling unless an extended delivery system (infusion pump or ALZET mini-osmotic pump) is used. Implantation of indwelling catheters or ALZET mini-osmotic pumps also results in minor stress from the surgical procedures needed to implant these extended-delivery devices, but these systems have a long history of successful use in developmental toxicity studies.

D. Carcinogenicity

There are three ICH guidelines that specifically address carcinogenicity testing. These guidelines discuss: (a) need for testing (5); (b) testing for carcinogenicity of pharmaceuticals (7); and (c) dose selection for carcinogenicity testing of pharmaceuticals (6). The ICH guidelines state that pharmaceuticals applied topically may need carcinogenicity testing (4). However, pharmaceuticals with little or no systemic exposure from dermal applications may not need oral carcinogenicity testing. These guidelines should be consulted for a complete description of carcinogenicity testing and discussions of special or unique situations. Those discussions will not be repeated here, but the following excerpt from the guideline discussing need for carcinogenicity testing highlights the major considerations that would necessitate testing.

> Several factors which could be considered may include: (1) previous demonstrations of carcinogenic potential in the product class that is considered relevant to humans; (2) structure–activity relationship suggesting carcinogenic risk; (3) evidence of preneoplastic lesions in repeated dose toxicity studies; and (4) long-term tissue retention of parent compound or metabolite(s) resulting in local tissue reactions or other pathophysiological responses.

Even if conducted, results of carcinogenicity studies are not required in support of clinical trials unless there are special concerns as mentioned in the ICH excerpt.

IPEC makes carcinogenicity testing conditional based on the excipient and the intended patient exposure. The need for dermal carcinogenicity testing of excipients by themselves is indeed rare. If an excipient were to be incorporated into a pharmaceutical formulation that was expected to be used by patients continuously or intermittently over extended periods, then final product testing might be warranted. As for all other repeated-dose studies, testing for an excipient intended for an iontophoretic transdermal delivery system may require tests of the entire iontophoretic delivery system to gain adequate exposure and examine interaction of current and excipient. Forbes has summarized methods for dermal carcinogenesis testing (143).

V. SPECIAL STUDIES

Topical studies that address the unique features of dermal application may be required for excipients intended for topical or transdermal formulations. In particular, the dermal route of delivery will often result in exposure of a new excipient to ultraviolet radiation from sun exposure. Because orally or parenterally administered materials are known to distribute into the skin and cause phototoxicity or photoallergy (144), dermal delivery would be anticipated to result in the same outcome for test materials with the propensity to produce similar toxicity.

A. Phototoxicity (Photoirritation)

Phototoxicity testing identifies chemicals (excipients) that react with ultraviolet (UV) light in the skin to cause local toxicity. Phototoxicity is not mediated by the immune system. Phototoxic compounds can reach the skin through direct skin penetration (topical or transdermal preparations) or systemic distribution through the bloodstream into the skin. The phototoxic compound may be either the parent compound or be a metabolite formed in the skin or, for systemically administered compounds, the GI tract or liver.

The ICH guidelines do not specifically address phototoxicity testing. However, if the test article is suspected to be a photoirritant owing to its chemical structure, ICH recommendations that apply to acute dermal toxicity and repeated-dose toxicity would also apply to phototoxicity testing. IPEC recommends phototoxicity testing as part of the base set of toxicity studies for topical and transdermal products.

Phototoxicity testing is not a routine test for most drugs or excipients. However, with the increasing number of compounds delivered by topical or transdermal routes, regulatory authorities are more cognizant of the need for this assay under specific circumstances. For example, topical formulations specifically intended for direct exposure to UV light (i.e., sunscreens) are obvious candidates for phototoxicity testing. Although many transdermal dosage forms are currently hidden by clothing or have a colored backing, these characteristics are changing owing to both technical and social changes. The trend in transdermal formulation is to make systems smaller, thinner, less conspicuous (e.g., clear or translucent) and capable of being worn on numerous skin locations. For example, the original Testoderm Transdermal Therapeutic System was applied exclusively to scrotal skin. A newer version of the Testoderm TTS can be applied to the arm, back, or upper buttocks—locations that are more likely to be exposed to sunlight.

Marzulli and Maibach (144) describe the fundamentals of phototoxicity; Lambert et al (145) discuss animal models and in vitro models for phototoxicity testing. The reader is directed to these two chapters for a thorough review of the salient features of phototoxicity testing.

B. Photocarcinogenicity

The ICH guidelines mention photocarcinogenicity testing in Sec. 4.6 (Extent of Systemic Exposure) in the guideline addressing the need to conduct carcinogenicity studies (5). The guideline states (in vague terms) that photocarcinogenicity studies should be conducted for compounds that cause concern.

IPEC recommends photocarcinogenicity testing as a conditional requirement (e.g., based on the nature of the compound and patient exposure). The FDA evaluates the need for photocarcinogenesis testing on a case-by-case basis (146). If there is concern for photocarcinogenicity, a mouse photocarcinogenicity assay may be required.

1. Photocarcinogenesis Background

Photocarcinogenesis testing methods do not directly identify chemicals that are carcinogens. The purpose of the assay is to identify chemicals that will hasten the development of skin neoplasms owing to exposure to the known carcinogen, UV radiation (UVR); (146). In other words, UVR is the carcinogen, the test article controls the prevalence of tumors. Because there are no validated short-terms assays to predict if a compound will increase the incidence of UVR-induced tumors, a long-term mouse assay is the current assay of choice.

2. Photocarcinogenesis Protocol Design

Because of the specialized nature of photocarcinogenesis testing, few pharmaceutical companies are willing to invest the time, effort, and resources necessary to conduct this type of assay ''in-house.'' However, the same is true for most toxicology contract research organizations. Sample protocols can be obtained on request from laboratories that have experience with this type of work.

An abbreviated summary of the protocol developed by Forbes et al. follows (143). Albino hairless Crl:SKH-hrBR mice (36 per sex per group) are topically dosed 5 days per week with test or control article (on the back and side) and are exposed to UVR on the days of dosing. On Mondays, Wednesdays, and Fridays, the test and control articles are applied to the mice before UVR exposure; on Tuesdays and Thursdays, test and control articles are applied after UVR exposure. The dose of UVR is measured with a Robertson-Berger detector which records both intensity and cumulative dose. Test and control mice receive approximately 600 Robertson-Berger units (RBU) per week. A ''higher control'' group receives approximately 1200 RBU per week. Mice are exposed to UVR for 40 weeks and then are monitored for an additional 12 weeks before sacrifice. Experimental endpoints include: clinical observations, tumor mapping, body weights, gross necropsy, tumor data, and other routine toxicological data. Tumor statistics are used to define the median onset (time at which one-half of group members have one or more qualifying tumors [>1 mm in diameter]); mortality-free prevalence; and tumor yield (mean tumors per mouse).

VI. CONCLUSION

Requirements and strategies for safety evaluation of new topical excipients as individual agents, separate from topical or transdermal formulations, have not been clearly delineated by regulatory agencies. ICH guidelines for pharmacologically active agents may be extrapolated to new excipients, but all aspects of active drug substance testing are not relevant to the safety evaluation of excipients. The initiative by IPEC to outline safety testing for excipients, separate from a clearly defined drug product, provides modified strategies that the toxicologist can consult in planning a safety program. Ultimately, the scientific acumen of the toxicologist, in collaboration with scientists from a variety of disciplines, including pharmaceutics, drug metabolism, and pharmacology, and the clinical pharmacologists who are responsible for the first introduction of a new excipient to humans, must judge if the nonclinical safety evaluation supports human clinical experimentation. Scientists within regulatory agencies should not be overlooked as valuable resources that can help a sponsor design a science-based safety assessment package for a new pharmaceutical excipient.

REFERENCES

1. M Steinberg, JF Borzelleca, EK Enters, FK Kinoshita, A Loper, DB Mitchell, CB Tamulinas, ML Weiner. The Safety Committee of the International Pharmaceutical Excipients Council. A new approach to the safety assessment of pharmaceutical excipients. Regul Toxicol Pharmacol 24:149–154, 1996.
2. The International Conference on Harmonisation of Technical Requirements for Registration of Pharmaceuticals for Human Use (ICH). Detection of toxicity to reproduction for medicinal products. 1993.
3. ICH. Toxicity to male fertility; an addendum to the ICH Tripartite Guideline on Detection of Toxicity to Reproduction for Medicinal Products. 1995.
4. ICH. Non-clinical safety studies for the conduct of human clinical trials for pharmaceuticals. 1997.
5. ICH. Guideline on the need for carcinogenicity studies of pharmaceuticals. 1995.
6. ICH. Dose selection for carcinogenicity studies of pharmaceuticals. 1994.
7. ICH. Testing for carcinogenicity of pharmaceuticals. 1997.
8. American Society for Testing and Materials (ASTM). Standard test method for determining subchronic dermal toxicity. Designation: E 1103-96. West Conshohocken: ASTM, 1996.
9. International Organization for Standardization (ISO). 10993-5 Test for cytotoxicity: *in vitro* methods. Geneva, Switzerland, 1992.
10. Organization for Economic Cooperation and Development (OECD). Guideline 406: skin sensitization, 1992.
11. Code of Federal Regulations (CFR). Washington DC: US Government Printing Office, Title 16, 1500.41, 1997.

12. CFR. Washington DC: US Government Printing Office, Title 21, 314.50 (d) (2) (ii), 1997.

13. CFR. Washington DC: US Government Printing Office, Title 21, 58, 1997.

14. Center for Drug Evaluation and Research, (CDER), U.S. Food and Drug Administration. Points to consider in the nonclinical pharmacology/toxicology development of topical drugs intended to prevent the transmission of sexually transmitted diseases (STD) and/or for the development of drugs intended to act as vaginal contraceptives. Informal communication under 21 CFR 10.90 (b) (9), 1996.

15. JC Liu, SJ Wisniewski. Recent advances in topical drug delivery systems. In: TK Ghosh, WR Pfister, SI Yum, eds. Transdermal and Topical Drug Delivery Systems. Buffalo Grove: Interpharm Press, 1997, pp 593–612.

16. WR Pfister. Transdermal and dermal therapeutic systems: current status. In: TK Ghosh, WR Pfister, SI Yum, eds. Transdermal and Topical Drug Delivery Systems. Buffalo Grove: Interpharm Press, 1997, pp 33–112.

17. B Berner, GC Mazzenga, JH Otte, RJ Steffens, RH Juang, CD Ebert. Ethanol: water mutually enhanced transdermal therapeutic system II: skin permeation of ethanol and nitroglycerin. J Pharm Sci 78:402–407, 1989.

18. JE Harrison, AC Watkinson, DM Green, J Hadgraft, K Brain. The relative effect of Azone and Transcutol on permeant diffusivity and solubility in human stratum corneum. Pharm Res 13:542–546, 1996.

19. K Knutson, SL Krill, J Zhang. Sovent-mediated alterations of the stratum corneum. J Controlled Release 11:93–103, 1990.

20. GB Kasting, WR Francis, GE Roberts. Skin penetration enhancement of triprolidine base by propylene glycol. J Pharm Sci 82:551–552, 1993.

21. WJ Lambert, WI Higuchi, K Knutson, SL Krill. Dose-dependent enhancement effects of Azone on skin permeability. Pharm Res 6:798–803, 1989.

22. LK Pershing, LD Lambert, K Knutson. Mechanism of ethanol-enhanced estradiol permeation across human skin *in vivo*. Pharm Res 7:170–175, 1990.

23. H Tanojo, JA Bouwstra, HE Junginger, HE Bodde. *In vitro* human skin barrier modulation by fatty acids: skin permeation and thermal analysis studies. Pharm Res 14:42–49, 1997.

24. M Yazdanian, E Chen. The effect of diethylene glycol monoethyl ether as a vehicle for topical delivery of ivermectin. Vet Res Commun 19:309–319, 1995.

25. DW Osborne, JJ Henke. Skin penetration enhancers cited in the technical literature. Pharm Technol Nov:58–66, 1997.

26. SC Chattaraj, RB Walker. Penetration enhancer classification. In: EW Smith, HI Maibach, eds. Percutaneous Penetration Enhancers. Boca Raton, FL: CRC Press, 1995, pp. 5–20.

27. MS Roberts, J Singh, N Yoshida, KI Currie. Ionotophoretic transport of selected solutes through human epidermis. In: RC Scott, RH Guy, J Hadgraft, eds. Prediction of Percutaneous Penetration. Methods, Measurements and Modelling. Proceedings of the Conference held in April, London: IBC Technical Services, 1989, pp 231–241.

28. BH Sage, JE Riviere. Model systems in iontophoresis—transport efficacy. Adv Drug Deliv Rev 9:265–287, 1992.

29. P Singh, HI Maibach. Transdermal iontophoresis pharmacokinetic considerations. Clin Pharmacokinet 26:327–334, 1994.

30. NG Turner, L Ferry, M Price, C Cullander, RH Guy. Iontophoresis of poly-L-lysines: the role of molecular weight? Pharm Res 14:1322–1331, 1997.

31. BR Meyer, W Kreis, J Eschbach, V O'Mara, S Rosen, D Sibalis. Successful transdermal administration of therapeutic doses of a polypeptide to normal human volunteers. Clin Pharmacol Ther 44:607–612, 1988.

32. BR Meyer, W Kreis, J Eschbach, V O'Mara, S Rosen, D Sibalis. Transdermal versus subcutaneous leuprolide: a comparison of acute pharmacodynamic effect. Clin Pharmacol Ther 48:340–345, 1990.

33. SB Ruddy, BW Hadzija. Iontophoretic permeability of polyethylene glycols through hairless rat skin: application of hydrodynamic theory for hindered transport through liquid-filled pores. Drug Design Discov 8:207–224, 1992.

34. PG Green, RS Hinz, A Kim, FC Szoka, RH Guy. Iontophoretic delivery of a series of tripeptides across the skin in vitro. Pharm Res 8:1121–1127, 1991.

35. MJ Pikal, S Shah. Transport mechanisms in iontophoresis. III. An experimental study of the contributions of electroosmotic flow and permeability change in the transport of low and high molecular weight solutes. Pharm Res 7:222–229, 1990.

36. J Hirvonen, RH Guy. Iontophoretic delivery across the skin: electroosmosis and its modulation by drug substances. Pharm Res 14:1258–1263, 1997.

37. SB Ruddy, BW Hadzija. The role of stratum corneum in electrically facilitated transdermal drug delivery. I. influence of hydration, tape-stripping and delipidization on the DC electrical properties of skin. J Controlled Release 37:225–238, 1995.

38. PM Elias, GK Menon. Structural and lipid biochemical correlates of the epidermal permeability barrier. Adv Lipid Res 24:1-26, 1991.

39. S Mitragotri, D Blankschtein, R Langer. Transdermal drug delivery using low-frequency sonophoresis. Pharm Res 13:411–420, 1996.

40. S Mitragotri, D Blankschtein, R Langer. Ultrasound-mediated transdermal protein delivery. Science 269:850–853, 1995.

41. R Vanbever, E Le Boulenge, V Preat. Transdermal delivery of fentanyl by electroporation I. Influence of electrical factors. Pharm Res 13:559–565, 1996.

42. DF Woodard, AL Nieves, LS Williams, CS Spada, SB Wawley, JL Duenes. A new hairless strain of guinea pig: characterization of the cutaneous morphology and pharmacology. In: HI Maibach, NJ Lowe, eds. Models in Dermatology. vol 4. Basels: Karger, 1989, pp 71–78.

43. EV Buehler, JJ Kreuzmann, A Sakr, XH Gu. Comparable sensitivity of hairless and Hartley strain guinea pigs to a primary irritant and a sensitizer. J Toxicol Cutan Ocul Toxicol 9:163–168, 1990.

44. AE Chester, TG Terrell, E Nave, AE Dorr, LR DePass. Dermal sensitization study in hairless guinea pigs with dinitrochlorobenzene and ethyl aminobenzoate. J Toxicol Cutan Ocul Toxicol 7:273–281, 1988.

45. J Ademola, HI Maibach. Safety assessment of transdermal and topical dermatological products. In: TK Ghosh, WR Pfister, SI Yum, eds. Transdermal and Topical Drug Delivery Systems. Buffalo Grove: Interpharm Press, 1997, pp 191–214.

46. EV Buehler. Prospective testing for delayed contact hypersensitivity in guinea pigs: the Buehler method. In: GR Burleson, JH Dean, AE Munson, eds. Methods in Immunotoxicol, vol 2. New York: Wiley-Liss, 1995, pp 343–356.

47. M Prevo, M Cormier, K Nichols. Predictive toxicology methods for transdermal

delivery systems. In: FN Marzulli, HI Maibach, eds. Dermatotoxicology. 5th ed. Washington, DC: Taylor & Francis, 1996, pp 397–410.

48. Guideline for industry toxicokinetics: the assessment of systemic exposure in toxicity studies. Federal Register, (60 FR 11264) March 1, 1995.

49. JC Keister, GB Kasting. The use of transient diffusion to investigate transport pathways through skin. J Controlled Release 4:111–117, 1986.

50. B Illel, H Schaefer, J Wepierre. Follicles play an important role in percutaneous absorption. J Pharm Sci 80:424–427, 1991.

51. J Kao, J Hall, G Helman. *In vitro* percutaneous absorption in mouse skin: influence of skin appendages. Toxicol Appl Pharmacol 94:93-103, 1988.

52. A Rolland, N Wagner, A Chatelus, B Shroot, H Schaefer. Site-specific drug delivery to pilosebaceous structures using polymeric microspheres. Pharm Res 10:1738–1744, 1993.

53. HJ Bidmon, JD Pitts, HF Solomon, JV Bondi, WE Stumpf. Estradiol distribution and penetration in rat skin after topical application, studied by high resolution autoradiography. Histochemistry 95:43–54, 1990.

54. AC Lauer, LM Lieb, C Ramachandran, GL Flynn, ND Weiner. Transfollicular drug delivery. Pharm Res 12:179–186, 1995.

55. AS Michaels, SK Chandrasekaran, JE Shaw. Drug permeation through human skin: theory and *in vitro* experimental measurement. AICHE J 21:985–996, 1975.

56. SK Chandrasekaran, AS Michaels, PS Campbell, JE Shaw. Scopolamine permeation through human skin *in vitro*. AICHE J 22:828–832, 1976.

57. RH Guy, EM Carlstrom, DAW Bucks, RS Hinz, HI Maibach. Percutaneous penetration of nicotinates: *in vivo* and *in vitro* measurements. J Pharm Sci 75:968–972, 1986.

58. DR Friend. *In vitro* skin permeation techniques. J Controlled Release 18:235–248, 1992.

59. TS Spencer. Preclinical assessment of transdermal drug delivery systems. In: TK Ghosh, WR Pfister, SI Yum, eds. Transdermal and Topical Drug Delivery Systems. Buffalo Grove: Interpharm Press, 1997, pp 167–179.

60. GB Kasting, TG Filloon, WR Francis, MP Meredith. Improving the sensitivity of *in vitro* skin penetration experiments. Pharm Res 11:1747–1754, 1994.

61. JP Skelly, VP Shah, HI Maibach, RH Guy, RC Wester, G Flynn, A Yacobi. FDA and AAPS report of the workshop on principles and practices of *in vitro* percutaneous penetration studies: relevance to bioavailability and bioequivalence. Pharm Res 4:265–267, 1987.

62. TJ Franz. Percutaneous absorption. On the relevance of *in vitro* data. J Invest Dermatol 64:190–195, 1975.

63. A Rougier, C Lotte, H Maibach. *In vivo* percutaneous penetration of some organic compounds related to anatomic site in humans: predictive assessment by the stripping method. J Pharm Sci 76:451–454, 1987.

64. AK Shah, G Wei, RC Lanman, VO Bhargava, SJ Weir. Percutaneous absorption of ketoprofen from different anatomical sites in man. Pharm Res 13:168–172, 1996.

65. GB Kasting, LA Bowman. Electrical analysis of fresh, excised human skin: a comparison with frozen skin. Pharm Res 7:1141–1146, 1990.

66. TJ Franz, PA Lehman. The use of water permeability as a means of validation for

skin integrity in *in vitro* percutaneous absorption studies. J Invest Dermatol 95: 525, 1990.

67. GB Kasting, LA Bowman. DC electrical properties of frozen, excised human skin. Pharm Res 7:134–143, 1990.
68. J Kao, FK Patterson, J Hall. Skin penetration and metabolism of topically applied chemicals in six mammalian species, including man: an *in vitro* study with benzo [*a*]pyrene and testosterone. Toxicol Appl Pharmacol 81:502-516, 1985.
69. RC Wester, J Christoffel, T Hartway, N Poblete, HI Maibach, J Forsell. Human cadaver skin viability for *in vitro* percutaneous absorption: storage and detrimental effects of heat-separation and freezing. Pharm Res 15:82–84, 1998.
70. F Raffali, A Rougier, R Roguet. Measurement and modulation of cytochrome-P450 dependent enzyme activity in cultured human keratinocytes. Skin Pharmacol 7: 345-354, 1994.
71. B Coulomb, C Lebreton, L Dubertret. The skin equivalent: A model for skin and general pharmacology. In: HI Maibach, NJ Lowe, eds. Models in Dermatology. vol. 4. Basel: Karger, 1989, pp. 20–29.
72. AE Loper, AS Michaels, G Bamopoulos, LL Ng, JV Bondi, EM Cohen. Drug transport through membranes from a reservoir with time-dependent diffusional resistance. Proceedings of the 11th International Symposium of Controlled Release of Bioactive Materials, 1984.
73. MT Hojyo-Tomoka, AM Kligman. Does cellophane tape stripping remove the horny layer? Arch Dermatol 106:767–768, 1972.
74. RL Bronaugh, RF Stewart. Methods for *in vitro* percutaneous absorption studies V: permeation through damaged skin. J Pharm Sci 74:1062–1066, 1985.
75. NG Turner, YN Kalia, RH Guy. The effect of current on skin barrier function *in vivo*: recovery kinetics post-iontophoresis. Pharm Res 14:1252–1257, 1997.
76. GL Flynn, SH Yalkowsky, TJ Roseman. Mass transport phenomena and models: theoretical concepts. J Pharm Sci 63:1276–1280, 1974.
77. J Palm, FG Ferguson. Fuzzy, a hypotrichotic mutant in linkage group I of the Norway rat. J Heredity 67:284–288, 1976.
78. AE Loper, LA Grabowski, EK Fong, MD Wiernik, JV Bondi. Fuzzy rat skin as a surrogate for cadaver skin for *in vitro* diffusion studies. Pharm Res 3:72S, 1986.
79. RL Bronaugh, RF Stewart. Methods for *in vitro* percutaneous absorption studies VI: preparation of the barrier layer. J Pharm Sci 75:487–491, 1986.
80. GG Krueger, ZJ Wojciechowski, SA Burton, A Gilhar, SE Huether, LG Leonard, UD Rohr, TJ Petelenz, WI Higuchi, LK Pershing. The development of a rat/human skin flap served by a defined and accessible vasculature on a congenitally athymic (nude) rat. Fundam Appl Toxicol 6:S112–S121, 1985.
81. JE Riviere, KF Bowman, NA Monteiro-Riviere, LP Dix, MP Carver. The isolated perfused porcine skin flap (IPPSF). Fundam Appl Toxicol 7:444-453, 1986.
82. E Cheriathundam, R Almirez, AP Alvares. Comparisons of hepatic and renal cytochrome P-450–dependent monooxygenases from Fuzzy and Sprague–Dawley rats. Drug Metab Dispos 20:19–22, 1992.
83. WM Muck, JD Henion. High-performance liquid chromatography/tandem mass spectrometry: its use for the identification of stanozolol and its major metabolites in human and equine urine. Biomed Environ Mass Spectrom 19:37–51, 1990.

84. TV Olah, DA McLoughlin, JD Gilbert. The simultaneous determination of mixtures of drug candidates by liquid chromatography/atmospheric pressure chemical ionization mass spectrometry as an *in vivo* drug screening procedure. Rapid Commun Mass Spectrom 11:17–23, 1997.

85. PK Noonan, RC Wester. Cutaneous metabolism of xenobiotics. In: RL Bronaugh, HI Maibach, eds. Percutaneous Absorption Mechanisms—Methodology—Drug Delivery. New York: Marcel Dekker, 1985, pp 65-85.

86. DAW Bucks. Skin structure and metabolism: relevance to the design of cutaneous therapeutics. Pharm Res 1:148–153, 1984.

87. RJ Martin, SP Denyer, J Hadgraft. Skin metabolism of topically applied compounds. Int J Pharm 39:23–32, 1987.

88. JC Loo, S Riegelman. New method for calculating the intrinsic absorption rate of drugs. J Pharm Sci 57:918–928, 1968.

89. A Yacobi, RA Baughman, DB Cosulich, G Nicolau. Method for determination of first-pass metabolism in human skin. J Pharm Sci 73:1499–1500, 1984.

90. WR Gillespie, P Veng Pedersen. A polyexponential deconvolution method. Evaluation of the gastrointestinal bioavailability and mean *in vivo* dissolution time of some ibuprofen dosage forms. J Pharmacokinet Biopharm 13:289–307, 1985.

91. JG Wagner. Pharmacokinetics for the Pharmaceutical Scientist. Lancaster: Technomic Publishing, 1993, pp 159–206.

92. Guideline for Industry Pharmacokinetics: Guidance for repeated dose tissue distribution studies. Federal Register 60(40):11274-11275, 1995.

93. WR Brown, MH Hardy. The asebia mouse: an animal model of psoriasiform disease. In: HI Maibach, NJ Lowe, eds. Models in Dermatology. vol 1. New York: S Karger, 1985, pp 220–227.

94. SM Puhvel. Animal models for comedolytic agents. In: HI Maibach, NJ Lowe, eds. Models in Dermatology. vol 2. New York: S Karger, 1985, pp 64–69.

95. VP Shah, GL Flynn, A Yacobi, HI Maibach, C Bon, NM Fleischer, TJ Franz, SA Kaplan, J Kawamoto, LJ Lesko, JP Marty, LK Pershing, H Schaefer, JA Sequeira, SP Shrivastava, J Wilkin, RL Williams. AAPS/FDA workshop report bioequivalence of topical dermatological dosage forms—methods of evaluation of bioequivalence. Pharm Technol Mar:140–148, 1998.

96. CFH Vickers. Stratum corneum reservoir for drugs. In: W Montagna, EJ Van Scott, RB Stoughton, eds. Advances in Biology of Skin, vol. 12. Pharmacology and the Skin. New York: Appleton-Century-Crofts, 1969, pp 177–189.

97. EK Fong, JD Frank, ME Cartwright, AE Loper. Improved method for rhesus monkey or rat skin preparation and cryosectioning for topical or transdermal skin distribution studies. Exp Dermatol 5:45–48, 1996.

98. JAM Neelissen. Visualization of percutaneous steroid transport. PhD dissertation, Leiden University, The Netherlands, 1962.

99. C Anderson, T Andersson, A Boman. Cutaneous microdialysis for human *in vivo* dermal absorption studies. In: MS Roberts, KA Walters, eds. Dermal Absorption and Toxicity Assessment. New York: Marcel Dekker, 1998, pp 231–244.

100. CS Auletta. Acute, subchronic and chronic toxicology. In: MJ Derelanko, MA Hollinger, eds. CRC Handbook of Toxicology. Boca Raton, FL: CRC Press, 1995, pp 51–104.

101. RE Rush, KL Bonnette, DA Douds, TN Merriman. Dermal Irritation and Sensitization. In: MJ Derelanko, MA Hollinger, eds. CRC Handbook of Toxicology. Boca Raton, FL: CRC Press, 1995, pp 105–162.

102. JN Lawrence. Application of *in vitro* human skin models to dermal irritancy: a brief overview and future prospects. Toxicol In Vitro 11:305–312, 1997.

103. JH Draize, G Woodard, HO Calvery. Methods for the study of irritation and toxicity of substances applied topically to the skin and mucous membranes. J Pharmacol Exp Ther 82:377–390, 1944.

104. AJP Klein-Szanto, CJ Conti, CM Aldaz. Skin and oral mucosa. In: WM Haschek, CG Rousseaux, eds. Handbook of Toxicologic Pathology. San Diego: Academic Press, 1991, pp 165–193.

105. O de Silva, DA Basketter, MD Barratt, E Corsini, MTD Cronin, PK Das, J Degwert, A Enk, JL Garrigue, C Hauser, I Kimber, J-P Lepoittevin, J Peguet, M Ponec. Alternative methods for skin sensitisation testing: the report and recommendations of ECVAM Workshop 19. ALTA 24:683–705, 1996.

106. H Bour, M Krasteva, J-F Nicolas. Allergic contact dermatitis. In: JD Bos, ed. Skin Immune System (SIS): Cutaneous Immunology and Clinical Immunodermatology. 2nd ed. Boca Raton, FL: CRC Press, 1997, pp 509-522.

107. SC Gad. Safety Evaluation of Medical Devices. New York: Marcel Dekker, 1997, pp 131–188.

108. I Kimber, RJ Dearman. Contact hypersensitivity: immunological mechanisms. In: I Kimber, T Maurer, eds. Toxicology of Contact Hypersensitivity. London: Taylor & Francis, 1996, pp 4–25.

109. E Patrick, H Maibach. Dermatotoxicology. In: Hayes AW, Principles and Methods of Toxicology. 3rd ed. New York: Raven Press, 1994, pp 767–803.

110. RJ Scheper, BME von Blomberg. Mechanisms of allergic contact dermatitis to chemicals. In: JG Vos, M Younes, E Smith, eds. Allergic Hypersensitivities Inducted by Chemicals: Recommendations for Prevention. Boca Raton, FL: CRC Press, 1996, pp 185–201.

111. G Klecak. Test methods for allergic contact dermatitis in animals. In: FN Marzulli, HI Maibach, eds. Dermatotoxicology Methods: The Laboratory Worker's *Vade Mecum*. Washington DC: Taylor & Francis, 1998, pp 121–143.

112. T Maurer. Guinea pig predictive tests. In: I Kimber, T Maurer, eds. Toxicology of Contact Hypersensitivity. London: Taylor & Francis, 1996, pp 107-126.

113. M Prevo, M Cormier, J Matriano. Developing a toxicology evaluation plan for transdermal delivery systems. In: FN Marzulli, HI Maibach, eds. Dermatotoxicology Methods: The Laboratory Worker's *Vade Mecum*. Washington DC: Taylor & Francis, 1998, pp 75–87.

114. HC Maguire. Estimation of the allergenicity of prospective human contact sensitizers in the guinea pig. In: HI Maibach, ed. Animal Models in Dermatology. New York: Churchill Livingstone, 1975, pp 67–75.

115. JF Griffith. Predictive and diagnostic testing for contact sensitization. Toxicol Appl Pharmacol 3:90–102, 1969.

116. EV Buehler. Delayed contact hypersensitivity in the guinea pig. Arch Dermatol 91:171–175, 1965.

117. DA Basketter, G Cookman, GF Gerberick, N Hamaide, M Potokar. Skin sensitiza-

tion thresholds: determination in predictive models. Food Chem Toxicol 35:417–425, 1997.

118. B Haeberlin, DR Friend. Anatomy and physiology of the gastrointestinal tract: implications for colonic drug delivery. In: DR Friend, ed. Oral Colon-Specific Drug Delivery. Boca Raton, FL: CRC Press, 1992, pp 1–43.

119. O Braun-Falco, HC Korting. Normal pH value of human skin. Hautarzt 37:126–129, 1986.

120. D Matsushima, ME Prevo, J Gorsline. Absorption and adverse effects following topical and oral administration of three transdermal nicotine products to dogs. J Pharm Sci 84:365–369, 1995.

121. United States Pharmacopoeia 23/National Formulary 18. Rockville, MD: United States Pharmacopoeia Convention, 1995, pp 1699–1703.

122. RI Freshney. Measurement of viability and cytotoxicity. In: RI Freshney, ed. Culture of Animal Cells: A Manual of Basic Technique. 3rd ed. New York: Wiley-Liss 1994, pp 287–307.

123. T Mosmann. Rapid colorimetric assay for cellular growth and survival: application to proliferation and cytotoxicity assays. J Immunol Methods 65:55–63, 1983.

124. SC Gad. Safety Evaluation of Medical Devices. New York: Marcel Dekker, 1997, pp 75–84.

125. United States Pharmacopeia 23/National Formulary 18. Rockville, MD: United States Pharmacopeial Convention, 1995, pp 1697–1699.

126. PJ Wier, LD Wise. Middle Atlantic Reproduction and Teratology Association: bibliography of reproductive and developmental toxicology test guidelines and related publications. In: RD Hood, ed. Developmental Toxicology. Boca Raton: CRC Press, 1997, pp 735–749.

127. DJ Ecobichon. Reproductive toxicology. In: DJ Ecobichon, ed. The Basis of Toxicity Testing. 2nd ed. Boca Raton, FL: CRC Press, 1997, pp 117–156.

128. DJ Ecobichon. Reproductive toxicology. In: M.J. Derelanko, M.A. Hollinger, eds. CRC Handbook of Toxicology. Boca Raton, FL: CRC Press, 1995, pp 379–402.

129. KM MacKenzie, RM Hoar. Developmental toxicology. In: MJ Derelanko, MA Hollinger, eds. CRC Handbook of Toxicology. Boca Raton, FL: CRC Press, 1995, pp 403–450.

130. JM Manson, YJ Kang. Test methods for assessing female reproductive and developmental toxicology. In: AW Hayes, ed. Principles and Methods of Toxicology. 3rd ed. New York: Raven Press, 1994, pp 989–1037.

131. JM Rogers, RJ Kavlock. Developmental toxicology. In: CD Klaassen, ed. Casarett & Doull's Toxicology, The Basic Science of Poisons. 5th ed. New York: McGraw-Hill, 1996, pp 301–331.

132. JA Thomas. Toxic responses of the reproductive system. In: CD Klaassen, ed. Casarett & Doull's Toxicology, The Basic Science of Poisons. 5th ed. New York: McGraw-Hill, 1996, pp 547–581.

133. CA Kimmel, EZ Francis. Proceedings of the workshop on the acceptability and interpretation of dermal developmental toxicity studies. Fundam Appl Toxicol 14:386–398, 1990.

134. JM Rogers, RJ Kavlock. Developmental toxicology. In: CD Klaassen ed. Casar-

ett & Doull's Toxicology: The Basic Science of Poisons. New York: McGraw-Hill, 1996, pp 301–331.

135. PJ Becci, MJ Knickerbocker, EL Reagan, RA Parent, LW Burnette. Teratogenicity study of *N*-methylpyrrolidone after dermal application to Sprague–Dawley rats. Fundam Appl Toxicol 2:73–76, 1982.

136. MH Feuston, CR Mackerer. Developmental toxicity study in rats exposed dermally to clarified slurry oil for a limited period of gestation. J Toxicol Environ Health 49:207–220, 1996.

137. MH Feuston, CR Mackerer, CA Schreiner, CE Hamilton. Systemic toxicity of dermally applied crude oils in rats. J Toxicol Environ Health 51:387–399, 1997.

138. MH Feuston, SL Kerstetter, EJ Singer, MA Mehlman. Developmental toxicity of clarified slurry oil applied dermally to rats. Toxicol Ind Health 5:587–599, 1989.

139. MH Feuston, SL Kerstetter, PD Wilson. Teratogenicity of 2-methoxyethanol applied as a single dermal dose to rats. Fundam Appl Toxicol 15:448–456, 1990.

140. AM Hoberman, MS Christian, S Lovre, R Roth, F Koschier. Developmental toxicity study of clarified slurry oil (CSO) in the rat. Fundam Appl Toxicol 1:34–40, 1995.

141. RE Seegmiller, MW Carter, WH Ford, RD White. Induction of maternal toxicity in the rat by dermal application of retinoic acid and its effect on fetal outcome. Reprod Toxicol 4:277–281, 1990.

142. RW Tyl, LC Fisher, MF Kubena, MA Vrbanic, R Gingell, D Guest, JR Hodgson, SR Murphy, TR Tyler, BD Astill. The developmental toxicity of 2-ethylhexanol applied dermally to pregnant Fischer 344 rats. Fundam Appl Toxicol 19:176–185, 1992.

143. PD Forbes, CP Sambuco, GE Dearlove, RM Parker, AL Kiorpes, JH Wedig. Sample protocols for carcinogenesis and photocarcinogenesis. In: FN Marzulli, HI Maibach, eds. Dermatotoxicology Methods: The Laboratory Worker's *Vade Mecum*. Washington DC: Taylor & Francis, 1998, pp 281–302.

144. FN Marzulli, HI Maibach. Photoirritation (phototoxicity, phototoxic dermatitis). In: FN Marzulli, HI Maibach, eds. Dermatotoxicology. 5th ed. Washington, DC: Taylor & Francis, 1996, pp 231–237.

145. LA Lambert, WG Wamer, A Kornhauser. Animal models for phototoxicity testing. In: FN Marzulli, HI Maibach, eds. Dermatotoxicology. 5th ed. Washington, DC: Taylor & Francis, 1996, pp 515–529.

146. PD Forbes. Carcinogenesis and photocarcinogenesis test methods. In: FN Marzulli, HI Maibach, eds. Dermatotoxicology. 5th ed. Washington, DC: Taylor & Francis, 1996, pp 535–544.

8
Routes of Exposure: Inhalation and Intranasal

Charmille B. Tamulinas and Chet L. Leach
3M Pharmaceuticals, St. Paul, Minnesota

I. EXCIPIENT TESTING BY THE INHALED ROUTE

For those excipients that are intended for use in inhaled or intranasal products, it is obviously necessary to conduct the bulk of the safety-testing program by the intended route of administration (e.g., inhalation and intranasal). Information by other routes can also be useful, but should be considered supplemental. For example, often those excipients that are given by inhalation have a significant oral component; therefore, oral toxicity data are useful and necessary in assessing safety. Similarly, toxicity information by the intravenous route can be useful when the excipient has a significant pulmonary exposure and crosses the lungs quickly and intact. Intratracheal studies can also be useful in preliminary screening studies, especially if test material quantities are limited or inhalation facilities are not available. However, the best and most relevant information is usually obtained from exposure systems that most closely mimic the intended route of exposure. This would include testing the product as a whole, including all excipients and drugs in one formulation delivered to the test species in the same manner as it will be with human exposure. However, this is not always possible and it may be desirable to test the excipient in a stand-alone situation to produce safety data on an excipient for use in many different products.

A. Physical Form of the Excipient

The term *excipient* can be broadly applied to any intentionally added substance to which humans may be exposed other than the active drug. The most common inhaled products are metered-dose inhalers (MDIs) used for the treatment of asthma. MDIs contain many excipients that have undergone extensive safety testing; hence; throughout this chapter examples will be drawn from MDI excipient testing programs. MDIs also contain all forms of excipients, including solid particles, volatile and nonvolatile liquids, and gases. Each of the physical forms of excipients presents a different challenge for conducting safety programs.

1.   Solid and Liquid Excipients

Examples of nonvolatile liquid excipients include surfactants, such as oleic acid, sorbitan trioleate, and lecithin. They are used to produce a well-suspended drug within the canister and as valve lubricants. They usually form an association with the drug and thus, are delivered to tissue along with the drug. Surfactant delivery may follow in direct proportion to the drug, or it may be skewed toward greater concentration in smaller or larger drug particles. Thus, it is useful to know the degree of association between surfactants and drugs. When formulated in high enough concentrations, surfactants can add enough mass to a drug to increase particle size, which can ultimately affect the site of drug deposition. Surfactants may also change drug deposition by altering the charge on the particle and by altering the hygroscopic properties of the drug. Thus, in excipient-testing programs it is important to perform tests on the pure surfactant as well as the final formulation of drug plus surfactant when feasible.

There are many other liquid excipients, with a variety of functions present in inhaled products. Examples of liquid excipients include water, chlorofluorocarbon-11, ethanol, phenylethylalcohol, and propylene glycol. It is important to know the ultimate fate of the liquid excipient to conduct relevant inhalation studies. For example, depending on the product configuration, ethanol can either deposit in the upper respiratory tract as liquid droplets, or it can be vaporized and deposited in all regions of the respiratory tract and pulmonary tissue. It is advisable to perform inhalation studies that assure exposure of all potential tissues where excipients may be deposited, but most excipient exposure should mimic the most probable human tissue sites of delivery. There are many ways to generate vapors of liquid aerosols that are volatile. The most direct way is by heating the liquid to produce vapors and then cooling the vapor atmosphere before animal inhalation (1). Pressure vessel generators can also be used whereby the liquid is placed in a pressure vessel, pressurized with air or nitrogen, and then simply metered through a spray nozzle into an airstream (2). Liquid droplet aerosols can be produced from Laskin-type generators, air-blast nebulizers, ultrasonic nebulizers, and spinning top generators.

Formulation aids for nasal sprays include benzalkonium chloride, edetate disodium, sodium chloride, sodium hydroxide, hydrochloric acid, citric acid, benzethonium chloride, cellulose, dextrose, and carboxymethylcellulose. Other solid excipients may be present as drug carriers, such as lactose in dry powder inhalers. These carriers add bulk to drugs and aid in the metering process, or they may be present to alter the final delivered particle size. Dry carriers may remain attached to the drug through final tissue deposition, or they may become separated during the aerosolization process. Laboratory methods, such as cascade impactor tests, are essential in determining particle size and drug association of excipients. Still other solid excipients, such as menthol, are used as taste enhancers to make a bitter drug more palatable.

There are a variety of methods for generating atmospheres of solid particulates. Common methods of aerosol generation include the Wright Dust Feeder (3), the Timbrell Dust Generator (4), fluidized bed generators (5), "hopper-type" generators, and dual-brush generators (6). The advantages and disadvantages of these generators have been reviewed elsewhere (1). Additional methods can be used by employing generators using solids dissolved or suspended in solvents that are then aerosolized through spacers to allow respirable-sized solid particle aerosols to be formed. Examples of these types of generators would include the many types of medicinal nebulizers available and some specialized generators, such as metered-dose inhalers and Laskin-type nebulizers (7). Spinning top generators have also been successfully employed (8).

In the future, additional solid excipients, such as protein stabilizers, sustained-release matrices, and more biocompatible surfactants, will be used in inhaled products.

2. Gas Excipients

Gas excipients are primarily used as aerosol propellants in metered-dose inhalers. The terms of the Montreal Protocol required the elimination of chlorofluorocarbons (CFCs) from general use, including use in MDIs (9). Two new propellants were developed by pharmaceutical consortia for use in pharmaceutical products including MDIs. The new propellants are hydrofluoroalkane-134a (HFA-134a) and HFA-227. They have been accepted as safe alternatives to CFCs, and products containing HFA-134a have been introduced in over 40 countries. Health authorities around the world required that these new propellant excipients be treated as new drug substances because of the extremely widespread, long-term use of products containing the propellants. As a result, full-safety programs were conducted on HFA-134a and HFA-227 as if they were active drugs. Additional propellants are being considered, and the question is whether new propellants structurally similar to HFA-134a and HFA-227 would have to undergo complete preclinical programs, taking 5 years and costing $20 million. Approaches to conducting full-safety programs on propellants have been previously outlined (10–13).

The generation of gases for inhalation is relatively simple compared with solid and liquid excipients. Usually, the gas can be placed into pressurized cylinders and then metered into an airstream which, in turn, is introduced into the exposure system. Excipient gases should by design be nontoxic; therefore, quite high concentrations are often required to produce any effects. These high concentrations can lead to very dense atmospheres, which can easily alter the exposure dynamics of an inhalation system. Caution is advised and a careful exposure system validation should be performed.

B. Conduct of Inhalation Studies

Before starting laboratory work on any program, pertinent information on the excipient must be assembled, starting with a complete literature review of the chemical, physical, pharmacological, and toxicological characteristics of the excipient itself as well as closely related excipients. This information will target possible adverse effects and provide guidance for the best design of studies, and thereby, eliminate the unnecessary repetition of studies. However, early studies may not have been conducted according to modern standards; therefore, they may need to be repeated. Studies essential to the safety issue must be conducted according to Good Laboratory Practices (GLP) requirements (14) and standard regulatory protocols (i.e., OECD requirements; 15). Subchronic, chronic, carcinogenicity, and reproductive inhalation studies are performed using the same study designs as used for the other routes of exposure (15,16). The safety evaluation program should be comprehensive enough to ensure adequate safety in use. The IPEC Guidelines provide a framework for designing the safety program for a new excipient to be used by the inhalation route (17). Table 1 summarizes the types of studies recommended for this route. The exception to the suggested tests is for the excipients that are gases, particularly propellants used in metered-dose inhalers. Gases cannot reasonably be tested by the oral, dermal, or parental routes, and the cold freon effects make eye irritation tests unrealistic. An indication of dermal toxicity and eye irritation can be partially inferred through standard inhalation tests because some degree of dermal and occular exposure occurs during inhalation exposure.

C. Acute Inhalation Study Types and Designs

Safety pharmacology–toxicology inhalation studies must be performed, the endpoints of which should include a general assessment of respiratory function, cardiac function (e.g., electrocardiography, blood pressure), renal function (e.g., urinary creatine clearance), pulmonary irritation, and sensitization, bronchospasm potential, and cardiac sensitization studies. Knowledge of potential adverse effects may lead the investigator to include more-specialized studies, such as central nervous system assessment studies. Excipient exposure concentrations in safety

Table 1 Summary of Excipient Guidelines Routes of Exposure for
Humans

Tests	Inhalation/intranasal[a]
Base set	
Acute oral toxicity	R
Acute dermal toxicity	R
Acute inhalation toxicity	R
Eye irritation	R
Skin irritation	R
Skin sensitization	R
Acute parenteral toxicity	—
Application site evaluation	R
Pulmonary sensitization	R
Phototoxicity/photoallergy	—
Ames test	R
Micronucleus test	R
ADME-intended route	R
28-day toxicity (2 species) intended route	R
Appendix 2	
90-day toxicity (most appropriate species)	R
Teratology (rat and rabbit)	R
Additional assays	C
Genotoxicity assays	R
Appendix 3	
Chronic toxicity (rodent, nonrodent)	C
Photocarcinogenicity	—
Carcinogenicity	C

[a] R, required; C, Conditionally required.
Source: Ref. 16.

studies are not usually performed at the overt toxic effect levels, but rather, at low
to intermediate doses that are smaller multiples of the human exposure situation.

1. Acute Toxicity

Acute inhalation studies are usually conducted using a single exposure of between
1- and 6-h duration, depending on the intended use of the excipient. Acute studies
provide information on single-dose product toxicity as well as accidental exposure situations. Increasingly, the use of satellite animals in all phases of the testing
program is becoming standard. The satellite animals can be used to determine
serum test article and metabolite levels, as well as bronchioalveolar lavage (BAL)
test article or metabolite levels and at times biological markers of exposure, such

as cytokines. The presence of test article or metabolite levels provides assurance that the animals did indeed receive an exposure. This is particularly important if there are no adverse effects of the excipients.

2. Special Acute Studies

Pulmonary irritation can be assessed in a single-exposure situation by placing animals, such as guinea pigs, in body plethysmographs adapted for use with inhalation chambers. Respiratory parameters are then measured under a variety of excipient exposure concentrations. Typical respiratory physiology parameters include tidal volume, breathing frequency, minute volume, maximum tidal inspiratory flow, inspiratory times, and expiratory times. Pulmonary irritation is characterized by rapid, shallow breaths with upper respiratory irritants, but slower pattern-shifted breaths for deep lung irritants (18,19). Pulmonary sensitization studies require a more extensive study. Typically, guinea pigs are exposed to test article by inhalation for 10-consecutive days, held for 5 days, and then challenged with an inhalation exposure of the excipient. The animals are housed in a body plethysomograph and the parameters described in the irritation test are evaluated. In addition, lung resistance and dynamic compliance are assessed. If a sensitization reaction happens during the exposure, then these parameters will be quite altered (20).

Bronchospastic activity can be assessed in a manner similar to pulmonary sensitization and irritation. Guinea pigs are placed in a plethysomograph and exposed to varying concentrations of methacholine or histamine, and a dose–response curve is generated. The animals are allowed a short recovery period, and they are then exposed by inhalation to a moderate concentration of test material. Varying concentrations of methacholine are then introduced to the animals concurrent with test material exposure, and a new methacholine dose–response curve is generated. A shift of the dose–response curve to the left indicates the degree of bronchospastic activity induced by the test material (21). The methacholine challenge test is useful in and of itself, and also because similar methacholine challenge studies are often performed in human clinical trials. Guinea pigs are usually more sensitive to bronchospasm than humans, except for possibly humans who have asthma.

Cardiac sensitization studies are indicated for those gas or liquid excipients that are halogenated or related to common solvents, such as benzene, heptane, or chloroform (22). The phrase ''cardiac sensitization'' is somewhat of a misnomer because it does not involve an immunological component. *Cardiac sensitization* is a measure of a chemical's ability to induce cardiac arrhythmias. Such arrhythmias, if severe enough, can cause serious adverse events, including death, such as in deaths caused by ''glue sniffing.'' No standard study design for conducting these studies has been promulgated. However, it appears that the beagle dog is the model of choice, although dogs are considered to be more sensitive to these effects than humans. An example of a study design involves injecting dogs with intravenous epinephrine using doses sufficient to cause a pharmacological response, such as a blood pressure change, yet the dose should not by itself cause cardiac arrhythmias. The animals are then exposed to graded concentrations of the test material concurrent with the epinephrine injection or infusion. The presence of cardiac arrhythmias is characterized by multiple ventricular premature complexes. If present, then maximum no-effect levels and minimum effect levels should be derived. A positive control such as chlorofluorocarbon-11 should also be employed.

D. Radiolabeled Excipient Deposition Studies

Eloquent and sensitive methods now exist for determining the actual site of inhaled excipient deposition within the respiratory tract and its subsequent tissue distribution. This is important because this information can confirm that the excipient reaches the same tissue in the preclinical studies as it does in the clinical application. Deposition studies combined with toxicokinetic studies can provide a clearer picture of the actual exposure and how it relates to the fate of the excipient. Classic methods include the use of beta-emitters, such as tritium and carbon 14, as radiotags covalently incorporated within the test material. This is primarily applicable to small animals that can be conveniently exposed to radiolabeled material by inhalation and then euthanized and either sectioned for whole-body autoradiography or individual tissues can be collected, homogenized, and counted using a liquid scintillation counter. Other ways of radioimaging include the use of positron emission tomography (PET) and magnetic resonance imaging (MRI). The test material used in PET must contain positron-emitting isotopes, such as carbon 11, nitrogen 13, oxygen 15, or fluorine 18. Positron emitters are short-lived, which makes them attractive from a safety perspective, but unfortunately facilities and staff are very expensive to set up and maintain (23). MRI facilities are more available than PET facilities. Nuclei useful for MRI include phosphorous 31, protons, sodium 23, potassium 39, carbon 13, and fluorine 19. If the test material contains fluorine then the use of fluorine 19 in MRI studies is particularly desirable because biological systems are almost completely devoid of natural fluorine 19, making its sensitive detection very easy (23).

   Another radioimaging technique involves radiolabeling the test material with a gamma-emitter, such as technetium 99m. The technetium is not covalently incorporated into the molecule of test material, but rather, is associated ionically or perhaps through colloidal interaction with the molecule. The close association of technetium and test material must be carefully validated for all sizes of particles to assure that the radioimages indeed match the test material deposition sites. These studies can be very useful in that facilities are much more available and

Table 2 Species Comparison of Pulmonary Parameters

Species	Body weight (kg)	Total lung capacity (mL)	Tidal volume (mL)	Frequency (breaths/min)	Minute volume (mL)
Man	70	5500	750	12	9000
Dog	10	800	200	20	4000
Monkey	3	200	20	40	800
Rabbit	4	60	16	40	640
Guinea Pig	0.5	15	2	90	180
Rat	0.4	6	1.5	160	240
Mouse	0.025	1	0.2	180	36

the imaging can safely and conveniently be performed in small animals, large animals, and human subjects (24). Only images obtained immediately after inhalation of the radiolabeled test material are useful, for the technetium can be rapidly dissociated from the test material once it has deposited in the tissue.

E. Physiological Parameters for Commonly Used Species

Table 2 lists pulmonary physiology parameters of species commonly used in inhalation studies. There are wide variations in the values presented for each species, depending on the strain and circumstances under which the measurements were obtained. However, the table provides a general guideline and starting point for study design aids. It is best to characterize the pulmonary physiology parameters for representative animals extensively used in the overall program.

F. Dose Selection for Inhalation Studies

Excipients are chosen primarily for their formulation improvement, but also for their safety. Thus, the expectation of an excipient is that there will be no serious toxicity, even at very high doses. The question for the toxicologist is how can a proper testing program be designed to characterize the toxicity of a nontoxic substance, or how can dose–response relations be characterized when there is no practical dose–response relation. Considering the example of HFA-134a, the acute and chronic inhalation toxicity of HFA-134a was so low that to see effects, several hundred thousand parts per million (ppm) concentration of HFA-134a in air had to be employed. This concentration was so high that oxygen levels were reduced and indeed most of the adverse effects recorded in the studies were attributed to oxygen deprivation. However, use of a concentration high enough to cause oxygen deprivation provides little useful safety data on the compound.

There are at least three possible answers to this dilemma. The most accepted historical approach to selecting inhalation doses is to expose animals to the highest dose concentration that causes adverse effects, irrespective of whether these effects are due to oxygen deprivation or direct excipient effects. This is the maximum tolerated dose (MTD) approach. Lower doses should include a no-effect concentration and at least one intermediate concentration for which adverse effects are moderate.

The second approach is to use a maximum feasible concentration as an upper limit. This upper limit has been proposed to be 5% of the atmosphere that results in an oxygen concentration of at least 19% (15). This concentration represents approximately 50,000 ppm of gas in air. This is still a relatively high concentration, yet it only reduces oxygen concentrations from about 20.5% in normal air to 19%, a level of oxygen depletion that is not toxic to any mammalian species. There is precedent to this approach because in some long-term feeding toxicity studies, it is acceptable to use a maximum level of 5% of test material in the diet. Higher values may reduce essential nutrients, such as protein, which could cause adverse effects not related to the test compound. The limit of 5% is not universally accepted among the regulatory authorities, but it can be defended. A valid criticism is that persons may be exposed instantaneously to concentrations of gases in air much greater than 5% so some acute studies may need to be conducted to mimic this condition.

A third approach is to produce an artificial atmosphere such that all doses have a 20% oxygen level, and the normal 80% nitrogen is proportionally replaced by the test gas. This approach is technically difficult to manage in a large inhalation program, and it is not recommended for many reasons, because these artificial atmospheres have no relevancy to the actual human exposure conditions.

There are similar dilemmas when choosing top doses for particulate aerosols. Ideally a particulate excipient should be free of adverse effects, even at high concentrations. However, it has been clearly shown that very high concentrations of inert particulate aerosols can be carcinogenic (25). These inert dusts are not directly carcinogenic, but rather, very high dust concentrations in the lungs cause an overload of clearance mechanisms, primarily those clearance paths involving macrophages. This was the case with diesel exhaust particles and carbon black particles. Thus, it is reasonable to set top doses at concentrations below those that would cause particle overloading of the lungs. Preliminary clearance studies may be necessary to estimate overload doses.

G. Inhalation Exposure Systems

There are a wide variety of inhalation systems in use, most of them custom fabricated by the testing facility. Most systems have been well validated, in general, but each must be revalidated and extensively characterized for the specific excipient being tested. One of the first choices to be made for small-animal exposures, such as mice, rats, hamsters, guinea pigs, and rabbits, is that of whole-body versus nose-only chambers. A more complete discussion of the advantages and disadvantages of each has been presented elsewhere (26), but the trend has generally been to conduct exposures in nose-only systems, especially for pharmaceutical applications. Although nose-only studies are more cumbersome and expensive, these exposures usually most closely mimic the human exposure situation and avoid the complications of whole-body contamination for a product intended exclusively for inhalation use. Undoubtedly, the biggest challenge in conducting a complete nose-only exposure program is the conduct of the reproductive toxicology studies, for which even minor stresses of confinement can have major effects on reproductive parameters. However, such reproductive nose-only studies have been successfully conducted where sham-exposed animals have incidences of reproductive effects similar to cage controls (27). Often nose-only exposure studies will include a sham-exposed control group in which animals are inserted into the exposure tubes, placed onto chambers, and subsequently exposed to filtered air only. At times a placebo group is used during which animals are exposed to inhaled formulations without the test excipient. When warranted, a cage control group may also be useful, but the addition of a cage control group is usually not necessary in an experienced laboratory with well-validated systems. The key to conducting such successful studies appears to be in the animal-handling techniques and gradual sham acclimation to the restraining tubes. Appropriate sham controls as well as caged controls should be used.

There are two fundamental types of nose-only systems. The flow-past nose-only chambers are more labor-intensive, but they usually provide for better dose consistency at each animal port. The flow-past chambers prohibit the animals from rebreathing another animal's expired breath, which may contain metabolites and carbon dioxide (28). The second nose-only system is the open cylinder design in which the animal tubes are placed on a large cylinder and atmospheres are drawn past the animal's noses out through the chamber bottom. These chambers are easy to set up and operate, but they do not always provide consistent doses to all animals.

Inhalation exposure systems for larger animals, such as dogs and monkeys, are less standardized than those for small animals. Often large animals are dosed individually with custom-made systems. However, there are systems described in the literature that allow the exposure of multiple large animals. A common design is to generate the desired atmosphere, pass it through a large, central chamber, and then draw it from the central chamber past the noses of the test animals. The transfer tubing usually employs a one-way valve to prevent back contamination of the central chamber. There is some debate over whether the test atmosphere should be delivered to the dog through the use of a mouth and nose muzzle or through a mouth tube, thereby bypassing unwanted nasal exposure. The mouth tube prevents the animal from raising its tongue and blocking inhalation exposure.

The decision should once again be based on the exposure situation most relevant to the eventual human use. For example, a mouth and nose muzzle system is appropriate for inert gas excipient exposure, but for solid and liquid excipients used in oral inhalers, the mouth tube exposure system is probably more relevant because humans will use inhalers exclusively through the mouth.

H. Test Atmosphere Monitoring

A significant advancement in inhalation science has been the employment of sensitive and sophisticated analytical techniques to better assess the test material concentrations to which the animals are exposed. The test material concentration along with particle size information for aerosols permits more accurate dose estimates to be calculated as well as relating results across many studies and several species, including humans. Typical monitoring procedures include continuous, real-time exposure monitoring along with more definitive, intermittent sampling. An example would be real-time monitoring of a gas atmosphere by in-line infrared analysis to detect changes and correct problems immediately. The more precise concentration measurement may then be determined by gas chromatography. For particles, the on-line concentration can be monitored by a variety of real-time aerosol monitors. The actual chemical-specific determination can be performed using filter collection of particles followed by high-performance liquid chromatography (HPLC) analysis. For aerosols, routine particle size analysis should be performed to assure that particle sizes are in the respirable range and are being consistently reproduced by the generation system. The most commonly used particle size analyzers are cascade impactors, such as the Andersen impactor and the Quartz Crystal Microbalance. Light-scattering particle size devices are also useful. An optimum particle size for deposition throughout the respiratory tract is approximately 2-μm–mass median aerodynamic diameter (MMAD) with a geometric standard deviation (GSD) of approximately 2.5.

Other parameters that must be measured and controlled include temperature, pressure, humidity, and airflow. It is vital and must be documented that control groups be kept free from test material exposure during exposure times as well as during off-exposure housing times. It is desirable to house control animals in separate rooms and assure that air-handling systems cannot be contaminated by airborne test material. More complete descriptions of general inhalation study conduct can be found in several texts (29–32).

II. EXCIPIENT TESTING BY THE INTRANASAL ROUTE

A. Studies to Support Use

When designing a toxicology program for an excipient used in an intranasal pharmaceutical formulation, it is very important to determine how the pharmaceutical preparation will be used. Considerations should include relevant concentrations, frequency of use, and the dose site or route. The nasal epithelium is in direct contact with inspired air and is the first tissue exposed to aerosol contaminants or toxicants; accordingly, the tissue of the nasal area is very susceptible to toxicity. Administration by the nasal route is essential to prevent the omission of any toxicity when extrapolating from other routes of exposure. The best approach to determine what will be needed to support intranasal use is to conduct a thorough literature search on what information is already available and what information is missing in profiling a new pharmaceutical excipient. The necessity of this approach is twofold: (a) It will prevent the conduct of redundant studies; and (b) it will also meet the requirements of the Animal Welfare Act (33) in assessing the need for the studies. Once this information is assembled, a complete strategy for further development can be established. Early in the development phase, meeting with the appropriate regulatory agencies about how to proceed is very helpful. Each regulatory agency may have specific ideas about the most appropriate way to continue development and may be able to provide insight from circumstances to which an industrial toxicologist may not have been exposed. The regulatory agencies may not divulge why they are requesting a specific study, but they can intimate if in the past a particular study provided needed information for safety assessment. It is recommended that regulatory agencies be consulted before starting a major testing program.

Table 1 summarizes study types considered to be necessary to adequately profile the toxicity of a new pharmaceutical excipient. All studies must be conducted in compliance with Good Laboratory Practices (21 CFR 58) (14). Other routes may be substituted in cases where exposure would be equivalent or better than the intended route.

Standard guidelines request that the acute toxicity be determined by several routes: by the intended exposure route as well as by a parenteral route (intravenous or intraperitoneal). Because of the potential for exposure to the skin during intranasal dosing and for worker safety, it is also wise to evaluate dermal toxicity. At least two species and both sexes should be used in the study. In intranasal studies, the rat and the dog are the two most commonly used species. These studies can be performed as escalating-dose studies, especially in the dog. This permits the use of a limited number of animals to determine upper bounds of drug exposure.

Repeat-dose studies are conducted to support the duration of the anticipated patient use period or to support the duration of clinical studies. These studies should be done by the intended exposure route. Occasionally, the nasal route cannot deliver sufficient test material to profile the toxicity. In those cases, other routes may be used to develop the toxicity or a combination of the exposure route with oral or IV, and so forth. If the toxicity has been profiled by another route, this information may be used in conjunction with additional supporting data by intranasal administration. Sufficient information should be available to determine local tolerance to the test material. These studies should include satellite groups for pharmacokinetic profiling. This will provide information on absorption, distribution, metabolism, and excretion. If the metabolic profile of this test material is substantially different from that of the nasal route, additional toxicology work might be required to support registration.

Current regulatory guidelines for active test materials recommend two studies to evaluate carcinogenic risk of new test materials. New ICH guidelines for pharmaceutical development have provided an alternative of using one species in a rodent carcinogenicity study and one short-term in vivo test (34). Examples being considered for short-term tests include the initiation–promotion model in the rodent, transgenic mouse assays (p53 ± deficient model, Tg.AC model, TgH*ras2* model, or other), and the neonatal rodent tumorigenicity model (34). Carcinogenicity studies should be conducted on any test material intended to be used in drug formulations clinically for at least 6 months or frequently in an intermittent manner. Other compounds that are administered infrequently or for short duration usually would not require carcinogenicity studies unless there is cause for concern. The studies should be conducted by the route of intended exposure whenever feasible. The carcinogenicity study may be conducted by other routes if pharmacokinetic–metabolic profiles show similar exposure levels, especially for the primary target site (i.e., nasal tissue). Existing carcinogenicity studies may also be supportive, and only one study by the intended route may be necessary. If there is little systemic exposure by nasal delivery, a carcinogenicity study by another route to enhance exposure may not be necessary. The pharmacokinetic study can be used to determine exposure and potentially justify not doing other routes of administration. The ICH guidelines on carcinogenicity studies provide information on choosing the appropriate dose levels.

Studies to evaluate reproductive outcomes should follow current ICH Guidelines for Detection of Toxicity to Reproduction for Medicinal Products (35). Teratology studies should be conducted in the rat and possibly the rabbit for determination of developmental effects during organogenesis. The studies may be conducted by the oral route, unless the pharmacokinetic–toxicokinetic profile is markedly different by intranasal administration. A second study with intranasal administration may be conducted to determine clinical relevance of the findings if effects are observed in the oral study. A further study to characterize effects on general reproductive development should be considered to determine effects on fertility, mating, delivery, and lactation.

Three types of genetic toxicity studies should be considered to fully characterize the test material. Evaluation of the material in a bacterial reverse mutation assay will have some predictive value of the genotoxicity in rodents. A second assay for the determination of test article effects in a mammalian in vitro assay for the detection of chromosomal damage including gene mutations and clastogenic effects (mouse lymphoma TK assay, CHO chromosome aberration, or human lymphocyte chromosome aberration assay) is recommended. The third assay to complete the standard battery is an in vivo test for chromosomal damage using rodent hematopoietic cells (bone marrow). These tests, provided they are negative, give sufficient certainty that the material is not genotoxic. Other assays may be included to clarify any tests that may produce positive results.

Studies to profile the potential for the test material to produce hypersensitivity responses are particularly important when dealing with inhalation or intranasal routes. These types of responses indicate potential of an adverse reaction through immune mechanisms. Delayed-type reactions at the dermal level (i.e., squamous epithelium of the nasal passages) as well as the pulmonary hypersensitivity response when exposed to a new test material are important to characterize. A hypersensitivity response in the airway may result in dypsnea and respiratory distress or tissue disruption, or death. The primary species used in these types of evaluations are the mouse and the guinea pig.

Delayed-type contact hypersensitivity can be evaluated in the mouse local lymph node assay (MLLN) which measures proliferation of lymphocytes using [^{3}H]thymidine incorporation. This assay provides an objective measurable endpoint and uses fewer animals than the traditional guinea pig assay. Some, but not all, regulatory authorities accept the MLLN assay. The other option is to use the traditional guinea pig assays for allergic contact sensitization. Regulatory authorities traditionally accept these assays for registration.

Pulmonary hypersensitivity should also be conducted in anticipation of inhalation of some of the intranasal dose. This methodology was discussed earlier in the chapter.

B. Special Considerations for Intranasal Delivery

It is important to mimic the human exposure not only in the form of the test material, but also with the use of a comparable volume. The test model then receives a similar ratio of volume to available surface area within the nasal cavity. The intranasal delivery to localized tissue presents different concerns when compared with pulmonary delivery. The surface area of the lung greatly exceeds that of the nasal turbinates. There is enhanced local exposure in a much reduced surface area following intranasal administration. This is characterized by the greater potential for local irritation and toxicity. Osmolarity, pH, and chemical irritation may produce effects on membrane integrity as well as ciliary transport. Therefore, the greater concentration at the site may produce more effects than those seen at the pulmonary sites.

Microscopic evaluation of nasal tissues will provide morphological confirmation of any toxic response. In addition, mucociliary transport will provide indications of any functional deficits. Measurement of ciliary beat frequency

(CBF) may be performed in vitro with canine trachea (36). However, in vitro methodology may not provide a clear interpretation of toxicity to inhaled irritants. Experiments investigating irritant CBF responses in single cells or tissue culture tend to find only modest stimulation followed by ciliary toxicity at higher concentrations (37). An in vivo evaluation is a better indication of actual toxicity. An in vivo method available for determining effects on ciliary beat frequency is the rabbit maxillary sinus model (38).

C. Delivery Systems

The approach taken in evaluating intranasal toxicity of a compound is very similar to that of inhalation or pulmonary delivery. The difference is the target site and the fact that it is much simpler to assure exposure of the intranasal target site. When delivering an aerosol formulation containing active and excipient, deposition may be quite different between the two molecules. Depending on the solubility and the particle size, these two different entities may deposit at varying sites and the toxicity associated with them may appear at differential sites as well.

Characterization of the form that the formulation will take and how the tissues will be exposed to the components in a patient use situation is also critical for accurate extrapolation or prediction of adverse effects. Will the patient be exposed to a solid particle or a solution? Will the formulation be administered in a powder, liquid, gas, or vapor? If dose administration in animal models is performed by aerosol generation and the delivery system for the patient will be an aqueous solution, exposure will be different with potentially dissimilar outcomes. Aerosol generation of the test material will deliver the exposure as a vapor or particle. In contrast, the patient may be exposed to the test material as a solution in water that may produce different conditions altogether. The solution should be tested as well, as aerosol generation would not provide a complete profile of potential adverse effects.

D. Animal Species and Physiology

At least two species should be used in evaluating the toxicity of a new excipient. There is no adequate substitute for human exposure, but the best model that would appropriately characterize human exposure should be considered. The species most commonly used are beagle dogs, mice (C57BL, Swiss CDI), rats (Wistar, Sprague–Dawley, Fisher-344), and New Zealand white rabbits for repeated-dose studies (see Table 2). The guinea pig is preferred for sensitization work. The primate is not as commonly used because of the difficulty and time involved in acclimating the animals to dosing as well as animal welfare issues. Septal windows are present in the guinea pig, rat, and mouse, which means that the two sides cannot be treated separately or one side cannot be used as a concurrent control. The nasal epithelium in all species is fairly similar. There are four distinct nasal epithelial types in all species; however, the distribution is slightly different (39). The major differences in nasal epithelium between species is the percentage of the nasal airway that is covered by olfactory epithelium. A greater percentage of the nasal cavity is lined by olfactory epithelium in the rat and mouse than in monkeys or humans (39). In primates (monkey and human), the olfactory region is in the upper reaches of the nasal cavity and is not subject to the direct effect of inspired air, and the cilia of the olfactory mucosa are nonmotile (40).

The factors that most influence on nasal absorption of compounds are anatomical. These include epithelial surface area, structure of the turbinates or conchae, presence of a septal window, and nasal cellular structure. In addition, the volume of material—solution or suspension—administered should not exceed that which is proportional to human exposure (i.e., 150 µL; 40).

Rodents are obligatory nose-breathers. The rodent nose is involved in several different functions, including filtering, humidifying, and warming air, and olfaction. The rodents also have a septal window that allows the flow of test materials from one side to the other, thus eliminating the use of one nostril as a concurrent control. They are one of the primary species for inhalation exposure as well as intranasal delivery for characterization of toxicity. Because of the highly developed turbinates or conchae in the nasal passages, the nose is a common area of particle deposition for inhaled materials.

Table 3 indicates the nasal volumes and comparative volumes for exposure in differenct species. The nasal volume of the mouse is about 30 µL. The volume of the rat nasal cavity is 400 µL. They can be dosed by nose-only aerosol exposure or by delivering liquids through pipettes.

Nose-only delivery systems can be used for exposure to rodents provided the particle size of the excipient is such that it deposits in the nasal area. Deposi-

Table 3 Species Comparison of Nasal Volumes and Comparative Volume for Exposure

Species	Nasal volume (µL)	Comparative volume for exposure (µL)
Man	20,000	150
Beagle dog	20,000	200
Rhesus monkey	8,000	58
Rabbit	6,000	58
Guinea pig	900	50
Rat	400	13
Mouse	30	3

Source: Ref. 40.

tion in the nasal area is enhanced by water solubility of the test material as well as particle size. Because of the turbulence produced as aerosolized material passes through the nose, highly water-soluble compounds can absorb moisture and quickly deposit in the moist tissues of the nasal area. Less water-soluble compounds are affected less by the humidity as by the turbulence and impact with tissues.

The design for aerosol studies involves an air control, vehicle control, and low-, mid-, and high-dose groups. Separate satellite groups would be needed to profile the toxicokinetics and demonstrate absorption or lack thereof. Exposure time is based on desired delivered dose and concentration. The exposure time should be equivalent for all groups. Similar groups including a sham control should be included when using a manual system of a pipette-applied dose. Dose volume should remain consistent (mouse approximately 3 μL and rat approximately 13 μL per nostril) while each group receives a different concentration. Rats and mice dosed in an upright position during manual dosing can then be held in a supine position for 1 min to cover the largest surface area.

The guinea pig also has a septal window that allows test article exposure to both sides of the nostrils (a comparative volume of about 25μL each). Dose the guinea pig manually in an upright position followed by holding it in a supine position for 1 min to ensure proper delivery.

Teratology studies and repeat dose studies are performed in rabbits. They can be dosed with an insufflator or by aerosol. Nasal volume is approximately 6 mL. A 58-μL volume per nostril approximates the human dose volume of 150 μL. Rabbits can be held during dosing or placed in a restraint box. For reproductive studies in the rabbit, oral or intravenous delivery of the test article may be sufficient to characterize developmental toxicity. If inhalation or intranasal provide better availability, then this route should be considered.

The dog is commonly used as a model for intranasal toxicology and drug delivery studies. The volume size and surface area of the nasal area is similar, but the anatomy is very different from the human (40). The volume that can be delivered to the dog and approximates human exposure, is about 200 μL per nostril. Dogs are relatively easy to work with, they can be dosed unanesthetized, and they can be dosed with the human delivery device. They do require acclimation to the delivery procedure to eliminate stress during dosing. Because of the large size of the dog, many procedures can be performed on one animal. All clinical pathology as well as toxicokinetics can be conducted on the toxicology test group. Again, air and vehicle controls should be included in the study design.

Intranasal studies have been conducted in primates, particularly in the area of delivery of intranasal hormones. Monkeys do require a significant amount of training before dosing can be accomplished in a relatively stress-free atmosphere. The nasal volume of the rhesus money is approximately 8000 μL, whereas a human equivalent dose would be approximately 58 μL. Anatomically, the monkey and the human are very similar. The nasal turbinates in both primates (monkey and human) are much less complex and have a reduced surface area as compared with the rat and the dog.

E. Dose Level Selection, Dose Frequency, and Exposure Duration

The high-dose level in repeat-dose toxicity studies is dictated either by the solubility of the test material in the limited volume if the vehicle is aqueous or the maximum feasible atmosphere that can be generated as an aerosol or alternatively by the relative toxicity of the test material. The maximum tolerated dose from intranasal delivery can be based on systemic toxicity resulting from absorption and systemic distribution or from local toxicity at the site of impact. An acceptable multiple of anticipated human exposure can also be used in choosing the high dose with comparable exposure based on deposition or serum AUC (area under the curve) or, where appropriate, C_{max} levels of test material. Multiples of 100 are acceptable if no signs of toxicity are observed.

When toxicity is observed, the safety factor or approximate therapeutic index is based on the benefit obtained from use of the test material versus the risk associated with its use. Under ordinary circumstances, the benefit of an excipient would be difficult to justify if its toxicity profile did not provide a significant safety factor. In certain cases, as in the inability to formulate a pharmaceutical product of great clinical benefit in the absence of the use of the excipient, then use of such an excipient might be justified. The acceptable safety factor is then a judgment based on the benefit of the combination of the two (new pharmaceutical excipient and active ingredient).

III. SUMMARY AND CONCLUSIONS

In comparing and contrasting the two delivery routes, the basic principles are very similar by either inhalation or intranasal. Particle size is important for both routes and can make the difference between whether the test material is delivered into the nasal cavity or into the lung. The particle size would be adjusted to deliver to the specific target site. The most difficult issue in inhalation delivery is to ensure that the target site is reached and that the intended site for treatment of the lung has received adequate exposure. With intranasal administration, exposure is easier to attain.

The inhalation and intranasal delivery systems for rodents are very similar. The same systems can be, although may not necessarily, the same. The rodent is not as relevant in extrapolating to human exposure from inhalation in comparison with intranasal exposure. In inhalation delivery of pharmaceuticals in patients, the drug is usually not delivered through the nares and, therefore, the large animal provides a more relevant exposure model for inhalation in humans. Species differences in anatomy are an additional hurdle in interpretation and extrapolation between humans and the test species following intranasal administration. The physical differences and surface area between species are significant in the nasal cavity. The basic anatomy of the lung varies less among the various test species and humans.

A consideration for intranasal administration is that the amount delivered must be adjusted to a specific volume to mimic the comparative delivery volume in humans. The potential for local toxicity and irritation is much greater with intranasal delivery because the test material can be delivered in a more concentrated form to a much-reduced surface area. Vehicle effects and formulation osmolarity can produce a greater effect at the target tissue. The comparative volumes are needed to address absorption and local effects between the target population and the test species. The primary goal with inhalation delivery is to deliver the test article at a specific dose and, although concentration of the aerosol is important, the volume is only needed to deliver the accurate dose.

In developing new excipients for use in pharmaceutical products, several general concepts must be considered. Those concepts include reviewing the relevant data available on the test material, ascertaining how the material will be used for an anticipated indication, and developing a plan based both on international guidelines and good scientific rationale. In the end, the primary goal is that the program should be scientifically valid and provide assurance of the safety in the proposed indication or exposure.

REFERENCES

1. JT Stevens, JD Green. Test article administration. In: H Salem, ed. Inhalation Toxicology. New York: Marcel Dekker, 1987, pp 59–92.
2. CL Leach, NS Hatoum, AD Ledbetter. Generation of gas and liquid hydrocarbon mixtures from a single pressurized cylinder for inhalation exposure. Am Ind Hyg Assoc J 46:97–99, 1985.
3. BM Wright. A new dust-feed mechanism. J Sci Instrum 27:12–15, 1950.
4. V Timbrell, AW Hyett, JW Skidmore. A simple dispenser for generating dust clouds of standard reference samples of asbestos. Ann Occup Hyg 11:273–281, 1968.
5. RT Drew, S Laskin. A new dust generating system for inhalation studies. Am Ind Hyg Assoc J 32:327–330, 1971.
6. EM Milliman, DPY Chang, OR Moss. A dual flexible brush dust-feed mechanism. Am Ind Hyg Assoc J 42:747–751, 1981.
7. RT Drew, DM Berstein, S Laskin. The Laskin aerosol generator. J Toxicol Environ Health 4:661–670, 1978.

8. M Lippmann, RE Albert. A compact electric-motor driven spinning disc aerosol generator. Am Ind Hyg Assoc J 28:501–507, 1967.

9. Montreal Protocol. Handbook for the International Treaties for the Protection of the Ozone Layer. Nairobi, Kenya: Ozone Secretariat, United Nations Environmental Program, 1996.

10. P Graepel, DJ Alexander. CFC replacements: safety testing, approval for use in metered dose inhalers. J Aerosol Med 4:193–199, 1991.

11. K Olejiniczak, R Bass. CFC and replacements in medicinal products. J Aerosol Med 4:205–209, 1991.

12. DJ Alexander. Safety of propellants. J Aerosol Med 8(suppl 1):S29–S34, 1995.

13. CL Leach. Approaches and challenges to use freon propellant replacements. Aerosol Sci Technol 22:328–334. 1995.

14. Code of Federal Regulations. Title 21, Part 58, Good Laboratory Practice for Non-clinical Laboratory Studies. Washington, DC: U.S. Government Printing Office, 1995.

15. OECD. OECD Guidelines for Testing of Chemicals. Guidelines 403 (Acute Inhalation Toxicity), 412 (Repeated Dose Inhalation Toxicity: 28-day or 14-day Study), 413 (Subchronic Inhalation Toxicity: 90-day Study), 451 (Carcinogenicity Studies). Paris: Organization for Economic Cooperation and Development, 1981.

16. JM Rogers, RJ Kavlock. Developmental toxicology. In: CD Klassen, ed. Casarett & Doull's Toxicology. New York: McGraw-Hill, 1996, pp 319–320.

17. M Steinberg, JF Borzelleca, EK Enters, FK Kinoshita, A Loper, DB Mitchell, CB Tamulinas, ML Weiner. A new approach to the safety assessment of pharmaceutical excipients. Regul Toxicol Pharmacol 24:149–154, 1996.

18. Y Alarie. Dose–response analysis in animal studies: prediction of human responses. Environ Health Perspect 42:9–13, 1981.

19. LA Buckley, XZ Jiang, RA James, KT Morgan, CS Barrow. Respiratory tract lesions induced by sensory irritants at the RD50 concentration. Toxicol Appl Pharmacol 74:417–429, 1984.

20. DL Costa, JS Tepper. Approaches to lung function assessment in small mammals. In: DE Gardner, JD Crapo, EJ Massaro, eds. Toxicology of the Lung. New York: Raven Press, 1988, pp 147–183.

21. M Schaper, J Kegerize, Y Alarie. Evaluation of concentration–response relationships for histamine and sulfuric acid aerosols in unanesthetized guinea pigs for their effects on ventilatory response to CO_2. Toxicol Appl Pharmacol 73:533–542, 1984.

22. CF Reinhardt, A Azar, ME Maxfield, PE Smith, LS Mullin. Cardiac arrhythmias and aerosol sniffing. Arch Environ Health 22:265–279, 1971.

23. WR Hendee, ER Ritenour. Medical Imaging Physics. St. Louis, MO: Mosby Year Book, 1992, pp 613–614.

24. CL Leach. Relevance of radiolabeled steroid inhalation studies to clinical outcomes. J Aerosol Med 11 (supplement 1), pp S29–S34, 1998.

25. PE Morrow. Contemporary issues in toxicology: dust overloading of the lungs: update and appraisal. Toxicol Appl Pharmacol 113:1–12, 1992.

26. CL Leach, SG Oberg, RP Sharma, DB Drown. A nose-only inhalation exposure system for the generation, treatment, and characterization of formaldehyde vapor. Am Ind Hyg Assoc J 45:269–273, 1984.

27. BM Ryan, C Aranyi, CL Leach. Nose-only inhalation developmental toxicity study of a salbutamol sulfate/HFA-134a metered dose inhaler in rats. Toxicologist 30:196, 1996.

28. WC Cannon, EF Blanton, KE McDonald. The flow-past chamber: an improved nose-only exposure system for rodents. Am Indus Hyg Assoc J 44:923–928, 1983.

29. RF Phalen. Inhalation Studies: Foundations and Techniques, Boca Raton, FL: CRC Press, 1994.

30. H Salem. Inhalation Toxicology: Research Methods, Applications, and Evaluation. New York: Marcel Dekker, 1987.

31. BK Leong. Inhalation Toxicology and Technology. Ann Arbor, MI: Science Publishers, 1981.

32. RO McClellan, RF Henderson. Concepts in Inhalation Toxicology. 2nd ed. Washington DC: Taylor & Francis, 1995.

33. Animal Welfare Act as Amended, U.S. Code of Federal Regulations, Title 7 USC, Sections 2131–2159, 1996.

34. Testing for carcinogenicity of pharmaceuticals. International Conference on Harmonization of Technical Requirements for Registration of Pharmaceuticals. 16 July 1997.

35. ICH Guidelines for Industry: Detection of Toxicity to Reproduction for Medicinal Products. International Conference on Harmonization for Technical Requirements for Registration of Pharmaceuticals. September 1994; ICH-S5A.

36. LB Wong, IF Miller, DB Yeates. Stimulation of tracheal beat frequency by capsaicin. J Appl Physiol 68: 2574–2580, 1990.

37. SW Clarke, DB Yeates. Deposition and clearance. In: JF Murray, JAM Nadel, eds. Textbook of Respiratory Medicine. Philadelphia: WB Saunders, 1994, pp 345–367.

38. A Reimer, C von Mecklenberg, NG Toremalm. The mucociliary activity of the upper respiratory trace III. A functional and morphological study on human and animal material with special reference to maxillary sinus diseases. Acta Otolaryngol 355(suppl 1–20), 1978.

39. JR Harkema, KT Morgan. Normal morphology of the nasal passages in laboratory rodents. In: TC Jones, DL Dungworth, U Mohr, eds. Respiratory System. Monographs on Pathology of Laboratory Animals. 2nd ed. Berlin: Springer–Verlag, 1996, pp 3–17.

40. S Gizurarson. Animal models for intranasal drug delivery studies. Acta Pharm Nord 2:105–122, 1990.

9

Routes of Exposure: Parenteral

David B. Mitchell
The Procter & Gamble Company, Mason, Ohio

I. INTRODUCTION

This chapter will review the safety evaluation strategy and conduct of toxicology tests for new pharmaceutical excipients administered by parenteral routes. Parenteral routes of administration to be discussed include intravenous (IV), intramuscular (IM), subcutaneous (SC), and intraperitoneal (IP). Less frequently used parenteral routes of administration for toxicology studies would include intrapleural, intra-arterial, intrathecal, and intracerebral. These routes of administration generally require anesthesia or surgical techniques to perform (1,2).

The term parenteral comes from the Greek (*para enteron* = ''beside the intestine'') and refers to the route of administration of drugs by injection under or through one or more layers of skin or mucous membrane. Parenteral administration is not the usual human route of administration for most drug substances, and is more frequently used in hospital situations. However, there are numerous advantages and disadvantages of parenteral routes of administration that must be understood to provide adequate toxicological information about a novel pharmaceutical excipient for use in parenteral dose forms (3,4).

Advantages: Parenteral administration generally produces a rapid onset of action, particularly from an intravenous dose, because there is immediate access of the medication and excipients to the systemic circulation. Thus, parenteral administrations are very useful if an immediate physiological or pharmacological action is required, as in emergency medical situations. Parenteral routes may be used to overcome a drug substance with poor oral absorption (because gastrointestinal absorption is circumvented) and may produce more predictable blood levels and lower variability. Intramuscular and subcutaneous administrations may provide a therapeutic advantage by enabling a long duration of drug delivery and drug action.

Disadvantages: One of the primary disadvantages of parenteral administration is that one cannot recover the dose once it has been administered. There is an increased risk of both local and systemic adverse effects. The formulation may produce irritation at the site of injection or infusion, or may be involved in systemically mediated reactions, such as allergic sensitization. More rigorous manufacturing procedures are required to ensure aseptic and pyrogen-free products. Normally, trained personnel are required for parenteral administrations, and this would be a disadvantage if the patient had to visit a professional for daily or frequent drug administrations.

II. BACKGROUND INFORMATION ON THE TEST MATERIAL

A. Physical and Chemical Properties

1. Test Material and Product Form

As general background to test material and product form, the *United States Pharmacopeia 23/The National Formulary 18* defines five general types of parenteral preparations (5):

> *Drug injection*: liquid preparations that are drug substances or solutions thereof
> *Drug for injection*: dry solids that, upon addition of suitable vehicles, yield solutions conforming in all aspects to the requirements for Injections
> *Drug injectable emulsion*: liquid preparations of drug substances dissolved or dispersed in a suitable emulsion medium
> *Drug injectable suspension*: liquid preparations of solids suspended in a suitable liquid medium
> *Drug for injectable suspension*: dry solids, that upon addition of suitable vehicles, yield preparations conforming in all aspects to the requirements for Injectable Suspensions.

The physical form of the test material and product formulation are important determinants of the parenteral dose routes available for administration, for both the clinical setting and in considerations for toxicological testing in animals (6,7). Table 1 summarizes the strengths and weaknesses of the different parenteral routes of administration. For example, although the intravenous dose route works well for even large-volume aqueous solutions, it is not a suitable dose route for water-insoluble substances or oily solutions. The subcutaneous route is a suitable dose route for some insoluble suspensions and can be used for implantation of

Table 1 Comparison of Various Parenteral Routes of Administration for Toxicity Testing

Parenteral route	Advantage/strength	Disadvantage/weakness
Intravenous	Able to dose large volumes	Cannot use water-insoluble substances, some emulsions acceptable
	Immediate exposure to circulation	
	Accurate delivery	Test material cannot be retrieved
		Extravasation at site of administration
Intraperitoneal	Relative ease of administration in rodents	Infrequent human dosing route
	Large absorptive surface	Sensitivity of peritoneum to chemical properties of dose solution
	Larger dose volumes than IM or SC routes	
Intramuscular	Can be used for both water-soluble and water insoluble substances	Potential for pain or inflammation at injection site
	Convenient/ease of injection	
Subcutaneous	Can be used for water-insoluble substances and for implantation of solid pellets	Only small-dose volumes can be injected
	Convenient/ease of injection	Potential for pain or inflammation at site of administration
	Injection need not be isotonic	

solid pellets. The intramuscular route of administration can use substantial volumes of either water-soluble or oily vehicles. The absorption pattern of subcutaneous and intramuscular dose forms can be purposefully altered by the formulator's use of aqueous solutions that produce rapid absorption or by use of repository preparations that provide slow and sustained absorption.

2. Relevant Information on Physicochemical Properties

The standard analytical data on test material properties that should be obtained as part of the background material for the excipient (8,9) are available on material safety data sheets or in *Handbook of Pharmaceutical Excipients* monographs (10). This information is considered important in the formulation of a stable, safe, and effective dose form. In addition, Good Laboratory Practice (GLP) regulations require that the test material be adequately characterized before conduct of toxicology studies. Some of the more typical background data that are relevant to parenteral dose forms are discussed in the following (2,3,6,7).

a. Description   Some of the more important factors that can help define the test material are general appearance, color, odor, and maybe even taste. These simple identifiers are rapid and inexpensive checks to uncover any obvious discrepancies between what was intended and what is proposed for testing.

b. Particle Size   Particle size can be important for a test material in a suspension because particle size has an effect on test material absorption and because sedimentation and flocculation rates in suspensions are partly governed by particle size.

c. Partition Coefficient   The partition coefficient is a measure of lipophilicity of the compound. Because biological membranes are lipoidal, they play a major role in test material transport. The ability of a test material molecule to cross a membrane at an absorption site can be related to the oil–water partition coefficient of the test material.

d. Solubility   Test material solubility can be a great determinant in the development of the parenteral dose form. For example, most intravenous dose forms require test materials with high water solubility that can be incorporated into aqueous solutions. Very lipophilic or water-insoluble materials may have to be incorporated into an oily vehicle that is more suitable for intramuscular or subcutaneous administration. Solubility of the drug in biological fluids at the injection site has a major effect on absorption of the drug. In general, solubility is a function of chemical structure: salts of acids or bases represent the test material class having the best chance of attaining the degree of water solubility desired. Other compound classes that cannot be solubilized in water within the desired pH range may require the use of nonaqueous solvents.

e. pK_a   The ionization constant (pK_a) provides information about the solubility dependence of the compound on the pH of the formulation. The pK_a of the test material at the neutral pH of the biological fluids at the injection site can greatly effect the ionized/nonionized ratio of the substance. It is the nonionized form that would have the greater absorption.

f. Chemical Properties

Stability of bulk.   The stability of the bulk excipient must be well characterized before the commencement of the toxicology program and the appropriate storage conditions defined. Various functional groups within a molecule may be prone to a specific type of reactivity under appropriate conditions. The conditions necessary for degradation are generally more pronounced when the test material is in solution or in suspension (9).

Stability of the Formulation.   Before beginning toxicology testing, the formulator must have an awareness of the potential for the new pharmaceutical excipient to undergo hydrolytic degradation or oxidation in the test formulation (9).

3. Compatibility with Blood

There are multiple mechanisms by which a pharmaceutical excipient may be incompatible with human or animal erythrocytes. Hemolysis appears to involve such factors as pH, lipid solubility, molecular and ionic sizes of solute particles, and certainly the tonicity of the solution (11). The specific chemical reactivity of the solute in solution is often more important in producing hemolysis than are the osmotic effects.

Many investigators test the reactivity and tonicity of injectable solutions by observing variations of red cell volume produced by the solutions (12). Such in vitro studies of the hemolysis of erythrocytes usually mix a large amount of solution with a small amount of blood, with a ratio of perhaps 100:1 (12,13).

4. pH of Solution

The pH of the pharmaceutical excipient solution is known to affect intramuscular or subcutaneous injection tolerance. For intravenous administration, a solution pH that is significantly different from neutrality might mean that the solution must be injected slowly to enable ''dilution'' of the irritating substance into the large blood volume.

5. Tonicity of Solution

A solution is isotonic with a living cell if there is no net gain or loss of water by the cell, or other change in the cell when it is in contact with that solution. Physiological solutions with an osmotic pressure lower than that of body fluids, or of 0.9% sodium chloride solution, are commonly referred to as being hypotonic. Physiological solutions having a greater osmotic pressure are termed as hypertonic.

Osmotonicity is of great importance in parenteral injections—the influence of osmotic effects depends on the degree of deviation from tonicity, the concentration, the location of the injection, the volume injected, the speed of the injection, the rapidity of dilution and diffusion, and other such (14). When formulating parenteral solutions, hypotonic solutions usually have their tonicity adjusted by addition of dextrose or sodium chloride. Hypertonic parenteral drug solutions cannot be adjusted. Hypertonic or hypotonic solutions are usually administered slowly in small volumes, in which dilution and distribution occur rapidly. Excessive infusion of hypotonic solutions may cause swelling of the red blood cells, hemolysis, and cellular uptake of water. Excessive infusion of hypertonic solutions may lead to intracellular dehydration, and osmotic diuresis, with loss of water and electrolytes. In addition, solutions that differ from the serum in tonicity are generally thought to cause tissue irritation, pain on injection, and electrolyte shifts (2,3,14).

B. Purity and Stability or Impurities

1. Importance of Specifications

Pharmaceutical excipient and product formulation specifications define the test material that is used in the toxicology program. Strict adherence to specifications will ensure that the entire toxicology program is conducted with the same test material. Quality control testing and evaluation is involved primarily with incoming raw materials, the manufacturing process, and the final product (2,3,6,7). Testing of incoming raw materials includes routine testing on all actives, chemicals, and packaging materials. While preparing parenteral dose forms, the formulator must also be aware that the *U.S. Pharmacopeia/National Formulary (USP/ NF)* limits the use of some "added substances" (5). Table 2 indicates the maximum amounts of added substances, as defined by the *USP* for injectable products.

2. Influence of Impurities on Toxicity Is Magnified by Parenteral Exposures

The parenteral routes of exposure can provide more complete and more rapid access to the systemic circulation and circumvent the highly efficient protective barriers of the human body. This enables the pharmaceutical excipient to achieve a very high blood level in a shorter time frame. The same would be true for impurities or contaminants present in either the bulk excipient or in the final parenteral formulation. Toxic agents generally produce their greatest effect and the most rapid response when given directly into the bloodstream. The intravenous route is the most "effective" in this respect (4,15), with intraperitoneal, subcutaneous, intramuscular, and intradermal being generally less efficient than

Table 2 Maximum Amounts of Added Substances Permitted in *USP* Injectable Products

Substance	Maximum (%)
Mercury compounds	0.01
Cationic surfactants	0.01
Chlorobutanol	0.5
Cresol	0.5
Phenol	0.5
Sulfur dioxide or	0.2
Sodium bisulfite equivalent or	0.2
Sodium sulfite equivalent	0.2

Source: Ref. 5.

intravenous (see Table 1) but more efficient that either oral or dermal routes of administration. The 50% lethal dose (LD50) values for common xenobiotics are often several orders of magnitude less for parenteral routes of administration than for oral or dermal routes (16).

3. Particular Emphasis on Residual Monomers, Heavy Metals, and Solvents

For specifications intended to demonstrate purity and impurities in the new pharmaceutical excipient, particular emphasis should rightly be placed on those materials, potentially common in synthetic or manufacturing processes, that historically have a high toxicological profile. Typically, those materials of concern have been associated in humans with cancer, neurotoxicity, or reproductive toxicity. For example, the International Conference on Harmonization (ICH) has drafted guidelines on residual solvents in pharmaceutical products that categorize solvents based on degree of human toxicity (17). In the *USP 23/NF 18*, several limit test methods are described for heavy metals (i.e., arsenic, iron, lead, and mercury), which are the heavy metals typically considered to have toxicological significance in pharmaceutical excipients and drug products.

4. Sterility and Pyrogenicity

Because the parenteral route of administration magnifies the potential for adverse effects, the *USP 23/NF 18* defines that parenteral products are prepared scrupulously by methods designed to ensure that they meet pharmacopeial requirements for sterility, pyrogens, particulate matter, and other contaminants.

Whenever possible, parenteral products should be sterilized as one of the end steps in the manufacturing and packaging process. The method of sterilization (autoclaving, dry heat, ionizing radiation, sterile filtration, or other) may be chosen based on the test material form and physicochemical properties (2,3,6,7). Sterility tests are either defined as part of the quality control in the manufacturing process or as part of the individual monograph requirements for sterility of the test article (19,20).

The presence of pyrogens in parenteral products is evaluated by a qualitative fever response test in rabbits (3). Rabbits are used as test animals because they show physiological response to pyrogenic substances that are similar to those of humans. The USP Pyrogen Test is described in *USP* Chapter ⟨151⟩ (21).

A relatively new test for pyrogens has been accepted, the limulus test, which is an in vitro test based on the geling or color development of a pyrogenic preparation in the presence of a lysate of the amebocytes of the horseshoe crab (*Limulus*). The bacterial endotoxins test (22) employs a limulus amebocyte lysate (LAL) reagent to estimate the concentration of bacterial endotoxin that may be present in a sample of the novel excipient. The unknown concentration in the sample is compared with a reference standard endotoxin that has a defined potency of 10,000 USP endotoxin units per vial. The limulus test provides greater sensitivity, and is simpler and more rapid to conduct because automated techniques have been developed.

Because of the increased concern for particulate matter in injectable preparations, there is often a step in the manufacturing process of liquids that employs a filtration (3,9). The primary objective of filtration is to clarify a solution. A high degree of clarification can be achieved when particulate matter, down to approximately 2 µm in size, is removed. A further filtration step, removing particulate matter down to 0.2 µm in size, would also eliminate microorganisms and would accomplish sterilization.

C. Background Biological Information

1. Structure–Activity Relationships to Similar Chemical Structures with Known Biological Effects

One of the first activities that the toxicologist should do before definition of the toxicology program is to compare the chemical structure of the new pharmaceutical excipient against computerized databases that have been established to show relations between functional groups–structural components and endpoints of toxicity, such as carcinogenicity, mutagenicity, and sensitization (23–25). This type of screening activity may highlight specific toxicology studies or safety evaluations on which to concentrate the toxicology program.

2. Known Pharmacological Effects, Nutritional Effects

Excipients, by definition, display either no pharmacological activity or very limited and directed activity (26). However, the toxicology program must be able to support safe exposure to the novel excipient over a wide potential range of exposure conditions. Although exposure to a new pharmaceutical excipient parallels the drug exposure categories, the overall exposure is less well controlled because patients can be exposed to a given excipient in several different dosage forms for different clinical conditions and for varying lengths of time.

There are several well-known types of experimental systems that can define the pharmacological or nutritional profile of the new pharmaceutical excipient. First, contract laboratories have established procedures for whole-animal pharmacology-screening capabilities that evaluate a wide range of potential pharmacological and nutritional actions of the test material. Second, there has been a proliferation of laboratories that offer receptor-binding assays that compare the binding affinity of novel compounds with ligands known to bind specific receptors that affect pharmacological activity.

III. BASE SET TESTING

The Safety Committee of the International Pharmaceutical Excipients Council (IPEC) has published a paper that outlines guidelines for suggested toxicology studies for novel pharmaceutical excipients (26). The guidelines provide for a tiered approach based on the chemical and physical properties of the excipient, review of the scientific literature, exposure conditions, and absence or presence of pharmacological activity. After a careful review of the results of the base set studies, single-dose studies should be conducted in humans. The data are critically evaluated and may support the use of the new excipient in a product with a short half-life that will not be given in a frequency what would provide for residual excipient in the body or in a product that may be used only once or twice in a lifetime (e.g., a diagnostic agent).

A. Base Set Tests Relevant to Parenteral Routes

For single or limited exposures in humans by parenteral dose route(s) (26) the following tests may apply.

1. Acute Toxicity Tests

Acute toxicity (i.e., limit test, approximate lethal dose, up-and-down procedure, and such) should be conducted to provide a comparison of toxicities between the oral and the intended parenteral route(s) of administration.

2. Mutagenicity Tests

The mutagenicity studies are conducted to ascertain if the novel excipient has any genotoxic potential. Typical studies that comprise a battery of tests might include the Ames test, the in vivo chromosomal aberration test, and the mouse micronucleus test.

3. 28-Day Repeat-Dosing Studies

Repeat-dosing studies should be conducted in two animal species (one rodent and one nonrodent) by appropriate parenteral dose route(s). Generally, if the intended product would use both intravenous and intramuscular–subcutaneous routes of administration, repeat-dose studies that evaluate both types of administration should be included. Important for parenteral products is a specific study designed to evaluate the injection site for the various parenteral route(s) of administration. It is extremely helpful to compare adverse effects delineated in toxicity studies conducted by the oral route versus the parenteral route(s).

4.   ADME–PK Studies

Such studies are designed to determine the absorption, distribution, metabolism, excretion (ADME), and pharmacokinetics (PK) of the novel excipient. These studies should compare ADME–PK from a single dose versus multiple doses. Multiple animal species, normally those employed in the repeat-dosing toxicity studies, should be tested and preferably by multiple dose routes.

5.   Other Appropriate Acute Toxicity Studies

Additional toxicity studies, such as dermal and eye irritation studies and acute inhalation toxicity studies, should be conducted as part of the generation of background toxicology information required to compile Material Safety Data Sheets (MSDS) that provide critical information on worker safety in manufacturing, processing, and handling the test material.

B.   Additional Animal Toxicity Testing

As the extent or duration of proposed human exposure increases, additional animal testing should be considered. All studies proposed should be based on previous findings in the toxicology program and the intended use of the new excipient. For limited and repeated exposure in humans (26) the following tests may apply.

1.   Base Toxicity Set (described in Sec. III.A)

2.   Subchronic Study

The subchronic study should be conducted in an appropriate animal species, as defined in the base toxicity set data. Normally 28–90 days of daily dosing is required.

3.   Teratology Study

The teratology study should be a Segment II teratology study conducted in rat or rabbit, or both, using the appropriate parenteral route(s) of administration.

4.   Additional Mutagenicity Data

As the duration of human exposure increases, additional mutagenicity data (e.g., mammalian cell gene mutation assay) should be added to complete the mutagenicity battery.

IV. ROUTE SPECIFIC REQUIREMENTS

A. Required Toxicity Tests: Parenteral

The IPEC Safety Committee safety assessment guidelines (26) indicate that the animal toxicology studies should be conducted using the route of expected human exposure (i.e., that valid animals studies must simulate the conditions of human exposure.) This is especially important for parenteral routes of administration because of the potential for adverse effects at the site of injection. In addition, the IPEC guidelines indicate that the extent of animal toxicity testing would be determined by the extent of human exposure.

Thus, the required parenteral toxicity tests should employ experimental designs that are based on the following principal considerations (1,2,9,27):

1. The appropriate route of administration, which is based on knowledge of the following:

    Desired rate and extent of systemic absorption.
    Physicochemical properties of the test material.
    The resultant dosage forms available.
    The total volume of the formulation to be injected.
    The frequency of the injections.
    The intended human route of administration.

2. The appropriate duration of treatment or dosing to mimic expected human dosing situations.

    Parenteral toxicity studies in animals usually require an exaggeration, both in dose and duration, of intravenous, intramuscular or subcutaneous treatment regimens.
    A frequent limiting factor in the conduct of multiple dose studies, especially for studies with subchronic duration of 2–4 weeks or longer, is local site intolerance (2,9,27).

3. The appropriate animal species, as influenced by several factors.

    Regulatory requirements may indicate the need for both rodent and nonrodent species to be tested.
    Study requirements in terms of practical considerations relative to required dosing volume, repetitive dosing, ease of dosing to minimize animal stress, study logistics, and such.

1. Acute Toxicity Evaluation by Intended Parenteral Dose Route

Acute toxicity studies in rodents that utilize parenteral dose routes are conducted in the same manner as their oral dosing route counterparts. For pharmaceutical excipients, in which the test material is not expected to have high pharmacological or toxicological activity, the limit test would be useful (28). In a limit-dose study, the dose that is used is considered high enough that if no mortality or significant toxicity is seen in animals receiving this dose, then no higher doses are required. The study would use a single dose on day 1 with necropsy on day 15; for rodent studies five males and five females would be employed. For nonrodent species, 2 to 5 animals per sex would be tested. This design would limit the number of animals on test, in accordance with IPEC guidelines for safety testing of pharmaceutical excipients.

If toxicity is expected, an alternative test method for acute toxicity evaluations would be the up- and down-test (29). In this design, a single animal is dosed, then undergoes an observation period along with clinical laboratory studies, then is sacrificed 14 days after dosing. If there is no effect seen from the first dose, a second animal is dosed at a multiple (usually between 1.5–3.0) of the first dose. If there is an effect seen from the first dose, then a second animal is dosed at a division (1.5–3) of the first dose. It is common to dose subsequent animals of different sexes and to dose animals of both sexes at both toxic and non-toxic levels.

2. Evaluation of Injection Site Irritation

There are two standard protocols suggested in Pharmaceutical Manufacturers Association (PMA) guidelines: the single-dose, intramuscular irritation study in the rabbit, and the multiple dose intramuscular study in the dog. (30).

a. Rabbit A single injection of the new pharmaceutical excipient is made with a 23-gauge needle in the midlumbar muscles of a New Zealand white rabbit, the injection is placed from 15- to 20-mm deep. Observations are made at 1, 3, and 7 days postinjection. A comparable number of control rabbits are injected with the equivalent volume of the vehicle only. Serum creatinine phosphokinase (CPK) values are used to evaluate muscle injury, and the serum CPK is determined pretest and at the 24-h postinjection observation to evaluate elevations close to the time of maximum serum activity. At termination, the lumbar muscles are perfused with a saline flush and fixed in formalin. Any injection site lesions are graded grossly according to its three-dimensional size and the degree of hemorrhage, degeneration, and necrosis present. Table 3 summarizes scoring of injection site lesions. Cumulative scores of 0 indicate no musculoirritation, 1–6 = "slight", 7–16 = "moderate", and scores higher than 16 indicate "severe" irritation. Histopathological examination is needed to establish the type of lesion (30).

b. Dog This special study is often conducted concomitantly with a subacute or subchronic toxicity study designed to determine systemic adverse effects of the test material. Thus, a typical design would include three treatment groups

Table 3 Evaluation and Scoring of Injection Site Lesions

Evaluation	Hemorrhage	Degeneration	Necrosis
None	0	0	0
Trace	1	2	4
Slight	2	4	12
Moderate	3	6	18
Severe	4	8	24

Source: Ref. 30.

and one control group of four animals each. The injection volumes should be kept small, in the range of 1.5 mL or less, either by adjusting the test material concentration or by dividing the daily dose among several sites in the muscle mass. Recommended injection sites are in the semitendonosis and semimembranosis muscles, but not close to the sciatic nerve. Serum glutamic oxaloacetic transaminase (GOT) is measured, as increases are characteristic of low-grade muscle injury. Normal histological examinations of the muscle at injection sites are made to delineate muscle fiber or bundle irritation (30). There should also be an evaluation of injection sites for test materials dosed by intravenous routes of administration, but there are no published standard methods to do this.

3. ADME–PK Studies

The conduct and experimental design of ADME–PK studies by the parenteral route is not significantly different from the conduct of these studies by the oral route of administration.

With intravenous administration, the test material is infused directly into the circulation, so there is no "absorption" phase to contend with. Significant differences in toxicity, reflecting availability, have been demonstrated in studies between oral and parenteral routes, even for readily absorbed compounds (1,16). Intravenous or intraperitoneal injections cause high blood level peaks followed by exponential declines (31). Test materials from intraperitoneal injections must undergo absorption, but the relative rate of absorption from this dose route is still faster than the oral route. The rate of absorption from intramuscular and subcutaneous injections is dependent on the physicochemical properties of the test material and the dose form (solutions are rapidly absorbed, but test materials in oily vehicles or suspensions–emulsions would be absorbed much more slowly).

The distribution of test materials once absorbed from parenteral routes of administration should be no different than from oral routes of administration.

The metabolic biotransformation of xenobiotics could be significantly altered, because the highest concentration of metabolizing enzymes exists in the liver (and a medium amount in the small intestine), and both of these tissues are circumvented from a "first-pass" effect by employing the parenteral dosing route (31).

Routes of excretion and mechanisms of excretion should not be significantly different between oral and parenteral dose administration, but because of the potential depot effect of subcutaneous and intramuscular dose forms, the rate of excretion from the body might be significantly longer in duration because the absorption phase has been delayed.

4. Subacute and Subchronic Toxicity Studies

The experimental design and conduct of subacute and subchronic toxicity studies by parenteral routes are not significantly different from similar studies that employ oral dosing as the route of administration (28,30). Standard in-life endpoints would include gross observations for signs of toxicity, measurements of body weight and food consumption, hematology and serum chemistry evaluations, and urinalysis. Necropsy measurements would include observations of gross lesions, organ weight measurements, and microscopic examination of tissues taken at necropsy.

B. Route-Specific Information

1. Considerations for Intravenous

The volume of test material that can be administered by intravenous injection and the rate at which it can be injected will vary considerably depending on the pH of the solution, its osmotic strength, and whether it is likely to have any physiological or pharmacological effects.

a. Blood Volume of Animal Species One of the primary considerations for intravenous administration will be the volume of the intended dose versus the blood volume of the test species. Table 4 will help illustrate how the various animal species differ in blood volume and plasma volume (27,28,32). If the proposed dose volume is large, the toxicologist may have to choose alternative routes of administration, may have to infuse the dose slowly, or may have to choose a larger species of animal in which to conduct the toxicity testing.

b. Recommended Site(s) of Injection The toxicology literature suggests numerous veins for intravenous injections in the various laboratory species, but some of them (such as the jugular veins) in smaller animals, necessitate sedation or anesthesia and surgical exposure (27,28,33,34). Thus, these veins may be considered difficult for routine acute toxicology studies, but may still have a place

Table 4 Information on Blood Volume and Plasma Volume Across Animal Species

Species	Body weight (kg)	Total blood volume (% of body weight)	Approximate total blood volume (mL)	Approximate plasma volume (mL/kg)
Mouse	0.030	7.5	2.4	48
Rat	0.30	5.6–7.5	21.0	31
Guinea pig	0.40	4.6–5.6	20.0	NA[a]
Rabbit	3.5	4.5–5.5	175.0	44
Dog	12.0	7.2–9.5	960.0	54
Monkey	7.0	7.5	525.0	45
Human	70.0	6.5–7.5	5,000	43

[a] NA, data are not available.
Source: Refs. 27, 28, 32.

when one considers repeated-dose studies in which indwelling catheters are surgically placed in each animal to facilitate repetitive injections or infusions. Table 5 summarizes preferred veins that can be used for intravenous administration of test materials for each animal species likely to be used in toxicology studies (34). Preferred sites of injection for other parenteral routes for each species are also included in this table.

c. Bolus Injection Versus Infusion Rapid intravenous injection (bolus) of a limited volume is possible if the dosing solution has a neutral pH and is isotonic without causing clinically significant cardiovascular disturbances (1,2). The simplest injection site would be through a tail vein in a mouse or rat or through a marginal ear vein of a rabbit. If the test solution cannot be administered by a bolus injection, slower infusion may still permit the choice of the intravenous route of administration. For example, a 1-ml bolus dose may be possible in a 200-g rat, and larger volumes of up to 2–3 mL may be infused over a 5-min period without harming the animal (27,33,34).

The basic components of an intravenous infusion system for laboratory animals are as follows (1,27,32–34):

A flexible plastic indwelling cannula
An extension line that is usually threaded subcutaneously to a convenient point of exit
An electronic infusion pump
A mechanical device to allow the animal to move and yet protect the infusion line and cannula

Table 5 Recommended Injection Sites and Needle Sizes for Parenteral Routes in Different Species

Species	IV	IP	SC	IM
Mouse	Tail vein, <25 G	Abdomen, <21 G	Scruff, <20 G	Quadriceps, caudal thigh, <23 G
Rat	Tail vein, <23 G	Abdomen, <21 G	Scruff, <20 G	Quadriceps, caudal thigh, <21 G
Hamster	Femoral or jugular vein, <25 G	Abdomen, <21 G	Scruff, <20 G	Quadriceps, caudal thigh, <21 G
Guinea pig	Ear vein, Saphenous vein	Abdomen, <23 G	Scruff, <20 G	Quadriceps, caudal thigh, <21 G
Rabbit	Marginal ear vein, <21 G	Abdomen, <21 G	Scruff, flank, <20 G	Quadriceps/caudal thigh, lumbar muscles, 20 G
Dog	Cephalic vein, <20 G	Abdomen, <20 G	Scruff, back, <20 G	Quadriceps, caudal thigh, <20 G
Monkey	Tail vein, <21 G	Abdomen, <21 G	Scruff, <20 G	Quadriceps, caudal thigh, <21 G

Source: Refs. 27, 32, 33, 35.

d. Use of Infusion Pumps Infusion pumps or syringe pumps can greatly facilitate slow infusion of test solutions. These pumps need to be calibrated for a specific syringe barrel size and test solution viscosity so that accurate in delivery rates milliliter per minute (mL/min) or milliliter per hour (mL/hr) can be determined. The resultant accuracy of delivery of syringe pumps is quite excellent and allows precise administration of test material solutions over a wide variety of intended infusion rates and times (1,27,32).

e. Indwelling Cannula or Catheters If repeated injections of test material need to be made over a relatively short time period, the toxicologist may choose to employ an indwelling cannula or catheter to facilitate dosing procedures (1,27,32–34). Depending on the species and the duration of infusion, indwelling cannulae can be surgically placed in the animal. "Butterfly" infusion needle sets can be attached to plastic cannulae and are very useful for intravenous infusions because the flexible butterfly extension set and catheter allows for some movement between the syringe and the needle without the needle becoming dislodged from the vein.

f. Use of Miniature Osmotic Pumps Another potential mechanism to infuse small volumes of test material over a long time period is to use osmotic minipumps to deliver the dosage. These pumps have an internal reservoir to contain the dosing solution that is filled before surgical implantation of the minipump subcutaneously in the test animal. A range of different pump sizes that provide infusion rates of 0.001–0.005 mL/h for 3–30 days are now available (e.g., Alzet minipump system (1,27)). The rate and duration of delivery are preset by the manufacturer. Although the osmotic minipump is typically implanted subcutaneously or intraperitoneally, it can be fitted with catheters for delivery of drug solution for intravenous infusions, into organs, or other areas.

2. Considerations for Intraperitoneal Administration

The intraperitoneal route of administration has only rare human applications, but can be a useful route of administration in animal toxicity studies. Injection by this route can be carried out quickly so it is often used in animal toxicology studies in which a parenteral dose route is to be compared with a oral dose route. Tables 6 and 7 indicate the suggested dose volumes (in mL/kg) and maximal dose volumes (in mL) for test material administration by various parenteral routes in different species, respectively. It can be noted from Tables 6 and 7 that large volumes of test material can be administered intraperitoneally (27,28,33,34). The injection is recommended to be made in the lower left quadrant of the abdomen, because there are no vital organs except the small intestine in this region. A primary concern for employing this dose route is the extreme sensitivity of the

Table 6 Suggested Dose Volumes (mL/kg) for Test Material Administration by Various Parenteral Routes in Different Species

Species	IV Target	IV Max	IP Target	IP Max	SC Target	SC Max	IM Target	IM Max
Mouse	5	25	5–10	50	1–5	20	0.1	1.0
Rat	1–5	20	5–10	20	1	20	0.1–1	10.0
Rabbit	1–3	10	NA[a]		1–2.5	10	0.1–0.5	1.0
Dog	1	10	3	5	0.5	2	0.1–0.2	1.0
Monkey	1	10	3	5	0.5	2	0.1–0.5	1.0

[a] NA, not available.
Source: Refs. 28, 34.

peritoneum, which may demonstrate irritation reactions to test materials or formulations that will confound the evaluation of systemic toxicity.

3. Considerations for Intramuscular Administration

The intramuscular route of administration is second only to the intravenous route in rapidity of onset of systemic action. Most injectable products can be given intramuscularly, so there are numerous dosage forms that can be used by this route of administration: solutions, oil-in-water or water-in-oil emulsions, suspensions (aqueous or oily base), colloidal suspensions, and others (3,6–9). Those product forms in which the drug is not fully dissolved generally result in slower, more gradual test material absorption, a slower onset of action, and sometimes

Table 7 Maximum Dose Volumes (in mL) for Test Material Administration by Various Parenteral Routes in Different Species

Species	Body weight	IV	IP	SC	IM
Mouse	20–30 g	0.5	1.0	0.5–1.0	0.05
Rat	100 g	1.0	2–5	2–5	0.1
Hamster	50 g	—	1–2	2.5	0.1
Guinea pig	250 g	—	2–5	5	0.25
Rabbit	2.5 kg	5–10	10–20	5–10	0.5
Dog	10 kg	5–10	10–20	5–10	1.0
Monkey	6 kg	10–20	10–30	5.0	3.0

Source: Refs. 27, 32, 34.

longer-lasting test material effects. Intramuscularly administered products typically form a depot in the muscle mass from which the drug is slowly absorbed. Generally, hypertonic dosage forms are contraindicated for intramuscular administration.

The intramuscular injection is easier to administer than the other parenteral injections, the main precaution is to avoid entering a blood vessel (27,32,33). This can be prevented by pulling back on the plunger of the syringe and if blood does not appear, then the needle is probably located properly in the muscle and not in a blood vessel.

a. Recommended Sites of Injection   The usual site for this route of injection is into the muscles of the hind limb. In rodents, the biceps femoris, the semitendinosus, and gluteus maximus muscles that make up the posterior aspect of the thigh and rump, or the quadriceps muscle group on the anterior thigh are most useful sites (27,33).

b. Alternative Sites for Repeat-Dosing Studies   For repeat-dosing studies, it is generally recommended to rotate injection sites on each animal daily to minimize the likelihood of local irritation or damage at the site of injection.

4. Considerations for Subcutaneous Administration

The subcutaneous injection is one into the loose connective tissue and adipose tissue beneath the skin. Drugs are more rapidly and more predictably absorbed by this route of administration than by oral routes, but absorption is generally slower and less predictable than by the intramuscular route (2). Test materials that are highly acidic, alkaline, or irritating that cause production of pain, inflammation, or necrosis of tissues should not be administered by this route of administration.

a. Recommended Sites of Injection   In rodents, this injection is usually made under the skin of the back and sides. The needle should be passed through the skin in an anterior direction and at a shallow angle to the skin surface (27,32,33,35). A successful subcutaneous injection made on the back or sides results in the formation of a bleb during delivery of 0.5 mL or more of solution from the syringe. The smallest size of hypodermic needle compatible with the type of material being injected should be used. Table 7 shows the recommended injection sites and needle sizes for parenteral routes in different species.

b. Alternate Sites for Repeat Dosing Studies   For repeat-dosing studies, it is generally recommended to rotate injection sites on each animal daily to minimize the likelihood of local irritation or damage at the site of injection.

C. Dose Route Selection and Feasibility

1. Maximum Injectable Volumes

Numerous reference materials are available to the toxicologist to provide estimates of maximum suggested dose volumes, listed by animal species and by
specific parenteral dosing route. Some tabulations are made on a milliliter per
kilogram volume basis (28) and some are in terms of maximum volume in milliliters (27,32,34; see Table 6).

2. Recommended Animal Species and Considerations

a. Practicality The IPEC Safety Committee guidelines for pharmaceutical excipients administered in parenteral dose forms are useful starting points for
the toxicologist to consider, but many of the specific choices about toxicology
study conduct will be dependent on the nature and physicochemical properties
of the test substance and the resulting product form. Larger animals, such as the
dog and monkey, are an easier test animal to manipulate for parenteral administration, and should be considered for intravenous administrations, in particular, because they have a larger blood volume to affect dilution of the test substance
and longer limbs in which to find multiple intravenous injection sites. Rodents,
because they are small, are easy to manipulate for all of the parenteral dosing
routes mentioned in this chapter, particularly for the intraperitoneal, intramuscular, and subcutaneous administrations. However, the toxicologist should pay close
attention to a rotation schedule for alternating sites of these injections in repeat-
dosing studies.

b. Regulatory Considerations Regulatory guidelines (36–38) for drug
products in the various countries usually require toxicology studies in at least
two species, one rodent and one nonrodent. Although no regulatory guidelines
for new pharmaceutical excipient toxicity testing currently exist, the IPEC Safety
Committee approach for safety assessment of pharmaceutical excipients (26) suggests that the same experimental design decisions, based on good scientific principles and the best science available, would hold true for a new pharmaceutical
excipient. The most specific guideline for parenteral dose toxicity studies comes
from the Pharmaceutical Manufacturers Association (30) in terms of describing
test methods for assessment of injection site irritation. This type of assessment
is also based on good scientific principles.

3. Options for Dealing with pH, Tonicity, Viscosity, and Compatibility with Blood Issues

If the toxicologist is faced with a test material or dosing preparation that has
properties that are not optimum (not isotonic, pH different from physiological,

high viscosity, or other), there are several options that can be employed in parenteral dosing studies.

For intravenous studies, the novel pharmaceutical excipient may be infused slowly, rather than injected by a bolus dose. The techniques for slow infusion (even for repeat-dose studies) and the availability of syringe pumps to accurately deliver the dose have improved to the point that this is a fairly common practice in toxicology laboratories.

For subcutaneous or intramuscular toxicity studies, the administration of the novel pharmaceutical excipient dosing material may be split so that several injection sites are used for one dose. It is also a common practice to have an injection site schedule for a repeat-dose study, so that the daily injection sites are rotated through various body parts or locations to decrease the chance for local adverse reactions.

V. CONCLUSIONS

The goal of this chapter was to review the safety evaluation strategy and conduct of toxicology tests for new pharmaceutical excipients intended to be administered by the most common parenteral dosing routes. The chapter provides an overview of specific considerations the toxicologist should assess as they define the toxicology program for a novel pharmaceutical excipient that will be employed in parenteral dose forms. Some of the details of general test methods for parenteral dosing studies have been discussed, but the specifics of test procedures and study protocols actually employed will be dependent upon several key considerations: (a) general toxicology information known from testing by other dose routes (if available): (b) specific considerations or limitations for parenteral dosing owing to the physicochemical properties of the new pharmaceutical excipient; (c) knowledge of the limitations and practicality of parenteral administration techniques, dose volumes, injection sites, and such, in common laboratory animals used in toxicology testing; (d) the intended use of the new pharmaceutical excipient in parenteral drug products; and (e) the use of test methods and study protocols that are generally recognized by toxicology experts and acceptable to regulatory agencies.

There cannot be enough emphasis placed on the raw material specifications and purity for the new pharmaceutical excipient relative to inclusion in parenteral dose forms. The adverse effect or influence of impurities (residual solvents, heavy metals, residual monomers, or others) on toxicity is magnified by parenteral exposures because body defense mechanisms and barriers are bypassed by these routes of exposure. There should also be great importance placed on preparation of final dose forms using the new pharmaceutical excipient that are sterile and free from pyrogens.

REFERENCES

1. D Walker. Parenteral toxicity. In: B Ballantyne, T Marrs, P Turner, eds. General and Applied Toxicology. vol 1. New York : Stockton Press, 1993, pp 387–412.
2. RJ Duma, MJ Akers, SJ Turco. Parenteral drug admininstration: routes, precautions, problems, complications, and drug delivery systems. In: KE Avis, HA Lieberman, L Lachman, eds. Pharmaceutical Dosage Forms: Parenteral Medications. vol 1. 3rd ed. New York: Marcel Dekker, 1992, pp 17–58.
3. KE Avis. Parenteral preparations. In: AR Gennaro, ed. Remington's Pharmaceutical Sciences. 18th ed. Easton, PA: Mack Publishing, 1990, pp. 1545–1569.
4. LZ Benet, DL Kroetz, LB Sheiner. Pharmacokinetics: the dynamics of drug absorption, distribution, and elimination. In: JG Hardman, LE Limbird, eds. Goodman and Gilman's The Pharmacological Basis of Therapeutics, 9th ed. New York: McGraw Hill, 1996, pp 3–27.
5. The United States Pharmacopoiea 23/The National Formulary. General Requirements for Tests and Assays, Injections, Chapter 1. Rockville, MD: United States Pharmacopoieal Convention, 1995, pp 1650–1652
6. JC Boylan, AL Fites. Parenteral products. In: GS Banker, CT Rhodes, eds. Modern Pharmaceutics. 2nd ed. New York: Marcel Dekker, 1990, pp 491–538.
7. HC Ansel, NG Popovich. Injections, sterile fluids, and products of biotechnology. In: Pharmaceutical Dosage Forms and Drug Delivery Systems. 5th ed. Philadelphia: Lea & Febiger, 1990, pp 255–306.
8. LJ Ravin, and GW Radebaugh. Preformulation. In: AR Gennaro, ed. Remington's Pharmaceutical Sciences. 18th ed. Easton, PA: Mack Publishing, 1990, pp 1435–1450.
9. S Motola, SN Agharkar. Preformulation research of parenteral medications. In: KE Avis, HA Lieberman, L Lachman, eds. Pharmaceutical Dosage Forms: Parenteral Medications. vol 1. 2nd ed. New York: Marcel Dekker, 1992, pp 115–172.
10. Handbook of Pharmaceutical Excipients. American Pharmaceutical Association/The Pharmaceutical Association of Great Britain, 1986.
11. CA Hall. Hemolysis. In: JA Halsted, CH Halsted, eds. The Laboratory in Clinical Medicine: Interpretation and Application. 2nd ed. Philadelphia: WB Saunders, 1981, pp 492–503.
12. DL Deardorff. Ophthalmic solutions. In: Remington's Pharmaceutical Sciences. 14th ed. Easton, PA: Mack Publishing, 1970, pp 1561–1562.
13. ER Hammarlund, K Pederson-Bjergaard. J Pharm Sci 50:24–30, 1961
14. FP Siegel. Tonicity, osmoticity, osmolality and osmolarity. In: AR Gennaro, ed. Remington's Pharmaceutical Sciences. 18th ed. Easton, PA: Mack Publishing, 1990, pp 1481–1498.
15. DL Eaton, CD Klaassen. Principles of toxicology. In: CD Klaassen, MO Amdur, J Doull, ed. Casarett and Doull's Toxicology: The Basic Science of Poisons. 5th ed. New York: McGraw-Hill, 1996, pp 13–33.
16. MJ Derelanko. LD_{50} values. Appendices: Tables of toxicological importance. In: MJ Derelanko, MA Hollinger, eds. CRC Handbook of Toxicology. Boca Raton, FL: CRC Press, 1995, pp 771–798.

17. International Conference on Harmonization. Residual Solvents Guideline, Q3C, Draft 6, 1996.

18. The United States Pharmacopiea 23/The National Formulary 18. General requirements for tests and assays, limit tests. Rockville, MD: United States Pharmacopieal Convention, 1995, pp 1724–1731

19. The United States Pharmacopoiea 23/The National Formulary 18. General Chapter ⟨61⟩, microbial limit tests. Rockville, MD: United States Pharmacopeial Convention, 1995, pp 1681–1686.

20. The United States Pharmacopoiea 23/The National Formulary 18. General Chapter ⟨71⟩ sterility tests. Rockville, MD: United States Pharmacopeial Convention, 1995, pp 1686–1690.

21. The United States Pharmacopoiea 23/The National Formulary 18. General Chapter ⟨151⟩ pyrogen test. Rockville, MD: United States Pharmacopeial Convention, 1995, pp 1718–1719.

22. The United States Pharmacopoiea 23/The National Formulary 18. General Chapter ⟨85⟩, bacterial endotoxins test. Rockville, MD: United States Pharmacopeial Convention, 1995, pp 1696–1697.

23. European Chemical Industry Ecology and Toxicology Center (ECETOC). Structure activity relationships in toxicology and ecotoxicology: an assessment. F Choplin, D Dugard, J Hermens, R Jaeckh, M Marsmann, DW Roberts, eds. ECETOC Monograph No 8, Brussels: 1986.

24. L Golberg, ed. In: Structure Activity Correlation as a Predictive Tool in Toxicology. Washington, DC: Hemisphere Publishing, 1983.

25. RE Rush, KL Bonnette, DA Douds, TN Merriman. Dermal irritation and sensitization. In: MJ Derelanko, MA Hollinger, eds. CRC Handbook of Toxicology. Boca Raton, FL: CRC Press, 1995, pp 105–162.

26. M Steinberg, JF Borzelleca, EK Enters, FK Kinoshita, A Loper, DB Mitchell, CB Tamulinas, ML Weiner. A new approach to the safety assessment of pharmaceutical excipients. Regul Toxicol Pharmacol 24:149–154, 1996.

27. HB Waynforth, PA Flecknell. Administration of substances. In: Experimental and Surgical Technique in the Rat. 2nd ed. New York: Academic Press, 1992, pp 1–67.

28. CS Auletta. Acute, subchronic and chronic toxicology. In: MJ Derelanko, MA Hollinger, eds. CRC Handbook of Toxicology. Boca Raton, FL: CRC Press, 1995, pp 51–78.

29. RD Bruce. An up-and-down procedure for acute toxicity testing. Fundam Appl Toxicol 5:151–157, 1985.

30. Pharmaceutical Manufacturers Association (PMA). Guideline for the Assessment of Drug and Medical Device Safety in Animals, 1977.

31. MB Abou-Donia. Metabolism and toxicokinetics of xenobiotics. In: MJ Derelanko, MA Hollinger, eds. CRC Handbook of Toxicology. Boca Raton, FL: CRC Press, 1995, pp 540–589.

32. WA Ritschel. In vivo animal models for bioavailability assessment. Sci Tech Pratiques Pharm (Paris) 3(2):125–141, 1987.

33. PA Flecknell. Nonsurgical experimental procedures. In: AA Tuffery, ed. Laboratory

Animals: An Introduction for New Experimenters. Chicester, UK: Wiley Press, 1987, pp 225–246.

34. CT Hawk, SL Leary. Needle sizes, sites, and recommended volumes for injection, Appendix IV A. In: Formulary for Laboratory Animals. Ames, IA: Iowa State University Press, 1995, pp 76–77.

35. YW Chien. Parenteral controlled-release drug administration. In: Novel Drug Delivery Systems. New York: Marcel Dekker, 1982, pp 219–310.

36. EI Goldenthal. Current views on safety evaluation of drugs. In: FDA Handbook of Total Drug Quality. 1970, pp 16–21.

37. W D'Aguanno. Drug Toxicity Evaluation–Preclinical Aspects. Food and Drug Administration, DHEW Publication No. (FDA) 74–30006, 1973, pp 35–40.

38. S Alder, G Zbinden. National guidelines: USA. In: National and International Drug Safety Guidelines. Zollikon, Switzerland: MTC Verlag Zollikon, 1988, pp 164–181.

10
Routes of Exposure: Other

Carol S. Auletta
Huntingdon Life Sciences, Inc., East Millstone, New Jersey

I. INTRODUCTION

This chapter will review the unique aspects of safety-testing strategies for excipients intended for ophthalmic, vaginal, rectal, and other mucosal routes of exposure. One obvious concern with materials intended for direct contact with mucosal surfaces is the delicate nature of these surfaces and the need to avoid inducing local irritation. Another feature is the rapid absorption that usually occurs for materials applied to mucosal surfaces. The feature that is probably most critical to establishing testing strategies for these mucosal formulations is the limited volume of material that can be administered by the intended routes and the consequent difficulties in establishing reasonable multiples of human exposure for safety assessment.

Table 1 summarizes proposed tests (required and conditionally required) for materials intended for use by these routes. This chapter discusses aspects of these tests and strategies specific to the ocular, vaginal, and other mucosal routes.

II. OPHTHALMIC EXCIPIENTS

A. Background

1. General

Excipients commonly used in ophthalmic preparations consist of carriers (usually saline or ointment bases), preservatives, and a wide variety of materials used in

Table 1 Testing Guidelines for Excipients Intended for Use in Ophthalmic and Mucosal Formulations

General requirements for materials intended for ophthalmic exposure
  Background information
    Define pH and tonicity of topical ocular dose form
    Base toxicity set
    Effects of acute exposure by ophthalmic route (include cytotoxicity test; e.g., agar overlay)
    Effects of repeated exposures by ophthalmic route (two species [one rodent, one mammalian nonrodent]; examination of anterior and posterior segments of the eye)
  Additional data
    Comparison of pharmacokinetic parameters of the route chosen for reproductive studies and the ophthalmic exposure is essential for extrapolation of potential toxicity by the ophthalmic route.

Summary of recommended testing;[a], materials intended for ocular and mucosal administration to humans

Appendix Exposure	1 Less than 2 wk		2 2–6 wk		3 More than 6 wk	
Base set						
	Acute oral toxicity	R	90-day toxicity (most appropriate species)	R	Chronic toxicity (rodent, nonrodent)	C
	Acute dermal toxicity	R	Teratology (rat and/or rabbit)	R	One generation reproduction	R
	Acute inhalation toxicity	C	Additional assays	C	Photocarcinogenicity	—
	Eye irritation	R	Genotoxicity assays	R	Carcinogenicity	C(M)
	Skin irritation	R				
	Skin sensitization	R				
	Acute parenteral toxicity	—				
	Application site evaluation	R(M)				
	Pulmonary sensitization	—				
	Phototoxicity/photoallergy	—				
	Bacterial gene mutation	R				
	Chromosomal damage	R				
	ADME-intended route	R				
	28-day toxicity (2 species) intended route	R				

[a] R, required; C, conditionally required; (M), required only for mucosal exposure.
Source: Ref. 1.

small amounts to optimize the final formulation. These materials may include buffers; pH- and tonicity-adjusting agents; and solubilizing, suspending, emulsifying, or wetting agents. A wide variety of materials have been tested and approved for ophthalmic use, either in over-the-counter (OTC) formulations or for prescription use. Information on these previously evaluated excipient ingredients can be found in several sources, including the *Physician's Desk Reference on Ophthalmic Preparations* (2) and such reference books as *Toxicology of the Eye* (3). Most ocular drug formulations contain several excipients. Testing guidelines discussed in the following refer to the final formulation of several excipients minus the active ingredient.

2. pH and Tonicity

Because these materials are intended for instillation directly into the eye, selection of nonirritating excipient formulations is critical. Isotonic formulations (pH 7) are ideal. Preservatives of high or low pH are frequently present in very small amounts; effects on pH of the final formulation must, therefore, be evaluated. Development strategies include the addition of buffers to adjust the final pH or the use of less-irritating ingredients. Although it is useful to understand the physical and chemical characteristics of all ingredients and to know their irritation potential, it is not realistic or useful to perform toxicity tests on individual components that will be present in very small amounts in the final formulation. Rather, effects of the complete final excipient formulation should be evaluated. If a range of concentrations of individual components is anticipated, it would be prudent to perform preliminary irritation evaluations over the proposed range or at the upper limit of the range.

B. Safety Evaluation Studies

1. Absorption, Distribution, Metabolism, and Excretion

Materials administered intraocularly are absorbed readily and very rapidly. An easily observed illustration is the presence in the urine of ocularly applied diagnostic fluoroscein dye within minutes after administration. Because of these absorption characteristics, any ophthalmic formulation would be expected to have systemic effects. Even the very small amounts administered routinely in an ocular irritation study can produce acute systemic toxicity and lethality when highly toxic materials (e.g., organophosphate insecticides) are administered. Pharmacokinetic profiles of excipient ingredients should, therefore, be well characterized and any excipient formulation component that is pharmacologically active should be considered carefully. Concentrations and total exposure estimates for excipient ingredients should be well below those that would be expected to produce unwanted activity or adverse effects. For materials intended for long-term use, studies with both the complete final drug formulation and the excipient formulation should be performed to evaluate pharmacokinetic behavior and any interactions between the excipient with the active ingredients. Because excipients generally contain multiple components, selection of the material for which analyses will be performed must be made carefully. This should generally be the material that constitutes the major portion of the excipient.

Pharmacokinetic studies should evaluate the initial profile of the excipient and, if repeated exposure is anticipated, the behavior under repeat-dose conditions. A single-dose absorption, distribution, metabolism, and excretion (ADME) study should be performed initially. The design of an ocular ADME study should be the same as one used for more traditional routes of exposure, with collection of blood, urine, feces and tissue sampling and use of radiolabeled materials when possible. Because of the small volume that can be administered, repeated administration over a short time period may be necessary. Care must be taken when administering materials into the eye to avoid loss of material from blinking the eye, shaking the head, or other movements of the animal. Manual restraint until no visible residual material is evident is recommended. If results are acceptable and further development of the formulation occurs, pharmacokinetic sampling to determine blood levels is easily incorporated into toxicity evaluations. If large blood volumes are required, it may be necessary to include satellite animals in rodent studies. However, if assays can be performed on small volumes of plasma, it is often possible to obtain samples from the test animals. A sparse sampling technique (4) can be used such that samples are obtained from as few as two animals per time point. Therefore, a group of ten animals sampled once or twice can provide data at five to ten post-dose intervals.

A route of administration other than ocular is sometimes selected for some of the toxicity studies, either because ocular administration is not feasible or because the drug may be intended for more than one route of exposure. In such cases, comparative pharmacokinetic data must be obtained to confirm the suitability of the alternative route and to provide a means of extrapolating effects to those that would be expected with ocular administration. A comparative study would be performed by administering the same doses of the excipient (or test material) formulation using the two routes and evaluating blood levels and pharmacokinetic parameters at various postdose intervals.

2. Animal Models: Species Selection

The traditional animal model for ocular evaluations is the albino (New Zealand White) rabbit. This is the species recommended in the earliest publication proposing ocular irritation and toxicity testing, *Methods for the Study of Irritation and Toxicity of Substances Applied Topically to the Skin and Mucous Membranes* (5) and is specified in the original Food and Drug Administration (FDA) guidelines for irritation screening (6). Although other tests were specified in this publication, the ocular irritation test in the rabbit is now commonly referred to as "the Draize test." The sensitivity of the albino rabbit to ocularly administered materials, the availability and ease of handling and husbandry, the relatively large size of the eye, and the lack of pigmentation make this model useful in performing studies predicting potential ocular irritation. Because of its extreme sensitivity to ocular irritants, the rationale is that this model will overpredict any hazards and that any material that is nonirritating to the rabbit eye is very unlikely to be irritating to the human eye. Studies in pigmented (Dutch-belted) rabbits have sometimes been performed with materials in which a pigmented model was crucial for evaluation of the intended use of a test material (7). However, these animals are less readily available and less hardy than albino rabbits. In addition, there is very little historical experience with this model.

A more realistic prediction of potential effects in humans may be obtained using a nonhuman primate model, generally the rhesus or cynomolgus monkey. The differences between the rabbit eye and primate eye in structure and function, as well as the absence of pigmentation in the albino rabbit, can result in markedly different responses to materials. Comprehensive comparisons of responses in albino rabbits and primates (8) demonstrated that responses in rabbits are generally more severe than those in monkeys. Comparisons of responses of rabbits, monkeys and humans to a soap solution (9) demonstrated that responses in humans and monkeys were similar to each other, but were different from those of rabbits. (In this study, corneal epithelial responses in the two primates were similar and more severe than those in rabbits.) However, nonhuman primates are expensive to purchase and maintain and dangerous to handle; their use in irritation screening studies is not practical, and is not often justifiable. They are probably best used in repeat-dose ocular toxicity studies.

The use of *any* animal model to predict ocular irritation has been the subject of much discussion, controversy, and research for more than a decade. In an effort precipitated by concerns of animal rights activists, the scientific community has devoted extensive amounts of time and money to research designed to develop nonanimal models that can be used as reliable predictors of ocular irritation in humans. Indeed, European regulatory agencies have legislated a ban, effective in the year 2000, on marketing (but not necessarily on performing) eye irritation tests in animals to predict the ocular irritation potential of cosmetic ingredients (10). The consensus, however, is that, even after all of the efforts to date, a satisfactory alternative has not yet been developed (11). Most researchers think that a battery of tests evaluating various structural and functional components of the eye will ultimately be developed (10). Many also argue that the "gold standard" for alternative methodologies should be their ability to predict human response, rather than their ability to duplicate responses in the Draize test. The species differences in responses are known. Differences between laboratories in interpreting responses to the same materials in rabbits have also been documented in a classic study (12) and continue to be a cause for concern.

When long-term studies in a rodent and nonrodent species are required, the use of the albino rabbit and the nonhuman primate (rhesus or cynomolgus monkey) is recommended. Although the rabbit is actually a lagomorph (of the order Lagomorpha) and, therefore, a nonrodent, some FDA regulators feel that it can and should serve as a "rodent" model for purposes of ocular testing. The albino rat can be and has been used in such studies. However, difficulties in administering adequate volumes (later discussion) are exacerbated because of the small size of the rat. The standard nonrodent models, the dog and the nonhuman primate, are considered acceptable for ocular testing. However, limited comparisons of responses in dogs, monkeys, rabbits, and humans (13) suggest that, not surprisingly, the monkey is a better predictor of human response than the dog. Thus, an extremely sensitive model (the albino rabbit) and a model with ocular structure, function, and pigmentation close to those in the human (the rhesus or cynomolgus monkey) are recommended for long-term testing of ophthalmic formulations. A note of caution and advice: It is always wise (and highly recommended) to review any proposed testing program with the regulatory scientists who will be reviewing the final drug applications to confirm that the program meets their expectations and requirements.

Classically, reproduction studies are performed in the rat and developmental toxicity (teratology) studies are performed in the rat and rabbit. These would be the test species of choice for evaluations of ophthalmic formulations and excipients. Similarly, carcinogenicity studies, if required, should be performed in species for which long-term experience and historical data are available (rats and mice). If dose volume limitations preclude ocular administration to rats and (most likely) mice, alternative routes (intravenous or oral) could be used, based on comparative pharmacokinetic studies.

3. Study Designs

a. In Vitro Tests The agar overlay cytotoxicity test, the *U.S. Pharmacopeia* (USP) test (14) used for evaluation of contact lenses is detailed in a number of guidelines for medical device testing (15–17) and is recommended for testing of ocular excipients (1). In this test, a culture of L929 cells (mouse connective tissue fibroblasts) is prepared and treated with neutral red (0.01%) until uptake of the stain is evident. Negative and positive control groups and a sample of the test material are applied to the surface of the cells and the culture is incubated for 24 h. The cells are then examined microscopically and scored for cytotoxicity. The study design and scoring system used by our laboratory are outlined in Table 2. This test is relatively insensitive because any potentially toxic material must diffuse through the agar layer to make contact with the cells, and the molecular

Table 2 Agar Overlay Test: Study Design and Scoring System

---

1. Negative control: medical or food-grade silicone rubber
2. Positive control: PVC disks, 1.3 cm in diameter, containing 0.57% dibutyl tin dimaleate
3. Test material: Liquid materials applied to sterile cellulose disks, 1.3 cm in diameter
4. Materials applied (in triplicate) to L929 cells in agar containing neutral red, cultures incubated (24 h) and areas under control and test materials evaluated for cytotoxicity by microscopic examination and scored as follows:

Scoring system

    0: No detectable zone around or under sample (no reactivity)
    1: Some malformed or degenerated cells under the sample (slight reactivity)
    2: Zone limited to the area under the sample (moderate reactivity)
    3: Zone extends from 0.5 to 1 cm from sample (moderate reactivity)
    4: Zone greater than 1 cm in extension from sample, but not involving entire dish (severe reactivity)

5. If in any test, a cytotoxic effect is observed for the negative contol, or no cytotoxic effect is elicited by the positive control, then the results for that test are considered invalid.
6. The sample meets the requirements of the test if the cultures treated with the sample show no greater than mild reactivity (grade 2).

Source: Ref. 18.

weight of the constituents will affect the rate of diffusion. As discussed earlier, there is general agreement that no acceptable *in vitro* alternative tests have yet been established for predicting ocular irritation. A review of various cytotoxicity tests and results of several materials tested in these assays, including the agar overlay test, found a poor correlation when compared with known *in vivo* effects (19) and concluded that the measurement of cytotoxicity using these assays is of limited value in predicting ocular irritation potential. However, the agar overlay cytotoxicity test is routinely used at the current time.

    b. Irritation and Acute Toxicity   The standard *in vivo* test for ocular irritation and acute toxicity (outlined in Table 3) is the primary eye irritation study and consists of administration of a small volume of the test material into the conjunctival sac of one eye of a test animal, most commonly the albino rabbit, followed by an observation period of 2–14 days. Evaluations for evidence of acute systemic effects are made immediately after dose administration and periodically thereafter. Ocular examinations to evaluate irritation may be performed shortly after test material administration (1 h postdose) and are routinely performed daily (approximately 24, 48, and 72 h postdose). If no irritation is evident after 72 h, observations are generally discontinued and the study is considered

Table 3 Study Design: Primary Ocular Irritation Study

Animals: Young adult albino rabbits (NZW), 2–3 kg, of either sex
Number: 6 (FSHA) 3 (OECD)
Dose: 0.1 mL per rabbit (one eye treated; other eye serves as control)
Duration: Single exposure; 3-day postdose observation
Irritation evaluations: 1 (OECD), 24, 48, 72 h OECD: additional observations if
  indicated

NZW, New Zealand white; FHSA, Federal Hazardous Substances Labeling Act; OECD, Organization
for Economic Cooperation and Development, Testing Guideline 405.
Source: Refs. 20 and 21.

completed. The observation period may be extended to evaluate time and extent of recovery from any irritation or if any potential for delayed irritation is expected.

Any potential ophthalmic formulation (or excipient) that produced irritation, especially irritation persisting for 72 h, would not be an acceptable candidate for further development. Similarly, any material with a potential for delayed irritation would not be developed for ophthalmological use. The number of animals tested has historically been six animals of either sex (5,6,20). However, current Organization for Economical and Coopertion Development (OECD) guidelines (21) consider three animals to be acceptable. Statistical evaluation of results for three versus six animals, using the standard Draize test (22) and the low-volume test (23) support the validity of performing ocular irritation tests in three animals. Young adult New Zealand white rabbits, weighing approximately 2–3 kg, are recommended for this study. In some cases, a nonhuman primate model (cynomolgus monkey) may be used for evaluation of acute effects, especially if subsequent repeat-dose studies in this species are anticipated.

c. Subchronic and Chronic Toxicity   Materials intended for repeated or long-term ocular administration should be tested for toxicity over a duration consistent with their intended use, as outlined in Table 1. Studies should include standard measurements of toxicity, in addition to special evaluation of ocular effects (irritation and structural alterations of the eye), as discussed later (see Sec. II.B.6). The excipient formulation would generally be tested concurrently with various dose levels of the final ophthalmic drug formulation, although a study could be performed on a new excipient alone by comparing effects with untreated or sham-treated animals or animals treated with a "control" material, such as physiological saline. A comparison with a known excipient formulation, used as a reference control, could also be performed to evaluate any differences between an old and a new formulation. A 28-day study is recommended for a material intended for repeated administration for less than 2 weeks and a 90-day study as well as developmental toxicity studies are recommended for a material intended for repeated administration for 2–6 weeks. Materials intended for longer-term administration should be evaluated for chronic and reproductive toxicity. Most regulatory agencies currently consider 6 months to be an adequate duration for a chronic study, although the U.S. FDA has requested a longer duration for studies in dogs, as discussed under Sec. II.D. Carcinogenicity evaluations may also be required, depending on the therapeutic use, duration, and anticipated human exposure. Pharmacokinetic evaluations are frequently conducted concurrently with such subchronic and chronic toxicity studies. Samples for analysis are usually collected after the first dose and near the end of the treatment period.

Reproduction and developmental toxicity studies should be performed by the ocular route of exposure when possible, although an alternative route, based on comparative pharmacokinetic studies, may be necessary, especially for the rat. Because these studies are designed specifically to evaluate reproductive function, ocular evaluations and pharmacokinetic evaluations are not generally incorporated. These data are obtained from the repeat-dose studies.

4. Dose Selection

A key issue in safety evaluation of materials intended for ocular administration is the limitation on dose volumes that can be administered by this route. The standard volume recommended for administration to rabbits for an ocular irritation study is 0.1 mL of the test material. The intent of this requirement was not to assure quantitative administration of this specific volume, but rather, to assure that all surfaces of the eye were exposed to the test material. Anyone who has performed such a study knows that this volume represents an ''overdose'' and that some material will invariably splash or be blinked from the eye immediately after administration. The ''low-volume'' test has been proposed (24), in which a volume of 100 µL (0.01 mL) is administered, as a more realistic approximation of the human exposure situation for the purposes of acute irritation testing and hazard evaluation.

Because of the small volume that can be administered, precise measurement of a specific volume for each animal, based on body weight, is not practical or realistic for ocular administration. It is more common to administer the same volume (1 or 2 drops, or a measured volume per eye) to each animal and calculate dose (exposure) based on an average body weight. The patient administering an ophthalmic formulation will generally be instructed to place ''a drop'' (or 2) in the eye; this appears to be a reasonable way in which to administer test materials for toxicity evaluations. A drop will vary according to the viscosity and density of the excipient (and the test material) and can be determined before study initiation to establish doses. Ideally, a medicine dropper calibrated specifically for the test material or excipient should be used. A drop of water administered by a pharmacopeial medicine dropper weighs between 45 and 55 mg (25).

Multiples of human exposure for systemic toxicity are achieved by virtue of the size of the test animal. A 2- to 3-kg rabbit or cynomolgus monkey that receives the same dose as a 60-kg human receives a 20–30 times higher dose based on body weight alone. (This is another advantage of the cynomolgus monkey over the 10-kg dog as a nonrodent species for safety evaluation of ophthalmic formulations). Because of the volume limitations, the only options to increase exposure are increasing either the concentration of the material or the frequency of application. However, increasing the concentration of excipient components to enhance exposure multiples may change the irritation potential and absorption characteristics and is often not an acceptable procedure. (If multiple concentrations are proposed for human use, the highest-proposed concentration should be tested.) Increasing the exposure multiple by administering multiple applications is a more viable option. Up to eight applications (one per hour) can be administered reasonably under laboratory conditions, especially in rabbits. However, the increased technical time required will result in a more costly study. A dose administration scheme of applications once, twice, and four times daily of the proposed human dose, assuming an average animal weight of 2.4 kg, would result in human exposure multiples of 25, 50, and 100, which should be acceptable to regulatory agencies. Testing of the excipient alone should be performed at the same volume and frequency as used for the highest dose of the drug formulation. The decision on whether to administer the material to one or both eyes obviously affects the exposure multiple and must be considered when developing the study design. It is common to dose one eye and leave the opposite eye untreated as a control for irritation and local effects. However, untreated control animals can serve this purpose and the primary goal of long-term studies is to evaluate systemic toxicity; the advantages of administration to both eyes to achieve the desired exposure multiples often make this dosing procedure preferable.

5. Administration Procedures

Administration should mimic the intended human use. Generally, this means holding the eyelids open and ''dropping'' the material onto the cornea. The eyelids may then be briefly held shut, if necessary, to assure exposure. Guidelines for single administration to assess irritation specify that the conjunctival sac of the lower eyelid of the rabbit be held open, the material administered into the sac, and the eyelids held shut for 1 s (6). However, this approach is unrealistic for human exposure and unnecessarily time-consuming for repeated administration. The primate eye does not have the loose folds of tissue that constitute a conjunctival sac in rabbits. Therefore, it is not possible to administer material into ''the conjunctival sac'' of a cynomolgus or rhesus monkey (or a human). An additional concern for very viscous formulations (e.g., ophthalmic ointments) is the tendency for materials to become trapped in the conjunctival sac and produce irritation that would not occur under human use conditions. It should be noted that guidelines for ocular irritation evaluations generally include a provision for rinsing the eyes of some animals immediately after administration to evaluate the effectiveness of such treatment in response to an accidental exposure. However, because exposure is intentional for ophthalmic pharmaceutical preparations, this procedure is not applicable here.

Any immediate response suggestive of pain is cause for concern. Should this happen, dosing should be discontinued and the cause of the response investigated. Current OECD guidelines suggest that local anesthetics may be administered before administration of materials that cause such responses (21). However, there is no uniform agreement on the wisdom or efficacy of this approach. In any case, an intended ophthalmic excipient that produces pain should be abandoned immediately. In some cases, a ''stinging'' sensation produced by a nonirritating material may result in a transient reaction in a laboratory animal. A prudent approach to any reaction would be to discontinue dosing and observe the reactor for 24 h for signs of irritation or other adverse effects. If no irritation is evident, additional animals may be dosed and observed. If consistent adverse responses are seen with additional animals, the components of the formulation should be reevaluated.

6. Evaluation Criteria

The Draize system for evaluating ocular irritation, introduced in 1944, remains the current industry standard for evaluation of acute irritation (5). Table 4 presents this system with some modifications and enhancements used by our laboratory. An evaluation system that is used less extensively, but that provides for a more thorough and objective evaluation, including use of a biomicroscopic slit-lamp evaluation, is the McDonald method (26), briefly summarized in Table 5. Although this procedure is more technically demanding than the Draize system and requires the use of a slit-lamp, it is recommended for acute irritation screening of materials intended for ophthalmic use. Use of either system must be performed by trained observers and interobserver variability must be minimized. A laboratory that performs ocular scoring should establish stringent requirements, including extensive training, practice, and testing, for the technical staff who are authorized to perform these evaluations. The *Illustrated Guide for Grading Eye Irritation by Hazardous Substances* (27) provides guidance, including color photographs, which are useful in training and standardizing Draize scoring. McDonald provides a complete description of his scoring system, as well as color photographs, in his chapter on eye irritation in *Dermatotoxicity and Pharmacology* (26). He and his colleagues also offer training courses in the technique.

The Draize and McDonald systems both employ fluorescein dye to evaluate any alterations in the corneal epithelium. Because this examination technique

Table 4 Grades for Ocular Lesions

Lesions	Grade
I. Cornea	
A. *Opacity—degree of density:*(area most dense taken for reading)	
No opacity	0
Slight dulling of normal luster	$+$[b]
Scattered or diffuse areas of opacity (other than slight dulling of normal luster), details of iris clearly visible	1[a]
Easily discernible translucent areas; details of iris slightly obscured	2[a]
Nacreous area, no details of iris visible, size of pupil barely discernible	3[a]
Opaque cornea, iris not discernible through opacity	4[a]
B. *Total area of cornea involved*: (Total area exhibiting any opacity, regardless of degree)[b]	
One quarter (or less) but not zero	1
Greater than one quarter, but less than half	2
Greater than half, but less than three quarters	3
Greater than three quarters, up to whole area	4
C. *Stippling*: (appearance of pinpoint roughening)[b]	
No stippling	0
One quarter (or less) but not zero	1
Greater than one quarter, but less than half	2
Greater than half, but less than three quarters	3
Greater than three quarters, up to whole area	4
D. *Ulceration*: (absence of a gross patch of corneal epithelium)[a]	
No ulceration	0
One quarter (or less) but not zero	1[b]
Greater than one quarter, but less than half	2[b]
Greater than half, but less than three quarters	3[b]
Greater than three quarters, up to whole area	4[b]
II. Iris	
A. *Values*:	
Normal	0
Slight deepening of the rugae or slight hyperemia of the circumcorneal blood vessels	$+$[b]
Markedly deepened folds (above normal), congestion, swelling, moderate circumcorneal hyperemia or injection (any or all of these or combination of thereof), iris still reacting to light (sluggish reaction is positive)	1[a]
No reaction to light, hemorrhage, gross destruction (any or all of these)	2[a]
III. Conjunctivae	
A. *Redness*: (refers to palpebral and bulbar conjunctivae)	
Vessels normal	0
Some vessels definitely hyperemic (injected above normal)	1
Diffuse, crimson red, individual vessels not easily discernible	2[a]
Diffuse beefy red	3[a]

Table 4 Continued

Lesions	Grade
B. *Chemosis*: (lids and/or nictitating membranes)	
No swelling	0
Any swelling above normal (includes nictitating membrane)	1
Obvious swelling with partial eversion of lids	2[a]
Swelling with lids about half closed	3[a]
Swelling with lids more than half closed	4[a]
C. *Discharge*:[b]	
No discharge	0
Any amount different from normal (does not include small amounts observed in inner canthus or normal animals)	1
Discharge with moistening of the lids and hairs just adjacent to lids	2
Discharge with moistening of the lids and hairs and considerable area around eye	3
D. *White tissue or ulcertion* of palpebral and bulbar conjunctivae or nictitating membrane[b]	
Not present	0
White tissue present	N
Ulceration present	U

[a] Score considered positive.
[b] Not included in Draize as OECD grading systems; values assigned if findings present.
Source: Ref. 5.

requires administration of a material onto the corneal surface, it should not be performed just before or after dose administration. A sterile fluorescein strip is held close to the eye until it becomes moistened with lacrimal fluid and releases a film of dye onto the corneal surface. The eye is then examined with a long-wave ultraviolet lamp (UV; "black light"). Any obvious stained areas or breaks in the film are indicative of missing areas of corneal epithelium. A slight pattern of stippling in the rabbit eye is occasionally seen. This is a reflection of the normal reepithelialization process that is continuously ongoing in the rabbit cornea and does not represent ocular irritation or an adverse effect.

Subchronic and chronic toxicity studies should incorporate routine observations for ocular irritation as well as periodic ophthalmological examinations for possible effects of the excipient or experimental formulation. A summary of recommended evaluations is presented in Table 6. Animals should be observed for grossly evident ocular irritation at the time of each dose administration. If administration occurs only once daily, additional postdose observations are suggested on the first day, perhaps 1 and 4 h after dose administration, to establish any unusual patterns or responses and to confirm that doses and concentrations have

Table 5 Summary: Ocular Examination and Slit-Lamp Scoring Procedure

Conjunctiva
  Conjunctival congestion
    0 = Normal. May appear blanched to reddish pink without perilimbal injection.
    +1 = A flushed, reddish color predominately confined to the palpebral conjunctiva with some perilimbal injection.
    +2 = Palpebral conjunctiva bright red with accompanying perilimbal injection covering at least 75% of the circumference of the perilimbal region.
    +3 = Dark, beefy red with congestion of both the bulbar and the palpebral conjunctiva along with pronounced perilimbal injection and the presence of petechia on the conjunctiva.
  Conjuntival swelling
    0 = Normal or no swelling of the conjunctival tissue.
    +1 = Swelling above normal without eversion of the lids.
    +2 = Swelling with misalignment of the normal approximation of the lower and upper eyelids; primarily confined to the upper eyelid.
    +3 = Swelling definite with partial eversion of the upper and lower eyelids essentially equivalent.
    +4 = Eversion of the upper eyelid is pronounced with less pronounced eversion of the lower eyelid.
  Conjunctival discharge
    0 = Normal. No discharge.
    +1 = Discharge above normal and present on the inner portion of the eye, but not on the lids or hairs of the eyelids.
    +2 = Discharge is abundant, easily observed, and has collected on the lids and around the hairs of the eyelids.
    +3 = Discharge has been flowing over the eyelids so as to wet the hairs substantially on the skin around the eye.
Aqueous flare
    0 = Absence of visible light beam light in the anterior chamber (no Tyndall effect).
    +1 = The Tyndall effect is barely discernible.
    +2 = The Tyndall beam in the anterior chamber is easily discernible and is equal in intensity to the slit beam as it passes through the lens.
    + 3 = The Tyndall beam in the anterior chamber is easily discernible; its intensity is greater than the intensity of the slit beam as it passes through the lens.
Iris involvement
    0 = Normal iris without any hyperemia of the iris vessels.
    +1 = Minimal injection of secondary but not tertiary vessels.
    +2 = Minimal injection of teriary vessels and minimal to moderate injection of the secondary vessels.
    +3 = Moderate injection of the secondary and tertiary vessels with slight swelling of the iris stroma
    +4 = Marked injection of the secondary and tertiary vessels with marked swelling of the iris stroma. The iris appears rugose; may be accompanied by hemorrhage (hyphema) in the anterior chamber.

Table 5 Continued

Cornea

    0 = Normal cornea.

    +1 = Some loss of transparency.

    +2 = Moderate loss of transparency.

    +3 = Involvement of the entire thickness of the stroma. With optical section, the endothelial surface is still visible.

    +4 = Involvement of the entire thickness of the stroma. Cannot clearly visualize the endothelium.

The surface area of the cornea relative to the area of cloudiness is divided into five grades from 0 to +4.

    0 = Normal cornea with no area of cloudiness.

    +1 = 1–25% area of stromal cloudiness.

    +2 = 26–50% area of stromal cloudiness.

    +3 = 51–75% area of stromal cloudiness.

    +4 = 76–100% area of stromal cloudiness.

Pannus

    0 = No pannus.

    +1 = Vascularization is present, but vessels have not invaded the entire corneal circumference.

    +2 = Vessels have invaded 2 mm or more around the entire corneal circumference.

Fluorescein staining

    0 = Absence of fluorescein staining.

    +1 = Slight fluorescein staining confined to a small focus.

    +2 = Moderate fluorescein staining confined to a small focus.

    +3 = Marked fluorescein staining.

    +4 = Extreme fluorescein staining.

Source: Ref. 26.

been selected appropriately. Any unusual irritation or response seen during the routine daily observations should be noted and recorded. However, recording irritation scores at the time of every dose, especially when little or no irritation is expected, appears unnecessarily burdensome. Therefore, it is recommended that scores be recorded periodically during the study (see Table 6). However, any unusual irritation observed during dose administration should be documented and followed up with appropriate observations and examinations on a case-by-case basis.

    Ophthalmological examinations should be performed before initiation of the study and any animals with ocular abnormalities that would interfere with interpretation of study results should be eliminated. Subsequent examinations should be performed periodically during the study (see Table 6) and at termina-

Table 6 Recommended Ocular Evaluations for Subchronic and Chronic Toxicity Studies of Ocularly Administered Material

Study duration	Irritation observation[a]	Irritation scoring[b]	Ophthalmological examination[c]
Subchronic			
28-day toxicity	Daily	Pretest, day 1, weekly	Pretest, termination, recovery
90-day toxicity	Daily	Pretest, day 1, weekly	Pretest, termination, recovery
Teratology	Daily	None	None
Chronic			
Chronic toxicity	Daily	Pretest, day 1, monthly	Pretest, every 3 months, termination, recovery
One-generation Reproduction	Daily	None	None

[a] General observations, notations of unusual/severe effects.
[b] McDonald or Draize systems.
[c] Direct and indirect ophthalmoscopy, intraocular pressure measurements if indicated.

tion of the study. If a postdose recovery period is included in the study design, examinations should be performed at termination of both the dosing and recovery period to evaluate reversibility of any effects as well as any potential for delayed effects. Examinations should include direct and indirect ophthalmological examinations and should be performed by a qualified individual, preferably a veterinary ophthalmologist. Because some variability is inherent in biological systems and because spontaneous ocular disease sometimes occurs, it is important for the examiner to have a thorough knowledge of laboratory animal ophthalmology and extensive experience in observing such animals. Measurements of intraocular pressure (IOP) can be made with a tonometer and should be incorporated into studies with materials with a potential to produce changes in intraocular pressure. Additional, more sophisticated, technologies that are available for ocular evaluations include specular microscopy, confocal microscopy, pachymetry (for corneal thickness measurements), and electroretinography. These may be useful in specific cases. However, the expense of the equipment and the need for specialized training, as well as the absence of background data in laboratory animals, do not make these practical or advisable for routine use.

C. Interpretation of Results

Both the Draize and McDonald systems have scoring systems that can assign a numerical irritation grade and allow the material to be characterized according to its irritation potential. An ocular irritation score ranging from 0 to 110 can be calculated using the Draize system (see Table 4). Scores ranging from 0 to +4 are assigned in several categories using the McDonald system, but a total score is not calculated. The use of such scoring systems for evaluation of irritant potential of ophthalmic excipients is limited. Systems that categorize the irritant potential of materials are designed to protect the public from accidental exposures and are not relevant to materials intended for ocular exposure. Any material that produces more than slight, transient irritation in the rabbit eye should probably not be considered for further development as an ophthalmic excipient. However, a material that has significant value in delivering a pharmaceutical, but causes some irritation in the rabbit eye, should be evaluated in another test species, preferably the nonhuman primate, and a judgment made about potential risk versus benefit. Thus, numerical ocular scoring systems are useful in documenting and evaluating effects and comparing various formulations, but any scores indicative of significant irritation would disqualify a potential ophthalmic excipient or formulation unsuitable for further development.

Abnormalities detected by ophthalmoscopic examinations that are considered by the ophthalmologist to represent an effect of the formulation administered would be cause to discontinue development of a proposed excipient. The experience of the ophthalmologist and testing laboratory are important in putting such abnormalities or changes in perspective and evaluating the potential for spontaneous occurrence of similar ocular lesions in the normal population of the species being tested.

D. International Issues and Harmonization

Efforts by the International Conference on Harmonization (ICH) of Technical Requirements for the Regulation of Pharmaceuticals for Human Use have established agreement by international regulatory agencies on the acceptability of many study designs. The OECD guidelines for ocular irritation testing (21) are considered acceptable and agreement has been reached on guidelines for developmental (teratology) evaluations (28). The duration of a ''chronic'' toxicity study in the nonrodent has been the subject of much discussion in the international arena, with the United States wanting a 12-month duration and other countries feeling that 6 months is acceptable; a recent compromise position states that a 9-month study is considered internationally acceptable (29).

The largest area of ongoing discussion and debate revolves around ocular irritation testing (the Draize test) and the acceptability of *in vitro* alternatives. As discussed previously, the current issues focus specifically on cosmetic testing, but advances and agreements reached in this area will surely affect future testing of ophthalmic formulations.

III. VAGINAL EXCIPIENTS

A. Background

Most animal studies performed using the vaginal route of administration are conducted to evaluate the local and systemic toxicity of materials intended for vaginal administration, primarily as contraceptives or gynecological therapeutic agents. Because of the difficulty in maintaining a drug product in the vagina for any length of time, excipients are used to enhance contact with the vaginal mucosa and, thereby, promote absorption of the active ingredient. Other excipients are of the type commonly used in other pharmaceutical preparations and include stabilizers, preservatives, and ingredients to enhance cosmetic appeal and ease of use.

B. Safety Evaluation Studies

1. Animal Models

The animal models most commonly used to evaluate safety of vaginally administered products are the rabbit and the ovariectomized rat. The rabbit is considered the more sensitive model, likely to overpredict irritation potential, and is used as an irritancy screening model. The response and the vaginal mucosa of the ovariectomized rat are closer to those of the human, and this is the model of choice for more definitive evaluation. In addition to the ocular irritation guidelines discussed previously, Draize's 1959 publication (6) suggested guidelines for evaluation of primary irritancy to vaginal mucous membranes and stated that the tissue tested should be the one to which the material would be applied for human use. In spite of Draize's reservations about the appropriateness of rabbits (because of the difficulty in dose administration and their relative sensitivity to irritants), they are currently the model used for irritation screening. Appropriate dosing techniques have been developed and the sensitivity of the rabbit is acknowledged when designing studies and interpreting data. Comparative studies using rabbits and monkeys (30) concluded that rabbits were indeed overly sensitive but interpreted this as a good reason to use them in screening studies. (This is the same reasoning for the use of rabbits in ocular irritation testing). For example, rabbits were clearly more sensitive than rats to administration of nonoxynol-9, the active ingredient in most spermicide preparations (31).

The vaginal epithelium of the rabbit consists of a single columnar cell layer (Figs. 1 and 2) that is especially responsive to topical irritants. The mature female is preferred over the prepubertal animal, in which the vaginal epithelium (Figs. 3 and 4) appears to be somewhat more resistant. A younger animal is sometimes selected based on a high body weight, which mistakenly suggests sexual maturity. The ovariectomized rat is recognized as exhibiting a response more similar to

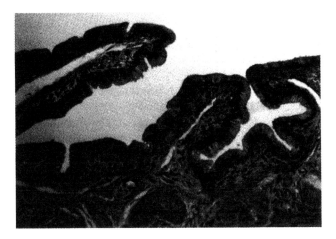

Figure 1 Normal vaginal epithelium: adult rabbit (50×). (From Ref. 34.)

that of the human and is the second species recommended for the 10-day repeat-dose vaginal irritation study when results using the more sensitive rabbit model are questionable (32). Ovariectomy in the rat induces a late diestrus-like condition characterized by a uniform, thin, noncornified epithelium (Fig. 5). Normal-cycling rats, with intact ovaries, will be in varying states of estrus at any given time and the structure of the vaginal epithelium will vary widely. Figure 6 illustrates the vaginal epithelium of a nonovariectomized rat in early diestrus.

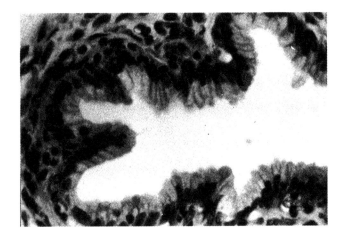

Figure 2 Normal vaginal epithelium: adult rabbit (125×). (From Ref. 34.)

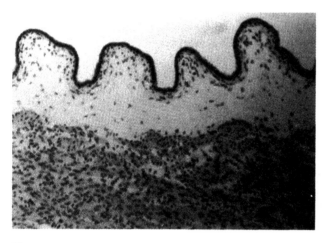

Figure 3   Normal vaginal epithelium: prepubertal rabbit (50×). (From Ref. 34.)

Although rabbits can be used for studies of up to 6–12 months in duration, rats are generally used for longer-term repeat-dose studies. Because the purpose of these studies is to evaluate systemic effects, rather than local toxicity, rats should not be ovariectomized to allow evaluation of any long-term effects on reproductive anatomy and physiology.

Larger species that may be and have been used include nonhuman primates

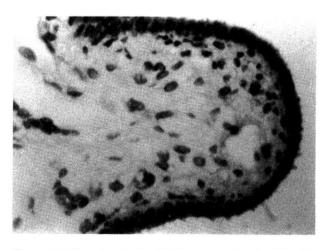

Figure 4   Normal vaginal epithelium: prepubertal rabbit (125×). (From Ref. 34.)

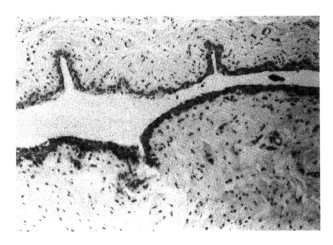

Figure 5   Vaginal epithelium: ovariectomized rat (50×). (From Ref. 34.)

and dogs. Selection of the appropriate species should be made on the basis of pharmacokinetic studies whenever possible. If human data are available, pharmacokinetic studies should be performed to compare metabolic profiles in the potential test species with the profile in humans. If human data cannot be obtained, anatomical and physiological aspects should be considered. Nonhuman primates

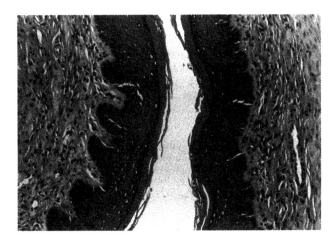

Figure 6   Vaginal epithelium: nonovariectomized rat (early diestrus)(50×). (From Ref. 34.)

(cynomolgus or rhesus monkeys), with anatomy and physiology similar to that of the human, are recommended. Although Draize (6) felt that the dog might be a suitable model, anatomical and physical differences between the dog and the human argue for the use of a primate species. (Dogs have estrus cycles, approximately every 6 months, whereas humans and nonhuman primates have monthly menstrual cycles). However, because of the higher cost of performing studies in non-human primates, dogs are sometimes used, especially when systemic toxicity is the primary concern. If alternative routes of administration are indicated by limited vaginal absorption, comparative pharmacokinetic studies should be performed to establish alternative routes, most likely intravenous or oral. If alternative routes of administration are used and human pharmacokinetic data cannot be obtained, either the dog or the monkey would be considered an acceptable species. Sheep have also been tested (33), but are not recommended.

Reproduction studies in the rat and developmental toxicity (teratology) studies in the rat and rabbit would be recommended for evaluations of vaginal formulations and excipients. If dose volume or absorption limitations preclude vaginal administration to these species, alternative routes (intravenous or oral) could be used. (Dose volumes of 1 mL for a rabbit and 0.2 mL for a rat are considered reasonable maximum doses.) Carcinogenicity studies, if required, should be performed in species for which long-term experience and historical data are available (rats and mice). The vaginal route would be preferred, but other routes may be necessary based on dose volume or absorption limitations.

2. Study Designs

a. Irritation and Acute Toxicity The standard primary vaginal irritation study, outlined in Table 7, consists of administration of 1.0 mL of the test material into the vaginas of six young adult female rabbits (8–12 weeks old) followed by observation and scoring for vaginal irritation (Table 8) at 30 min and 24, 48, and 72 h postdose. Details of administration procedures follow. Macroscopic examination of vaginal tissue for signs of irritation is generally performed at necropsy, but tissues are not preserved for microscopic examination. For materi-

Table 7   Study Design: Primary Vaginal Irritation Study

Animals	Young adult female albino rabbits (NZW), 2–3 kg
Number	6
Dose	1 mL/rabbit
Duration	Single exposure; 3-day postdose observation
Irritation evaluations	30 min; 24, 48, 72 h
Postmortem	Macroscopic examination of vaginal tissue

Source: Ref. 34.

Table 8 Scoring System for Evaluation of Response:
Vaginal Irritation

Response	Score
Erythema	
No reaction	0
Very slight (barely perceptible) erythema	±
Slight (well-defined) erythema	1
Moderate erythema	2
Severe erythema, with or without necrosis or eschar formation	3
Edema	
No reaction	0
Very slight (barely perceptible) edema	±
Slight (well-defined) edema	1
Moderate edema	2
Severe edema	3
Discharge	
No discharge	0
Very slight discharge	±
Slight discharge	1
Moderate discharge	2
Severe discharge	3
Other	
Necrosis	N
Eschar formation	E

Source: Ref. 34.

als intended for repeated administration, a 10-day repeat-dose vaginal irritation study, based on the recommendations of Eckstein et al. (30) should be performed. Study designs in rabbits and ovariectomized rats are outlined in Tables 9 and 10, respectively. These studies incorporate negative (sham-treated) and positive controls as well as a vehicle (excipient) control group and a test–material-treated group. A study conducted to evaluate the excipient alone should include the other two control groups. The sham-treated group provides information on background changes or changes related to handling. The positive control group provides a level of "acceptable" irritation produced by a material approved for vaginal use (the spermicide nonoxynol-9), at the concentration used in many contraceptive preparations (2%). This material, along with octoxynol and menfegal, was cited as acceptable for human use in a 1980 FDA OTC monograph (35) on vaginal contraceptives. Because spermicides act by disrupting cell membranes, some irritation is expected with these products. This is considered acceptable in the ab-

Table 9 Study Design: Repeat-Dose Vaginal Irritation
Study: Rabbits

Animals	Young adult female albino rabbits (NZW), 2.0–3.0 kg
Number	20 (5 per group)
Groups	Negative control (sham-treated); vehicle control; positive control (2% nonoxynol-9); test material 1.0 mL/rabbit per day
Dose	1.0 mL/rabbit per day
Duration	10 days
Irritation evaluations	Daily
Postmortem	Macroscopic vaginal examination; microscopic vaginal examination/ scoring

Source: Ref. 30.

sence of any human (clinical or epidemiological) evidence of hazard or developmental effects. (The authors did comment on the lack of animal data for these materials).

Treatment and observations are the same as for the primary irritation study, with observations performed before each daily dose. Animals are also checked shortly after dose administration for any unusual signs, including loss of administered material, signs of toxicity and evidence of bleeding or other vaginal discharge. After 10 days of treatment, animals are necropsied and macroscopic and microscopic examinations of vaginal tissue are performed.

b. Subchronic and Chronic Toxicity Excipients intended for use in pharmaceutical formulations to be used for repeated or long-term vaginal administration should be tested for toxicity over a duration consistent with their intended use, as outlined in Table 1. Studies should include standard measurements of toxicity in addition to evaluation of vaginal irritation. The excipient formulation would generally be tested concurrently with various dose levels of the test material (final formulation), although a study could be performed on a new excipient alone by comparing effects with untreated or sham-treated animals or animals treated with a negative or positive control material with known effects and acceptability criteria. The OTC monograph on vaginal contraceptives (35) suggests 30- to 180-day studies, with shorter studies for reformulations and longer studies for new formulations. This appears to be reasonable guidance for testing of excipient formulations and is consistent with the recommendations in Table 1. The rabbit

Table 10 Study Design: Repeat-Dose Vaginal Irritation Study: Rats

Animals	Young adult ovariectomized female albino rats (Sprague–Dawley), 200–300 g
Number	40 (10 per group)
Groups	Negative control (sham-treated); vehicle control; positive control (2% nonoxynol-9); test material
Dose	0.2 mL/rat per day
Duration	10 days
Irritation evaluations	Daily
Postmortem	Macroscopic vaginal examination; microscopic vaginal examination/scoring

Source: Ref. 30.

is recommended for the short-term study and the dog or nonhuman primate for studies of longer duration. However, if human metabolism data is available, pharmacokinetic studies should be performed to compare metabolic profiles in the potential test species with the profile in humans and select the most appropriate species. A 28-day study is recommended for a material intended for repeated administration for less than 2 weeks and a 90-day study as well as developmental toxicity studies are recommended for a material intended for repeated administration for 2–6 weeks. Materials intended for longer-term administration should be evaluated for chronic toxicity (6 months in rats and 9 months in nonrodents). Carcinogenicity evaluations may also be required, depending on the therapeutic use, genotoxicity potential, duration, and anticipated human exposure. These studies may be performed by an alternative route, especially in mice, depending on absorption and comparative pharmacokinetics. Pharmacokinetic evaluations are frequently conducted concurrently with subchronic and chronic toxicity studies. Samples for analysis are usually collected after the first dose and near the end of the treatment period.

Reproduction and developmental toxicity studies should be performed by the vaginal route of exposure, whenever possible, although an alternative route, based on comparative pharmacokinetic studies, may be necessary, especially for the rat. The use of the vaginal route of administration for studies designed to evaluate the reproductive system may be cause for concern. However, experience in our laboratory with vaginal dose administration for such studies has not demonstrated any unexpected adverse reproductive or developmental effects. The dose administration schedule used is the same as for an oral gavage study. During the mating phase, females should be dosed several hours before cohabitation. An exception to this procedure should be made for materials intended for contraceptive use, which should not be administered during the mating period. Because these studies are designed specifically to evaluate reproductive function, specific vaginal examinations or evaluations for irritation and pharmacokinetic evaluations are not generally incorporated. These data are obtained from the repeat-dose studies.

3. Dose Selection

Because of the limited size of the vaginal mucosal surface, dose limitations exist for the vaginal route of administration. Volumes of 1.0 mL for a rabbit and 0.2 mL for a rat are considered reasonable maximum doses. The maximum dose volume is 1.0 mL for a nonhuman primate; larger volumes can be used with dogs. As with ocular dosing, the intent of irritation testing is to expose the vaginal surface to the administered material, rather than to exactly quantify the dose. For toxicity evaluations, multiples of human use are achieved by virtue of the small size of the test animals versus the size of the human. Higher multiples can also be achieved by administering two or more doses over the course of a day. Concentrations of active ingredients should be those intended for therapeutic use. Increasing the concentration to increase the use-multiple can increase irritation or alter absorption. A significant issue with this route of administration is retaining the dose once it is administered. Animals are generally held briefly after administration, but leakage of materials, especially those with aqueous vehicles, can occur. Therefore, the dose selection process must account for this possibility. Accurate single-dose exposures can be achieved in restrained or anesthetized animals or by clipping or suturing the vaginal opening. However, this is not reasonable for repeat administration. If vaginal dose administration is to be used for repeat-dose toxicity studies, it is important to determine received versus administered dose. This can be accomplished by preliminary pilot studies to evaluate the potential for loss of material and to compare absorption of the same dose administered to restrained and free-moving animals. Based on this information, a decision can be made on the feasibility of vaginal dose administration and the need to consider alternative routes (based on comparative pharmacokinetic studies).

4. Administration Procedures

Vaginal instillation of nonviscous materials is accomplished by the use of a French catheter (size 18) lubricated with a vaginal lubricant such as (nonspermicidal) K-Y jelly. The catheter is attached to a syringe which is used to measure and deliver the material. Placement of the catheter is critical, especially in the rabbit, to assure that the material is administered intravaginally and that the urinary tract is avoided. Animals are cradled on their backs and held by a technician or animal handler while a second technician administers the dose. A similar procedure is used for larger species, with appropriate modifications for animal size and technician safety (nonhuman primates). For rabbits, the catheter is inserted deep into the vagina (approximately 7.5 cm); for rats, the catheter is inserted approximately 2.5 cm deep. After administration, the catheter is gently withdrawn and the animal is returned to its cage. Occasionally, suppository-type materials, which cannot be readily drawn into a syringe, are formulated for vaginal use. These materials can be packed into the syringe barrel, often with gentle warming in a water bath to facilitate handling. Administration is accomplished by use of a 1 mL tuberculin syringe, which is packed with the material to be tested, lubricated, and inserted into the vagina, using the procedure just described. The material is deposited by slowly withdrawing the syringe while depressing the plunger, leaving a cylindrical section of material that will melt when warmed to body temperature.

5. Evaluation Criteria and Interpretation of Results

Observations of vaginal irritation are limited by the nature of the target organ. Scoring of signs of irritation is performed by careful examination of the vaginal orifice and any surrounding tissue. Use of a speculum or other manipulation should be avoided because of the potential to induce mechanical irritation. The scoring system outlined in Table 8 is used during in-life observations. Any direct injury to the vaginal mucosa is evaluated by macroscopic examination at necropsy. Criteria similar to those used for in-life examinations are used for macroscopic postmortem examinations; any evidence of erythema, edema, discharge, or tissue damage is recorded. In general, severe irritation is considered cause for concern; slight to moderate irritation, especially in the rabbit, occurs with materials accepted for human administration (nonoxynol-9) and is not considered significant for human risk evaluation. The use of fluoroscein dye to assist in evaluating mucosal epithelial damage at necropsy has proved to be of limited usefulness (34) and is not recommended.

Results of microscopic examination of vaginal tissue obtained at necropsy after the completion of the 10-day repeat-dose vaginal irritation study (see Tables 9 and 10) is considered the definitive predictor of irritation potential. Three levels of vagina (upper, middle, and lower) are examined and graded, using the system presented in Table 11, for epithelial integrity and evidence of injection, edema, and leukocyte infiltration. Each section is scored separately, and an average score, which can range from 0 to 16, is calculated. Scores of less than 12 in the rabbit are considered "acceptable." The fact that a high degree of irritation (scores of up to 11 on a 16-point scale) is considered acceptable in this test is an acknowledgment that the rabbit is a sensitive model that will overpredict the human response. Materials such as 2% nonoxynol-9, which produce high scores in this

Table 11 Scoring Procedure: Microscopic Examination of Vaginal
Sections: 10-Day Vaginal Irritation Study

Reaction	Score[a]
Epithelium	
Inact, normal	0
Cell degeneration or flattening of the epithelium	1
Metaplasia	2
Focal erosion	3
Erosion or ulceration, generalized	4
Leukocytes	
Minimal: <25/HPF	1
Mild: 25–50/HPF	2
Moderate: 50–100/HPF	3
Marked: >100/HPF	4
Injection	
Absent	0
Minimal	1
Mild	2
Moderate	3
Marked with disruption of vessels	4
Edema	
Absent	0
Minimal	1
Mild	2
Moderate	3
Marked	4

[a] A total score is calculated for each section. Irritation ratings for the various total
scores are as follows: minimal irritation 1–4; mild irritation 5–8; moderate irritation 9–11; marked irritation 12–16.
Based on these ratings, materials are given acceptability ratings as follows: acceptable 0–8; marginal 9–10; unacceptable 11–16.
HPF, high-power field.
Source: Ref. 30.

test (and are used as positive controls), are considered acceptable for human use
(35).

IV. RECTAL EXCIPIENTS

A. Background

The vast majority of pharmaceuticals that are administered rectally are designed
for systemic delivery when oral administration is poorly tolerated or ineffective.

Examples are anticonvulsants (diazepam), antinausea preparations (prochlorperazine), pain relievers for migraine headaches (codeine preparations), and anti-inflammatory, antiarthritic medications that are frequently associated with gastrointestinal complications (indomethacin). A great deal of research has been dedicated to the development of alternative routes of administration for materials that are poorly tolerated orally because of nausea, vomiting, or severe epigastric reflux (36), for proteins and peptides that are destroyed by digestive enzymes (insulin); and for materials that are inactivated by first-pass metabolism. Although much of this research has focused on nasal delivery systems, rectal administration appears to be the route of choice for some types of materials (37,38). A comparison of drug delivery from gastrointestinal and rectal mucosa is presented in Table 12.

The advantages of rectal administration of systemic drugs are the avoidance of first-pass hepatic metabolism, the elimination of contact with digestive enzymes, and the ability to treat patients with nausea and convulsions. Disadvantages include poor patient acceptability and volume and absorption limitations.

Little work has been performed to evaluate the effects of rectal excipients in laboratory animals. Some studies with various suppository bases demonstrated that both polyethylene glycol- and triglyceride-based materials produced irritation of the rectal mucosa, whereas monoglyceride- or fatty acid-based materials were less irritating (39). The need to enhance absorption for rectally administered materials mandates the use of several unique excipient components in rectal formulations, such as protease inhibitors and penetration enhancers (40). Because penetration enhancers exert their effects by compromising the integrity of mucous membranes, acceptable limits of irritation must be established and extent and rate of repair of mucosal damage must be assessed. Although this has been done successfully in the rabbit model for vaginal contraceptive materials, little comparable work has been performed for rectal penetration enhancers. Systemic effects of the enhancer as well as the potential for nonselective absorption of other components must also be considered.

Table 12 Comparison of Drug Delivery from Gastrointestinal and Rectal Mucosa

	Gastrointestinal	Rectal
Absorption area	Very large	Small
First-pass metabolism	Yes	No—lower Yes—upper
Affected by gastric contents	Yes	No
Dose lost upon defecation	No	Yes

B. Safety Evaluation Studies

1. Animal Models

Studies have been performed in several standard laboratory species (rats, rabbits, dogs, and minipigs). Little information is available for mice, presumably because of size limitations, or for nonhuman primates. Some scientists believe that the rabbit is an appropriate model because of the size of the rectum, which is similar to that of a 6-month-old human, and its histological characteristics, which are similar to those of the human rectum (39). The dog or the minipig would be appropriate larger nonrodent species because of the relatively comparable size of the rectum in these species and that of the human (41).

2. Study Designs

Specific study designs for rectally administered materials have not been developed. Studies conducted using this route would most likely be limited to evaluations of local irritation, with observations for acute toxic effects, in which a study design similar to those used for primary irritancy of other mucous membranes is suggested (Table 13). Because many rectal formulations are developed as new delivery systems for known drugs, it is probable that repeat-dose and long-term studies of drug effect would not be required if blood levels comparable with those seen by more conventional routes of administration can be established. If repeat-dose systemic toxicity studies are required, they would most likely be conducted by other routes of administration (probably intravenous) based on comparative pharmacokinetics. Studies (as recommended in Table 1), similar to those used with other routes of administration, are suggested.

3. Dose Selection

Dose selection for rectal administration is severely limited by the poor accessibility of the rectal mucosa. For acute irritation evaluations, a volume that provides

Table 13 Study Design: Primary Mucous Membrane Irritation Study

Animals	Young adult albino rabbits (NZW), 2–3 kg
Sex/number	6 males (penile irritation)
	3 males, 3 females (other mucous membranes)
Dose	1 mL/rabbit (or as appropriate)
Duration	Single exposure; 3-day postdose observation
Irritation evaluations	30 min; 24, 48, 72 h
Postmortem	Macroscopic examination of mucosal surface

exposure of the rectal surface to the material should be administered. Dose volumes of 0.2–0.5 mL for rabbits, 0.1–0.2 mL for rats, and 0.5–1.0 mL for dogs and minipigs are suggested. The dose limitations inherent in this route of administration will usually necessitate use of an alternative route with similar absorption characteristics to achieve adequate multiples of human use.

4. Administration Procedures

Challenges in developing appropriate dosing procedures arise from volume limitations and retention difficulties. Materials intended for local use, such as creams and ointments, can be administered perianally. Suppositories can also be administered to larger species, such as dogs and minipigs and, possibly, rabbits. However, materials intended for systemic delivery must be administered at the appropriate site, the lower rectum (42). This area, measuring approximately 2 cm in the rat (43) is drained by the lower and middle hemorrhoidal veins, which enter the systemic circulation. Because the upper portion of the rectum is drained by vessels that enter the portal circulation, it is not an appropriate site for systemic delivery.

Much of the animal work reported in the literature consists of acute (up to 4-h) exposures using such methods as perianal purse-string sutures in anesthetized rats (44) or closure of the anal orifice with surgical adhesive to retain the administered dose (45). Suppositories, which are the most widely used rectal drug delivery system in humans, have been successfully administered to animals by some researchers. DeMuynck *et al*, (39) report success in administering suppositories to rabbits, followed by a 30-minute closure of the anus (using clothes pins) every 8 h for 14 days. (Because of the high interanimal variability, these authors recommend the use of 20 animals per group). Other potential delivery systems reported in the literature for research or therapeutic use include microenemas, hydrogels, rectal capsules, and osmotic pumps (45,46). Other researchers have successfully administered materials to rats using rectal cannulation and minipumps (47), or in rabbits using multiple actuations of a metered-dose spray pump (48). The latter technique appears to have the best potential for a repeat-dose study.

5. Evaluation Criteria

Histologically, the rectal mucosa has many characteristics in common with that of the vaginal mucosa. A thorough microscopic examination and a histological scoring system similar to that described in Table 11 should be considered. However, further work is needed to establish limits of acceptability.

V. OTHER MUCOSAL EXCIPIENTS (SUBLINGUAL, BUCCAL, PENILE)

Some materials are readily absorbed sublingually and this route has been an acceptable choice for human administration. Because of the difficulties in administering materials to laboratory animals by this route, testing is generally limited to irritation evaluations of the buccal mucosa and comparative pharmacokinetic studies to develop an alternative route for further testing (frequently subcutaneous or intravenous). Dose administration is accomplished by placing the desired amount of material, usually in tablet form, beneath the animal's tongue and holding the mouth closed until the material has dissolved. A general study design for mucosal irritation, based on those used for the more standard mucosal routes, is outlined in Table 13.

The penile route of exposure may be applicable for materials used to treat urological or erectile dysfunctions. Issues are similar to those discussed in the foregoing; testing is generally limited to irritation evaluation and pharmacokinetic studies to determine a more appropriate alternative route for more extensive safety evaluations.

VI. CONCLUSION

Evaluation of the safety of excipients intended for use in formulations to be administered to humans by the ocular, vaginal, and rectal routes, and by other nonstandard routes involving mucosal exposure, presents unique challenges. Standard study designs, evaluation procedures, and laboratory animal models have been developed to meet the specific needs of some of these evaluations. In many instances, the most appropriate approach is to select an alternative, more standard, route of exposure based on comparative pharmacokinetic data. Although local testing for irritation potential will continue to be needed, active searches for nonanimal alternatives to irritancy testing are likely to alter future testing strategies. The rapid growth of increasingly sophisticated bioanalytical technology has contributed to the development of comparative pharmacokinetics and will continue to provide reliable data on which to base decisions about acceptable alternate routes of administration.

ACKNOWLEDGMENT

Photomicrographs were provided by R. F. McConnell, D.V.M., Flemington, New Jersey and were previously published in the *Journal of the American College of Toxicology* (34).

REFERENCES

1. M Steinberg, JF Borzelleca, EK Enters, FK Kinoshita, A Loper, DB Mitchell, CB Tamulinus, ML Weiner. A new approach to the safety assessment of pharmaceutical excipients. Regul Toxicol Pharmicol 24:149–154, 1996.
2. Physician's Desk Reference for Ophthalmology. Montvale, NJ: Medical Economics, 1998.
3. WM Grant. Toxicology of the Eye. 2nd ed. Springfield, IL: Charles C Thomas, 1974.
4. JR Nedelman. Serial versus sparse sampling in toxicokinetic studies. Pharm Res 13: 1105–1108, 1993.
5. JH Draize, G Woodard, HO Calvary. Methods for the study of irritation and toxicity of substances applied topically to the skin and mucous membranes. J Pharmacol Exp Ther 82:377–390, 1944.
6. JH Draize. Appraisal of the Safety of Chemicals in Foods, Drugs and Cosmetics. Austin, TX: Association of Food and Drug Officials of the United States, 1959.
7. CS Auletta, LF Rubin, IW Daly, WR Richter, K Hosoi, H Suda, T Ikuse. 26-Week ocular toxicity studies in pigmented rabbits treated topically with bunazosin hydrochloride ophthalmic solution. Atarishii Ganka (J Eye) 12(3), 1995.
8. EV Buehler, EA Newmann. A comparison of eye irritation in monkeys and rabbits. Toxicol Appl Pharmcol 6:701–710, 1964.
9. JH Beckley, TJ Russell, LF Rubin. Use of the rhesus monkey for predicting human response to eye irritants. Toxicol Appl Pharmacol 15:1–9, 1969.
10. Toxicology Forum. The search for *in vitro* alternatives for prediction of ocular toxicity, Washington, DC, Feb 3, 1998.
11. T Herzinger, HC Korting, HI Maibach. Assessment of cutaneous and ocular irritancy: a decade of research on alternatives to animal experimentation. Fundam Appl Toxicol 24:29–41, 1995.
12. CS Weil, RA Scala. Study of intra- and interlaboratory variability in the results of rabbit eye and skin irritation tests. Toxicol Appl Pharmacol 19:276–360, 1971.
13. JH Beckley. Comparative eye testing: man vs. animal. Toxicol Appl Pharmacol 7: 93–101, 1965.
14. United States Pharmacopeia (USP 23). Biological Tests ⟨87⟩ Biological Reactivity Tests *In Vitro*. Rockville, MD: U.S. Pharmacopeial Convention, 1995, p. 1697.
15. International Standard ISO 10993-5. Biological Evaluation of Medical Devices—Part 5: Tests for Cytotoxicity: *In Vitro* Methods. 1992.
16. ASTM F 895-84. Standard test method for agar diffusion cell culture screening for cytotoxicity. In: Annual Book of ASTM Standards. vol 13.01. Philadelphia: ASTM, 1984.
17. British/European Standard BS EN 30993-5. Biological Evaluation of Medical Devices—Part 5. Tests for Cytotoxicity: *In Vitro* Methods. 1994.
18. P Uphill. Standard Protocol CTX/CBI/AGO–L929/USP (VI.I): Cytotoxicity test, agar overlay method *(USP 23). Cambridgeshire, UK: Huntingdon Life Sciences, 1998.*
19. JF Sina, GJ Ward, MA Lazek, PD Gautheron. Assessment of cytotoxicity assays as

predictors of ocular irritation of pharmaceuticals. Fundam Appl Toxicol 18:515–521, 1992.

20. United States Consumer Product Safety Commission, Federal Hazardous Substances Act Regulations Federal Register 38(187):27014–27019, 1973.

21. Organization for Economic Cooperation and Development. (OECD) Guideline for Testing of Chemicals No. 405 Acute Eye Irritation/Corrosion, 1987.

22. DJ De Sousa, AA Rouse, WJ Smolen. Statistical consequences of reducing the number of rabbits utilized in eye irritation testing: data on 67 petrochemicals. Toxicol Appl Pharmacol. 76:234–242, 1984.

23. LH Bruner, RD Parker, RD Bruce. Reducing the number of rabbits in the low-volume eye irritation test. Fundam Appl Toxicol 19:330–335, 1992.

24. JF Griffith, GA Nixon, RD Bruce, PJ Reer, EA Bannan. Dose–response studies with chemical irritants in the albino rabbit eye as a basis for selecting optimum testing conditions for predicting hazard to the human eye. Toxicol Appl Pharmacol 55:501–513, 1980.

25. The United States Pharmacopeia. 20th Rev. The National Formulary 15th ed. Rockville, MD. United States Pharmacopeial Convention. USP XX:1101, 1980, p. 1019.

26. TO McDonald, JA Shadduck. Eye irritation. In: FN Marzulli, HI Maibach, eds. Dermatotoxicity and Pharmacology. Washington DC: Hemisphere Publishing, 1977, pp 139–191.

27. Food and Drug Administration. (FDA) Illustrated Guide for Grading Eye Irritation by Hazardous Substances. Washington DC: U.S. Government Printing Office, 1965.

28. International Conference on Harmonization (ICH) of Technical Requirements for the Regulation of Pharmaceuticals for Human Use. Harmonized Tripartite Guidelines: Detection of Toxicity to Reproduction for Medicinal Products. Study Design 4.1.1. study of fertility and early embryonic development to implantation. Federal Register 59(183), Sept 22, 1994.

29. FDA. Draft Guidance. S4A duration of chronic toxicity testing in animals (rodent and nonrodent toxicity testing. Federal Register, Nov 18, 1997.

30. P Eckstein, MCN Jackson, N Millman, AJ Sobrero. Comparison of vaginal tolerance tests of spermicidal preparations in rabbits and monkeys. J Reprod Fertil 20:85–93, 1969.

31. M Kaminsky, MM Sivos, KR Brown, DA Willigan. Comparison of the sensitivity of the vaginal mucous membranes of the albino rabbit and laboratory rat to nonoxynol-9. Food Chem Toxicol 23:705–708, 1985.

32. RJ Staab, CS Auletta, DL Blaszcak, RF McConnell. A relevant vaginal irritation/subacute toxicity model in the rabbit and ovariectomized rat [abstr 1069]. Toxicologist 7, 1987.

33. RJ Staab, CS Auletta, RF McConnell. Evaluation of toxicity of four superabsorbant laminates when exposed to the vaginal mucosa of rats, rabbits and sheep [Abstr 1229] Toxicologist 6:1986.

34. CS Auletta. Vaginal and rectal administration. J Am Coll Toxicol 3:48–63, 1994.

35. FDA. 21 CFR Part 351, Department of Health and Human Services, Part V. Vaginal contraceptive drug products for over-the-counter human use; establishment of a monograph; proposed rule-making. Federal Register 45:1990.

36. SR Pollack, LS Olanoff. Clinical pharmacology approaches to the assessment of novel drug delivery concepts. J Controlled Release 11:331–341, 1990.

37. XH Zhou, A Po, W Li. Comparison of enzymatic activities of tissues lining portals of drug absorption, using the rat as a model. Int J Pharm 62:259–267, 1990.

38. Y Katagiri, T Itakura, K Naora, Y Kanba, K Iwomoto. Enhanced bioavailability of morphine after rectal administration in rats. J Pharm Pharmacol 40:879–881, 1988.

39. C DeMuynck, C Cuvelier, D Van Steenkiste, L Bonnarens, JP Remon. Rectal mucosa damage in rabbits after subchronical application of suppository bases. Pharm Res 8:945–950, 1981.

40. VHL Lee. Protease inhibitors and penetration enhancers as approaches to modify peptide absorption. J Controlled Release 13:213–223, 1990.

41. RD Franderson, EH Whitten. Anatomy and Physiology of Farm Animals. 3rd ed. Philadelphia: Lea & Febiger, 1981, pp 323; 507.

42. A Kamiya, H Ogata, HL Fung. Rectal absorption of nitroglycerine in the rat: avoidance of first pass metabolism as a function of rectal length exposure. J Pharm Sci 71:621–624, 1982.

43. BJ Aungst, G Lam, E Sheeter. Oral and rectal nalbuphine bioavailability: first-pass metabolism in rats and dogs. Biopharm Drug Dispos 6:413–421, 1985.

44. WA Behrendt, J Cserepes. Acute toxicity and analgesic action of a combination of buclizine, codeine and paracetamol ("Migraleve") in tablet and suppository form in rats. Pharmatherapeutics 4:322–331, 1985.

45. H Okada, I Yamakazi, I Ogawa, S Hirai, T Yashiki, H Mima. Vaginal absorption of a potent luteinizing hormone-releasing hormone analog (Leuprolide) in rats. I. Absorption by various routes and absorption enhancements. J Pharm Sci 71:1367–1371, 1982.

46. K Morimoto, K Morisaka, H Akasutchi, A Kamada. Effect of non-ionic surfactants in a polyacrylic gel base on the rectal absorption of [Asau^{1,7}]-eel calcitonin in rats. J Pharm Pharmacol 37:759–760, 1985.

47. EJ Van Hoogdalem, C Vermeig-Kerrs, AG DeBoer, DD Breimer. Topical effects of absorption-enhancing agents on the rectal mucosa of rats *in vivo*. J Pharm Sci 79:866–870, 1990.

48. DC Corbo, J-C Liu, YW Chen. Drug absorption through mucosal membranes: effect of mucosal route and penetrant hydrophilicity. Pharm Res 6:848–852, 1989.

11
Toxicokinetics and Hazard Identification

Frank M. Sullivan and Susan M. Barlow
Harrington House, Brighton, East Sussex, United Kingdom

I. INTRODUCTION

The goal of this chapter is to review the hazard identification process in relation to excipients. Just as it is expected that every new active ingredient has to be tested to identify the potential hazards from its use in patients, so all the excipients used in the manufacture of the final drug product have to be shown to be safe for the purpose intended. A major difference between excipients and active ingredients is that the excipient is expected to be toxicologically inert. Thus, it would be expected that the excipient would have few or virtually no hazardous effects at the doses used. On the other hand, it is becoming more common for active ingredients to be discovered that present major problems in formulation to provide good pharmacokinetic profiles, and in these cases novel excipients may form a crucial function in permitting use of the drug, so that some hazard from the excipient may be identified.

The term *hazard* is defined as the intrinsic properties of a chemical that produce adverse or toxicological effects. These effects may be manifest at any dose level, or by any route of administration. Hazard should be distinguished from *risk*, which relates to the adverse effects that may be observed under specified exposure scenarios (i.e., at specified doses and routes of exposure).

The first step in hazard identification is to examine the chemical structure and physical and chemical properties of the test material, as these may indicate areas of toxicological concern. The chemical structure may suggest structure-activity relations to known toxicants and so provide an alert to particular effects that should be investigated. Physicochemical properties information, such as a high vapor pressure suggesting the possibility of loss of test material by vaporization, may also be important to the evaluation of hazard. Similarly, a high octanol/water partition coefficient (Po/w) indicates the possibility of bioaccumulation owing to high lipid solubility and possibly poor excretion.

The process of hazard identification involves reviewing all the available toxicological studies, identifying all adverse effects, the dose levels at which these are observed, and the dose–response relations for each. Interpretation of the significance of the identified hazards for human exposures ideally requires knowledge of the mechanisms by which such effects are induced; the possible interactions of the different toxic responses, and any species specificity of the toxic responses. These are all relevant to the interpretation of animal data and their extrapolation to humans.

II. IMPORTANCE OF TOXICOKINETICS

Knowledge of how a chemical behaves when introduced into the body is fundamental to interpretation of any adverse effects induced. *Toxicokinetics* is the study of absorption, distribution, metabolism, and excretion (ADME) of chemicals in relation to their toxic effects. This term is identical with *pharmacokinetics*, which is used to describe the same information in relation to pharmacological effects of drugs.

Bioavailability is another term that is commonly used to describe the amount of the drug or other chemical that is absorbed and is available for action at target sites. This includes not just whether the chemical is absorbed and enters the bloodstream, but also whether it is free in the plasma to diffuse to the target sites or whether it may be tightly bound to plasma proteins and thus unavailable for diffusion to the target sites. It also includes whether the chemical is able to diffuse to all body compartments, intracellular as well as extracellular, into the cerebrospinal fluid (CSF); and across the placental and testes barriers. Studies on toxicokinetics should ideally include all the information necessary to estimate the bioavailability of the chemical at all sites, but frequently, the information available does not permit this.

Because some pharmaceutical excipients are high molecular weight materials, such as polymers or indigestible polysaccharides, very little excipient may be absorbed following oral administration; thus no systemic toxicity would be expected. However, it is difficult to prove lack of absorption, and it is rare that toxicological studies can be avoided on this basis. The fact that 100% of an orally administered dose appears in the feces is not proof of lack of absorption, for material may be absorbed and then excreted through the bile and undergo enterohepatic recirculation. Additionally, polymers usually contain low molecular weight monomers and oligomers that may be absorbed and may have toxicological import.

Age is another important variable in toxicokinetics. Absorption, metabolism, and excretion of specific chemicals may differ in the fetus, neonate, immature, adult, and elderly subjects, and these differences need to be taken into account in assessing safety. For example, maturation of the liver with development of metabolizing enzymes occurs late in gestation or in the neonatal period. Rat fetuses for example, are unable to metabolize carcinogens into their active forms until the last third of gestation, which accounts for their relative insensitivity to transplacental carcinogenesis until the late fetal period. In newborn animals the small intestine has the ability to absorb large molecular weight compounds, such as proteins and antibodies, and this persists in rats until about day 20 of age, when "closure" as it is termed, occurs (see Ref. 1 for discussion in relation to absorption of high molecular weight PVP).

Differences in toxicokinetics are a major cause of species differences in toxic response to chemicals. Differing rates of absorption, different metabolic pathways, and differing rates of excretion, all contribute to species differences in toxicity. Target tissues may not be exposed to the same concentrations of chemical or active metabolites in different species. In addition to toxicokinetic profiles, differences in end-organ sensitivities are the other important factor in species differences.

The various components of ADME are discussed separately in the following sections.

A. Absorption

In reviewing toxicological data for hazard identification, it is important that studies in animals should be performed using the same routes of exposure as will be used in the clinical application of the final drug product. Because many excipients have already been accepted for use as food additives, oral toxicity data will often be available. It is important to realize that chemicals that are safe following oral administration may not be safe for other routes of administration. The most obvious example is that insoluble substances may be safe orally but may be fatal if injected intravenously. Oral studies on food additives are usually, but not always, carried out by dietary administration, whereas drugs are usually investigated using gavage administration in which a single bolus dose is administered through a tube inserted into the stomach of the animal. For larger animals the use of capsules is possible. The blood levels achieved using dietary versus gavage methods of oral administration may differ by an order of magnitude, and the area under the blood concentration–time curve (the AUC) may differ by two orders of magnitude. Thus, applying data derived from a chemical investigated for food use to its use as a pharmaceutical excipient requires examination of the kinetics at different dose levels to ensure that they are adequate to assess safety and to provide an adequate safety margin under the new conditions of use.

On the other hand, oral absorption of chemicals is usually much greater than the absorption that would be achieved following other routes of exposure commonly used for drugs, with the exception of injection route. Good oral studies with demonstrated high bioavailability of the excipient, therefore, may remove the need for studies to be carried out to support some other routes of exposure, such as topical, ocular, vaginal, and transdermal, at least as far as systemic toxicity is concerned, although topical toxicity, such as irritation or sensitization, must also be considered. This is of importance in reducing the need for studies to support every separate use of excipients, with savings both in costs and numbers of animals. Each chemical should be evaluated on a case-by-case basis to determine the testing needs by various routes of exposure to be used commercially.

For example, absorption of chemicals by the dermal route is usually less than by the oral route, especially with water-soluble chemicals, for which absorption dermally is usually less than 10% of the administered dose and frequently less than 1%. Even oil-soluble chemicals, which are well absorbed dermally, rarely have bioavailabilities equal to oral administration. An obvious exception to this is when drugs are formulated for transdermal application (see Chapter 7), for which the route is chosen to avoid first-pass metabolism in the liver. The difference in metabolism when given in this way may be critical for the active ingredient, but may not make any difference for the excipient when often neither the parent substance nor the metabolites are very toxic.

All of the foregoing discussion refers to investigation of systemic toxicity only. For each separate route of exposure, topical adverse effects at the site of application also have to be considered, such as irritation and sensitization. Some consideration also must be given to clearance from the site to ensure that normal mechanisms, such as ciliary clearance from the respiratory tract, are not overloaded. These considerations may require only short-term bridging studies to permit use of an excipient by a route not envisaged in original studies.

For excipients used for parenteral administration by injection, special care is required if extrapolating from results of oral studies. Extrapolation will normally be valid only for chemicals of relatively low molecular weight. For low molecular weight substances below about 1000 Da, absorption from the gastrointestinal tract is good and rapid, provided other factors, such as solubility, degree of ionization, and lipid solubility, are favorable. Higher molecular weight substances can be absorbed to only a small degree, by pinocytosis that allows high molecular weight (MW) chemicals to gain entry to cells of the gastrointestinal tract, especially the lymphoid tissues of Peyer's patches and local draining lymph glands (persorption). For example, less than 0.1% of administered polyvinylpyrrolidone (PVP) at 8000 MW is absorbed by this process (see Ref. 1 for a discussion of the kinetics of the excipient, PVP, of different molecular weights). For chemicals that are well absorbed orally, comparison of the kinetic data, especially the peak plasma levels (C_{max}) and the AUC, with kinetic data from subcutaneous, intramuscular, or local injections will indicate whether the oral data can be reasonably used for hazard evaluation of these other routes of exposure.

Intravenous administration will usually permit a higher C_{max} to be attained than any other route of administration, and also carries other risks relating to the sudden rate of rise of blood level that can be achieved. However, most of the special risks of intravenous administration are likely to be transient effects and may often be addressed by a series of short-term bridging studies.

B. Distribution

Analysis of the volume of distribution, into the vascular space, extracellular space, total body water, or selective concentration at particular sites, will be helpful in assessing the toxicological effects of the chemical. Autoradiography is particularly helpful in defining sites of accumulation, which may, for example, be localized to individual organs, such as the eye, adrenal gland, or thyroid, or localized in particular areas of the brain. In such cases care should be taken to examine for specific toxicity at these sites that might not be revealed by standard toxicological investigations. The so-called brain, placental, and testis barriers are, in fact, never absolute barriers, and the results from short-term studies may not represent what may happen following repeated administration. The rate of entry through these barriers may be very slow, but the rate of exit may be even slower. There is clear evidence in animals and humans that chemicals may accumulate in these sites following repeated exposure. Transfer of chemicals into breast milk is common, but levels rarely exceed plasma levels, and the doses transferred to the baby are usually small. Important exceptions to this rule are highly fat-soluble chemicals (high *P*o/w), which are poorly metabolized and tend to be stored in body fat owing to poor excretion by the kidney. Such substances will be transferred to milk during lactation in amounts that may be of toxicological significance for the infant, and this should be taken into account in assessment of chemicals with these physicochemical properties.

C. Metabolism

Most if not all guidelines for the conduct of toxicology studies, have stated in the past that the test species used should metabolize the test chemicals in a manner similar to humans. In practice, however, studies are normally carried out in a rodent (usually the rat) and a nonrodent (usually the dog). Marked differences in metabolism of chemicals do occur, however, between species, and even between strains. Thus, a knowledge of the metabolism of the test chemical in the species used for the toxicology-testing program is essential for adequate interpretation of the results of studies. Information should be available on the routes of metabolism and the rates of conversion and elimination of the different metabolites. For extrapolation of animal test results to humans, knowledge of the kinetics and metabolic fate in humans is also necessary so that comparison can be made of the AUC, the peak plasma levels, and the rates of elimination for the parent substance and for each of the major metabolites. If major differences exist between animals and humans in metabolism, then a judgment has to be made of the relevance of the animal studies for assessing human safety. Clearly, satisfactory kinetic and metabolic parameters underlie all the toxicological tests, so it is normal for the metabolic studies to be carried out at an early stage of toxicological testing. Only with this information can the subacute, chronic, reproductive, and other repeat-dose toxicology tests be adequately designed.

The metabolic factors to be considered in assessing the relevance of the toxicology studies for hazard assessment for humans demonstrate that the administered dose is only one of the important considerations. A high margin between the dose administered to animals and that consumed by humans may be of little relevance if the substance is poorly absorbed in animals and well absorbed in humans. Toxicity will normally depend on the plasma and tissue levels of the parent substance and any active metabolites, and the difference in these levels between animals and humans at the same administered dose levels can differ by orders of magnitude. Careful examination of the plasma concentration data may show that a standard protocol study design will not be satisfactory, although it may be possible to make modifications to improve the amount and type of information obtained. For example, although it is recommended that substances used as food additives should normally be tested by administration in the diet, there may be metabolic or other reasons that would make it necessary to administer the chemicals by gavage or some other method to overcome rapid metabolism or clearance (2). Metabolism following long-term administration may differ from that following brief administration owing to enzyme induction. It is, therefore, important to test for such changes if the excipient is to be used in drug products for prolonged administration.

If humans metabolize a chemical to a major metabolite that is not produced by the test species, then consideration has to be given to whether an alternative species should be used. Sometimes it may be possible to synthesize the metabolite in amounts sufficient to administer it to animals for separate hazard evaluation. Care must be taken here to ensure that the metabolites, which are usually polar, water-soluble compounds, are actually absorbed and pass to the appropriate tissue site. For example, in a reproductive study, it may be impossible to feed or even inject a metabolite that will pass through the placenta to the fetus to mimic the situation in which the metabolite is generated in the fetal compartment when the parent compound is given.

In vitro techniques have proved very useful in the study of metabolism of foreign compounds using hepatocytes or other single cells, including gut flora, or using intact organs, such as gut or liver perfusion. Direct comparison in vitro of animal and human tissues is possible. It must be remembered, however, that the animal gastrointestinal tract differs in many ways from that of the human and that the gastrointestinal flora are also metabolically active. The artificial sweetener cyclamate is an interesting example of a substance that can be metabolized by human gut flora to the toxic metabolite cyclohexylamine, which is not produced in the rodent gut, and separate studies on the toxicology of the metabolite have had to be performed for hazard assessment (3).

Dose-dependent kinetics, in which plasma concentrations are not linearly related to the dose administered, are frequently observed in toxicology studies when high-dose levels are administered. This is usually related to saturation of metabolic processes by overloading the enzyme-binding capacity. Two possible consequences are either marked disproportionate increases in plasma levels of parent substance and associated toxicity, or alternatively diversion to a different metabolic pathway with production of a new and more toxic metabolite. Such a situation arises, for example, with acetaminophen (paracetamol) overdose when exhaustion of a cofactor, glutathione, alters the metabolic pathway to produce a new reactive metabolite not found with the normal low doses of acetaminophen (4). In such situations the results obtained in the high-dose groups in the toxicology studies may not be relevant to the normal use of the substance, and hazard evaluation has to take this into account.

An important factor in the choice of test species for toxicology testing is frequently the metabolic profile of the test material in different species. This is likely to affect the choice of the nonrodent species more than the choice of the rat, because there is such an extensive experience of rat toxicology that it is normally considered to be an obligate species for testing, unless overwhelming evidence to the contrary is provided. For acute toxicity, mutagenicity, and carcinogenicity studies, the mouse is also a widely used species and has a metabolic profile for chemicals that is often very different from the rat. For the nonrodent species, the dog or the primate have traditionally been used. The primate is not always closest to humans in metabolism and quite marked differences in metabolic pathways exist even between different primates. There is also an increasing reluctance to use primates for research for humane and conservation reasons. Even if a species can be identified that has a metabolic profile similar to humans, if little toxicological experience and background information are available about that species, then the disadvantages may outweigh the metabolic advantages. However, various other species have been investigated. For developmental toxicity testing the rabbit is the most commonly used nonrodent species. The ferret, guinea pig, pig, and cat have also been used, although never widely accepted.

D. Excretion

Most chemicals that are absorbed into the bloodstream are metabolized to more polar water-soluble substances that are excreted in the urine, usually by glomerular filtration, but organic acids may be secreted by the tubules. For pharmaceutical excipients that are thought not to be absorbed after oral ingestion, presence of the chemical in the urine provides clear evidence for absorption, although the converse is not true. The other major route of elimination, especially for chemicals with molecular weight close to 300–400 and those that are protein bound, is in the bile. After biliary excretion into the gut, reabsorption may occur with further cycles of excretion. A second peak plasma level 3–4 h after the first peak may be a clue to enterohepatic recirculation. Other routes of excretion, such as expired air (except for gases) and sweat, rarely account for more than 1% of the absorbed dose.

The influences of molecular weight and molecular size on excretion through the kidney of the rat, rabbit, dog, and human have been studied using intravenous infusions of the excipient PVP of different molecular distributions (1). Molecules with a radius up to 2.4 nm are cleared as readily as inulin (100% of the glomelular filtration rate; GFR), but as radius increases, clearance decreases up to a radius of 6 nm, at which total exclusion occurs. These data show that PVP with molecular weights up to about 25,000 are rapidly cleared by glomerular filtration. At higher molecular weights, some of the material may pass through postglomerular capillaries into the renal interstitium and be reabsorbed. Molecules with a weight greater than about 100,000 are not cleared by the kidney at all. Estimates of pore size using PVP do not apply to all molecules, as the clearance of albumin of similar size to a specific PVP is cleared at about 0.01% of the rate of the PVP. Thus, the effective pore size of the kidney for a particular molecule depends not only on the molecular radius, but also on the physical characteristics of charge, shape, rigidity, and such, of the macromolecules. The overall conclusion of studies using PVP suggest that any substance that can be absorbed by the gastrointestinal tract into the bloodstream can be easily cleared by the kidney.

When materials with a higher molecular weight than can be easily cleared by the kidney are introduced into the body by injection then, in the absence of metabolism, the only method of clearance from the blood is by the reticuloendothelial system (RES); hence, the Kupffer cells in the liver and the RES cells elsewhere, especially in the spleen, bone marrow, bone and kidney, become filled with the material, leading to the production of "foam cells" in these tissues. This is not pathologically important by itself unless the system becomes overloaded, although the phagocytosis of other materials may be reduced. High molecular weight excipients injected into areas of low vascular perfusion, such as subcutaneously, intramuscularly, joints, and the breast, have led to accumulations of material at the site of injection, with development of so-called pseudotumors (5). This happened when high molecular weight PVP (>25,000) was used as an excipient in slow-release drugs for repeated injection and when the total amount injected exceeded 200 g. In Europe the use of PVP as an excipient for intramuscular injections is restricted to PVP with an average molecular weight of less than

8000–10,000 and in amounts of 50 mg per injection, which can be rapidly cleared by the kidney (6).

III. INTERPRETATION OF TOXICOLOGICAL DATA

A. General Concerns

The first task when starting to assess and interpret toxicological studies on excipients is to review the overall quality of each study. The aims and design of the study should be clearly stated and the protocol for the study should be included in the report. The report should be clear and easy to follow. Each and every animal should be accounted for during the study, and none should have disappeared. It is expected that studies carried out after 1979 will conform to the principles of Good Laboratory Practice (GLP; 7, 8), and any deviations from this should be mentioned in the report of the study. Was assignment to test groups done in an acceptable manner, and were all the animals treated concurrently, or were some added at a later date? Were the animals generally in good health? Good studies require healthy animals, and only a limited number of intercurrent infections or deaths (not more than 10% in the controls) may occur without jeopardizing the quality and statistical power of the study. Were intercurrent deaths dose-related, and what is the meaning of that? Were the numbers of animals for each test parameter, such as body weight, hematology and clinical chemistry, clinical examination, pathology, and such, correctly conducted according to the protocol? How many animals were lost before final pathological evaluation from cannibalism or autolysis, which might suggest poor animal husbandry? Where appropriate, animals should have been dosed 7 days/week. It is surprising that studies are still being reported in which 5-days/week dosing has been performed. It requires a good deal of experience of report reading to be able to carry out this initial assessment of a study because it involves a certain amount of ''detective'' work, but sometimes a study may be rejected at this early stage because of fundamental flaws in design or execution.

It would be expected that studies should be carried out according to some guidelines, as discussed fully in Chapter 5, but excipients are a special case and are neither foods nor drugs, although they may be used in both. No regulatory guidelines exist specifically for excipients. When an excipient has been tested as a food additive then the package of data may be quite different from what would be expected for a drug submission. Because of the worldwide pressure to reduce the numbers of animals used for experiments, as well as for commercial reasons, careful examination of the data package should be made to determine whether the proposed use in a drug product is adequately covered by the data available. For example, the route of exposure or mode of administration may not be exactly equivalent to the intended new use. However, if there are good toxicokinetic and metabolic data, it may be possible to show that the blood levels achieved are adequate to cover the safety of the new intended use. It may be possible with short-term–bridging studies to safely extend the use in various ways.

The major aim of toxicity testing is to identify the types of toxicity that a substance can produce; the target organs affected, and the dose–response relations for each toxic effect, including the no-observed effect levels (NOEL, NEL) and no-observed adverse effect levels (NOAEL). A study that fails to produce any detectable adverse effect is essentially a failed study for most chemical substances. However, excipients are intended to fulfill a pharmaceutical function, not a pharmacological function, and ideally they should be biologically inert. In this situation it is generally accepted that there is a limit to the highest dose that should be used in toxicity studies to avoid nonspecific effects from excessive amounts of chemical. Unfortunately, there is no worldwide agreement on what the limit dose should be. In Europe, different regulatory authorities, and even different regulations under the same European Union (EU) directive variably consider 1 or 2 g/kg body weight as a limit for substances given by gavage or parenterally, and 5% when the chemical is added to the diet. For dermal application 2 g/kg body weight, for inhalation of aerosols or particulates 5 mg/L for 4 h, and for gases and vapors 20 mg/L for 4 h are the limit levels (9). These European limits are essentially identical with those of the OECD (10). The FDA in the United States consider 5g/kg body weight as a limit for gavage dosing and do not specify a maximum in the diet although higher than 10% would not normally be expected (2); (see also Table 1 in Chapter 5).

When toxic effects are observed in studies, it is valuable to know whether these are reversible because more weight is given to irreversible effects in setting margins of safety. Reversibility may sometimes be inferred from the type of pathology produced, but the addition of satellite groups to study reversibility is helpful. The severity of the effects is also important in setting safety margins, although there is often disagreement between experts on the severity of specific observations. A particular example is at the lowest end of the hazard scale, for which there may be debate over whether an effect is toxic, or perhaps is merely a physiological adaptation to exposure to a xenobiotic. A common example is increase in liver weight, which may be due to enzyme induction consequent on the increased metabolic load from the administered substance. Sometimes, there is a change in absolute organ weight when compared with the control group, but no change in the relative organ/body weight ratio compared with the controls. Some regulatory authorities consider all differences from controls as "adverse," whereas others accept some differences as of no toxicological significance. This accounts for the different NOAELs that may be identified from the same data in different countries.

Another factor that contributes to differences in interpretation is the lack of agreement between experts on the methods and importance attached to statistical analyses of the results of toxicity studies. Because of the wide variability that is normally observed in biological systems in response to chemical exposures, it is not always easy to decide if statistically significant differences between groups are of biological importance. These issues are reflected in the problems of risk assessment discussed in Chapter 12. Dose–response relations are important in interpretation of studies. In general, statistically significant findings that are not dose-dependent may not be biologically significant. Examples include changes in clinical chemistry or hematology that occur without dose–response in some groups.

B. Specific Toxicity Studies

1. Acute Toxicity

Although not normally required for food additives, acute toxicity tests are required for drug applications and for excipients for drug use. The aim of the acute study is to give some idea of the lethal dose when given by a variety of routes, and how quickly the effect is observed. Careful clinical observation of the animals in such studies can reveal a good deal about the activity of the test substance. The studies are also useful for advice on poisoning, as required by some regulatory authorities, and for accidental and workplace exposure. In the past, acute LD_{50} studies were used for the bioassay and standardization of drugs for which chemical assays were not available and very elaborate designs and methods of statistical analyses were published that permitted an accurate LD_{50} with confidence limits to be calculated. Nowadays such accurate studies are not required as part of a toxicity package, and various much simpler designs, using the minimum number of animals, are available (2,10,11).

Examples of the type of information that can be obtained from a series of acute tests include an indication of oral absorption by comparison of the oral and parenteral LD_{50}s, and an indication of dermal absorption if animals can actually be killed by dermal application. If animals die very rapidly within the first few hours, then an action on the central nervous or cardiovascular systems might be suspected, and it is unlikely that any histological changes would be observed. On the other hand, if the animals die after a few days, or at intervals over a few days, then some action on vital organs is probable, and careful examination, including histopathology, may be worthwhile. The slope of the mortality–dose response curve indicates the extent of the variability of response and is valuable in initial risk assessment and may suggest different endpoints of organ toxicity in different animals. Examination of the animals surviving to the end of the 14-day observation period may indicate the target organs and may give some idea of the reversibility of the toxic actions. Thus, a well-conducted acute toxicity package can give a great deal of useful information, much more than the mortality data in each group, which is all that some reports contain.

2. Repeat-Dose Toxicity Studies

The repeat-dose studies include short-term (14–28 day) studies, subchronic (90 day) studies, and long-term (6 month to 2 years) studies. The ICH guidelines for testing of new drugs (12), which tend to be more up to date than the guidelines for food additive testing, now consider 6–9 months duration as adequate for chronic toxicity testing in rodents and dogs, with little to be gained from prolonging the test. This does not include carcinogenicity testing, for which lifetime exposure is required, usually in two species, although the use of short-term tests for carcinogenicity using transplacental exposure or transgenic strains of rodents, or other newer tests are acceptable in place of the second species (13). Whereas most food additives will have been tested for carcinogenicity, drugs and excipients would normally require carcinogenicity testing only if they are to be used for prolonged periods, usually 6 months or more, or used repeatedly.

The choice of the most appropriate repeat-dose studies and the key endpoints of toxicity are discussed in Chapter 5 in relation to the expected use of the excipient. The aim of these tests is to indicate the target organs on repeated exposure; to determine whether cumulative effects occur either systemically or in specific target organs; and whether the appearance of toxicity is delayed. For adequate interpretation, the package of tests should be evaluated as a whole and should demonstrate consistency of results. Effects seen in the shorter-term studies should also be seen in the longer-term studies, unless transient or not treatment-related. The target organs should be identified to identify potential hazards or precautions for future human studies. For example, detailed liver or renal function or hematological studies for early detection of toxic effects may be indicated in the first human studies. A clear no-effect level should also be identified by the studies, and if the repeat-dose studies fail to show a no-effect level for a potentially serious toxic effect, then further studies will be required at lower doses. In general it would be expected that excipients should have high no-effect levels if they are to be generally useful, although exceptions will occur.

3. Reproduction and Developmental Toxicity Studies

There are many guidelines for these studies, as discussed in Chapter 5, the most recent being the ICH guidelines for drug testing (14). Guidelines for both food additives and drugs will include tests for effects on fertility in males and females, and on development pre- and postnatally up to the age when the reproductive function of the offspring can be tested. The major difference between food additive and drug guidelines is that for food additives it is normal to test for multigenerational effects, whereas these are not normally evaluated for drugs. When assessing a new excipient, consideration should be given to the possible uses of the excipient to decide whether the package of reproductive tests is adequate.

In a recent review (15) of 117 chemicals and chemical classes tested for effects on male fertility by a variety of different testing methods, it was found that histopathology after 4 weeks of treatment together with organ weight analysis provided the best general-purpose means of detecting substances with potential to affect male fertility. If actual mating studies were added, then little improvement in the detection rate was achieved by adding other test systems, such as elaborate methods of semen and sperm analysis. Thus, the results of the 28 to 90-day repeat-dose toxicity tests provide good information on whether there are likely to be effects on male fertility. The assessment of the adequacy of the conduct and interpretation of developmental toxicity studies is an expert task and should not be undertaken by a general toxicologist; therefore, it will not be discussed further here (see Refs. 16,17 for reviews of reproductive and developmental toxicology).

4. Genotoxicity Studies

Genotoxicity studies form an essential component of any toxicology package, and their function is primarily to detect mutagens and to screen for potential carcinogens. The various studies are designed to detect mutagenic effects or structural or numerical changes in the chromosomes. Studies can be performed both in vitro, with and without metabolic activation, and in vivo. A wide variety of tests are available, each of which has special characteristics, and some of which are of particular value, or of no value, with certain classes of chemicals. For example, the in vitro testing of azo dye food colors requires the use of special strains of bacteria that possess azoreductase activity. The design of a battery of tests, therefore, may depend on the class of compound being studied. The testing of insoluble compounds, which would include many excipients, presents problems with many of the test systems. There is presently no international agreement on whether chemicals that produce negative results in a variety of in vitro systems, with and without metabolic activation, should also be tested in vivo. There is also no international agreement on which of the many tests should be included in a genotoxicity-testing battery. The assessment of the appropriateness of the battery of tests performed on specific chemicals and the interpretation of the results is an expert task and should not be undertaken by the general toxicologist; therefore, we will not discuss it further here (see Ref. 17 for further discussion, and Refs. 18, 19 for detailed discussions of the types of studies and methods).

C. Special Studies

Depending on the applications of the excipient, or the structural–activity relations of the excipient, other types of toxicity tests may be required. Consideration should be given to their necessity for each new application. For example, for excipients to be applied topically in drug formulations or in cosmetics, there are tests for skin and eye irritancy and skin irritation that are appropriate. There are in vitro tests available for irritancy, but these are really designed for humane testing of industrial chemicals or pesticides that may be corrosive, and are unlikely to be of value with inert excipients. For humane reasons, it is generally not necessary to test for eye irritancy those substances that have been shown to be skin irritants, for it can be assumed that they will also be eye irritants.

In recent years there has been an increasing interest in developing batteries of tests for immunotoxicity and neurotoxicity. All contemporary developmental toxicity studies include neurotoxicity and behavioral toxicity parameters. There are many discussion documents published by the World Health Organization (20), U.S. FDA (2), and the OECD (10) on the methods for these batteries of tests, but none are generally agreed on internationally. Many regulatory authorities do not require special tests for immunotoxicity and neurotoxicity beyond what can be detected in well-performed subchronic and chronic toxicity studies. Again, these are specialist areas beyond the scope of this chapter.

IV. CONCLUSION

Excipients used in formulation of final drug products will frequently have already been examined toxicologically for use in foods. Careful examination of the data available, particularly the kinetic and metabolism data will show whether the tests already performed are adequate for the new uses that are envisaged. If they are not, then additional tests may be required. Often short-term–bridging tests may be sufficient, thereby providing savings in costs and animals. The decision on the scope of the safety data should be made by trained toxicologists.

The key aspects to be assessed in the toxicology studies for hazard identification purposes are the identification of the most sensitive target organs; assessment of the observed severity, reversibility, and dose–response relations; and a decision on their relevance for humans. This information can then be used in the risk assessment process. The hazard identification process involves expert knowledge of many different areas of toxicology; thus, a team effort is essential if the best analysis is to be made.

REFERENCES

1. BV Robinson, FM Sullivan, JF Borzelleca, SL Schwartz. PVP: A Critical Review of the Kinetics and Toxicology of Polyvinylpyrrolidone (Povidone). Ann Arbor, MI: Lewis Publishers, 1990.

2. U.S. Food and Drug Administration (FDA). Toxicological Principles for the Safety Assessment of Direct Food Additives and Color Additives Used in Food. Redbook II (draft). Rockville, MD: FDA, 1993.
3. BA Bopp, RC Sonders, JW Kesterson. Toxicological aspects of cyclamate and cyclohexylamine. CRC Crit Rev Toxicol 16:213–306, 1986.
4. JR Mitchell, DJ Jollow, WZ Potter, DC Davis, JR Gillette, BB Brodie. Acetaminophen-induced hepatic necrosis. I. Role of drug metabolism. J Pharmacol Exp Ther 187:185–194, 1973.
5. F Cabanne, R Michiels, P Dusserre, H Bastien, E Justrabo. La maladie polyvinylique. Ann Anat Pathol 14:419–439, 1969.
6. Bundesanzeiger, No. 123/83, July 7, Bundesgesundheitsamt, p. 6666, 1983.
7. OECD. Good Laboratory Practice in the testing of Chemicals. ISBN 92-64-12367-9, Paris 1982. Republished in OECD Environ Monogr 45, 1992.
8. FDA: Good Laboratory Practice for Nonclinical Laboratory Studies, 21 C.F.R. Part 58 (1994).
9. Council Directive 92/32/EEC of 30 April 1992. Amending for the seventh time Directive 67/548/EEC on the approximation of laws, regulations and administrative provisions relating to the classification, packaging and labelling of dangerous substances, Annex V and VI. Official J Eur Communities L 154/1, 5.6.1992.
10. Organization for Economic Co-Operation and Development (OECD). OECD Guidelines for the Testing of Chemicals. Section 4: Health Effects. Paris: OECD Publications, 1993.
11. MJ van den Heuvel, DG Clark, RJ Fielder, PP Koundakjian, GJA Oliver, D Pelling, NJ Tomlinson, AP Walker. The international validation of a fixed-dose procedure as an alternative to the classical LD$_{50}$ test. Food Chem Toxicol 28:469–482, 1990.
12. ICH Note for guidance on duration of chronic toxicity testing in animals (rodent and nonrodent toxicity testing), Draft 3. London: EMEA, 16 July 1997.
13. ICH Note for guidance on carcinogenicity: testing for carcinogenicity of pharmaceuticals. London: EMEA, 16 July 1997.
14. ICH Harmonised Tripartite Guideline. Detection of toxicity to reproduction for medicinal products. London: EMEA, 24 June 1993.
15. B Ulbrich, AK Palmer. Detection of effects on male reproduction–a literature survey. J Am Coll Toxicol 14:293–327, 1995.
16. CD Klaassen, ed. Casarett and Doull's Toxicology: The Basic Science of Poisons. 5th ed. New York: McGraw-Hill, 1996.
17. CA Kimmel, J Buelke-Sam, eds. Developmental Toxicology. 2nd ed. New York: Raven Press, 1994.
18. DJ Kirkland, ed. Basic Mutagenicity Tests: UKEMS Recommended Procedures. Cambridge, UK: Cambridge University Press, 1990.
19. DJ Kirkland, M Fox, eds. Supplementary Mutagenicity Tests: UKEMS Recommended Procedures. Cambridge, UK: Cambridge University Press, 1993.
20. International Programme on Chemical Safety (IPCS). Environmental Health Criteria 180. Immunotoxicity associated with exposure to chemicals, principles and methods for assessment. Geneva: World Health Organization, 1996.

12
Exposure Assessment

David J. George
Whitehall–Robins Healthcare, Madison, New Jersey

Annette M. Shipp
ICF Kaiser, The K. S. Crump Group, Inc., Ruston, Louisiana

I. INTRODUCTION

Excipients embrace a diverse group of chemicals that are incorporated into pharmaceutical dosage forms for a variety of purposes. Typically, products contain multiple excipients (Table 1). Each excipient has been selected for its functionality, and it has been deemed safe in the particular application and compatible with other components of the formulation.

Pharmaceutical excipients are, by design, devoid of significant pharmacological or toxicological activity at the doses used in drug product formulations, and they are completely lacking predictable teratogenic and carcinogenic potential. The safety of individual excipients is paramount. Compounds with narrow safety margins are seldom, if ever, employed as excipients. Thus, excipients are exceptionally safe, but similar to all chemicals, cannot be assumed to be toxicologically inert in every situation (1–4). Examples of the potential toxicity of representative excipients are provided in Table 2.

The dose of an excipient administered to humans is usually selected initially as some fraction of a safe animal dose. However, the dose levels of an excipient used in chronic animal toxicity studies are selected based partly on the projected human dose and dosage schedule. Accordingly, beyond acute animal toxicology studies, the testing of new excipients usually occurs concurrently in humans and animals; the results from animal studies help guide the design of human studies

Table 1 Excipients Contained in a Currently
Marketed Ibuprofen Tablet[a]

Acetylated monoglyceride	Propylparaben
Carbuna wax	Silicon dioxide
Croscarmellose sodium	Simethicone
Iron oxides	Sodium benzoate
Lecithin	Sodium lauryl sulfate
Methylparaben	Starch
Microcrystalline cellulose	Stearic acid
Pharmaceutical glaze	Sucrose
Pharmaceutical ink	Titanium dioxide
Povidone	

[a] Excipients comprise approximately 60% of each tablet.
Source: Ref. 30.

Table 2 Potential Toxicity of Representative Excipients from Excessive Exposure or
Administration to Unusually Sensitive Patients

Excipient	Intended function	Potential toxicity[a]
Benzoic acid	Preservative	Hypersensitivity reactions
Benzyl alcohol	Preservative	Neonatal respiratory and metabolic abnormalities
Butylated hydroxytoluene (BHT)	Antioxidant	Hypersensitivity reactions
Chlorbutol	Preservative	Hypotension, somnolence
Fluorinated hydrocarbons	Aerosol propellant	Cardiac arrhythmias
Polyethoxylated castor oil	Surfactant	Anaphylaxis
Propylene glycol	Solvent	Contact dermatitis, hyperosmolality, thrombophlebitis
Sodium metabisulfite	Antioxidant	Hypersensitivity reactions
Sodium lauryl sulfate	Surfactant	Skin irritant
Tartrazine (FD&C Yellow No. 5)	Enhanced appearance	Hypersensitivity reactions

[a] References for these rate toxic actions as well as those encountered with other excipients can be
found in recent reviews (1–4).

and vice versa (5,6). Therefore, in the early stages of development, a provisional human exposure assessment is usually undertaken. This assessment undergoes refinement, as additional information becomes available. As illustrated in Table 3, the "safety package" for a new excipient to support its use in a new product development program commonly contains not only extensive and comprehensive toxicity study results, but also the results from initial human safety studies. This table also provides an example of the toxicology information, discussed in Chapter 8, which might be required for a new excipient for products administered by the inhalation route of exposure.

Exposure is defined by individual dose, dosage regimen, duration of therapy, and usually, the concentration of the excipient within specific dosage forms. For products administered topically, parenterally, or by inhalation, concentration becomes an important consideration for minimizing local toxicity (e.g., irritation), and for ensuring the compatibility of injectable products with body fluids. When the potential for producing systemic toxicity is of interest, the proposed dose of an excipient that is to be administered should be defined, as well as the proportion of this dose that may reach the systemic circulation and, thereby, the potential sites of pharmacological or toxicological action (i.e., the bioavailability).

Quantitative aspects of dosage must be combined with the characterization of the population expected to receive the product containing the excipient in question. Typical usage patterns for a consumer product or an over-the-counter (OTC) or prescription drug should be identified for the population of interest. Other population characteristics, such as age, sex, and weight, as well as concurrent disease states (e.g., liver or kidney disease, diabetes), may also be important.

Conceptually, exposure assessment is a relatively straightforward exercise that focuses on the proposed dosage schedule for the particular dosage form under consideration. This exposure is then compared with available safety data (animal or human, or both), and the suitability of the specified application is assessed. This comparison is discussed in Chapter 13. Quantitative estimates of exposure can, however, become complex owing to the physicochemical properties of an excipient, physiological characteristics of biological barriers, or influential pharmacokinetic characteristics of absorption, metabolism, distribution, and elimination (AMDE). In this chapter we will provide an overview of basic concepts for exposure assessment and general methods to quantify exposure for the major routes of administration. We will also review dose and some considerations for different dosage forms. Some situations that may require elaborate assessment approaches are mentioned so that when encountered, they are given proper consideration; the reader is referred to the references cited for more in-depth discussions.

Table 3 Studies on a New Ozone-Sparing Aerosol Propellant Recently Submitted to the U.S. Food and Drug Administration to Support a New Drug Application for a Therapeutic Agent Administered by Metered-Dose Inhaler[a]

I. Safety pharmacology
   Cardiac sensitization (dog)
   Cardiac, respiratory, renal, and endocrine function (rat, dog)
   Pharmacokinetics and metabolism (mouse, rat, dog)
II. General toxicology
   Acute (mouse, rat, dog)
   28-day (mouse, rat)
   90-day (mouse, rat, dog)
   12-month (dog)
III. Reproductive toxicology
   Fertility (rat)
   Teratogenicity (rat, rabbit)
   Embryotoxicity (rat)
   Pre- and postnatal toxicity (rat)
IV. Genetic toxicology
   Salmonella (with and without microsomes)
   Reverse mutation (bacteria)
   Human lymphocytes
   Mammalian cells
   Cytogenetic study (rat)
   Micronucleus (mouse)
   Unscheduled hepatocyte DNA synthesis (rat)
   Dominant lethal (mouse)
   Ovary mutation (hamster)
V. Oncogenicity
   2-yr (mouse, rat)
VI. Human safety studies
   Ascending single dose tolerance plus pharmacokinetics
   7-day multiple dose tolerance plus pharmacokinetics
   In vitro binding to plasma proteins and blood cells

[a] This is a summary of major studies only, and not meant to be an exact or complete listing, which would include dose range-finding studies, validation studies for analytical methodology, and others.
Source: Ref. 31.

II. GENERAL CONSIDERATIONS

A. Dose

The *dose* of an excipient is the amount administered (exposure) per unit of body weight by a specified route, in a specified time period (e.g., mg/kg per day, per os). Frequently, one finds the amount of excipient administered expressed only as a percentage of the total composition of a given dosage form, but for quantitative exposure assessments, this requires conversion to conventional units of dosage. In some cases, the dose administered must be distinguished from the biologically active dose (see following paragraph). In formal exposure assessments, *exposure* is typically expressed as the average daily dose (ADD), which is the total amount of exposure averaged over the length of time that the actual exposure occurs.

Bioavailability, expressed as a percentage, refers to the extent to which an administered dose of an excipient becomes available within the systemic circulation and, therefore, available to sites of biological action. Thus, the bioavailability of the administered dose essentially defines the biologically active dose. Technically, bioavailability also encompasses the rate of systemic availability, but rate is seldom, if ever, an important consideration for excipients and can generally be ignored. Except for intravascular administration, absorption is usually the primary determinant of bioavailability. In some cases, however, only a fraction of the absorbed amount of an excipient may reach the systemic circulation. For example, a portion of the absorbed excipient may be cleared (metabolized) by the liver before reaching the general circulation. The concept of bioavailability and its application to exposure assessment has recently been reviewed (7).

Although the bioavailability of individual excipients can vary widely between 0 and 100% (Table 4), a simplifying assumption often made in exposure

Table 4 Examples of Human Oral Bioavailability of Some Pharmaceutical Excipients

Excipient	Intended function	Oral bioavailability[a] (%)
Aluminum salts	Buffer component	4
Boric acid	Preservative	100
Ethylenediamine	Stabilizer	34
Lactose	Bulking agent	100
Microcrystalline cellulose	Suspending agent	0
Phenol	Preservative	90

[a] These values were obtained from various published sources having been determined under a variety of study conditions. They can be considered only as approximations and are provided to illustrate the significant differences that may exist among excipients.

assessment is that a substance is 100% bioavailable. This is usually done for one of two reasons: either there is no scientifically valid bioavailability data available for the case at hand, or one wishes to develop a conservative assessment that is more easily justified than employing assumptions based on little data. Consequently, the administered dose is assumed to be the bioavailable dose.

The bioavailable dose of an excipient is generally preferable over the administered dose when extrapolating human potential toxicity from animal data, or extrapolating from different routes of exposure or temporal dosing patterns. When required, the determination of the bioavailable dose can be complex and require extensive clinical data. Information on target tissue exposure doses may be estimated using advanced pharmacokinetic models (8–10).

B. General Approach to Estimating Dose

A simplified excipient exposure assessment checklist is provided in Table 5. Estimates of dose by any route of exposure can be calculated using the general equation found in Table 6. The parameters employed in this general equation can be divided into three broad categories: product-specific data, product usage data, and general default values. These are defined in the following paragraphs.

1. Product-Specific Data

The *concentration* (C) of excipient in a product is typically expressed as the amount per unit of dosage. The percentage *absorption* (AB) is typically assumed to be representative of bioavailability. Absorption is a product-specific value that can vary with the route of exposure, the matrix of the dosage form, and the dosing

Table 5 Excipient Exposure Assessment Checklist

---

1. Identify intended route of administration
2. Identify intended dosage form
3. Identify intended dose
    Maximal single dose
    Maximal daily dose
    Average daily dose
4. Identify intended duration of therapy
    2 wks or less: continuous or intermittent
    Longer than 2 wks: continuous or intermittent
5. Identify bioavailability by the intended route of administration
6. Identify enhanced susceptibiltiy issues for targeted patient population
7. Identify limiting concentration issues for administration by topical, parenteral or
    inhalation routes

Table 6 General Exposure Assessment Equation

$$ADD = \frac{C \times CR \times EF \times ED \times AB}{BW \times AT}$$

ADD = average daily dose (mg/kg/day)
C   = Concentration of excipient in dosage form (e.g., mg/mL; mg/tablet; mg/m^3)
CR   = Contact rate, amount of drug dosage form contacted per unit time (e.g., mL/ day, tablets/day)
EF   = Exposure frequency (e.g., days/yr)
ED   = Exposure duration (e.g., yr)
AB   = Amount absorbed (%)
BW   = Body weight (kg)
AT   = Average time over which the excipient has been used (e.g., yr)

regimen. A simplifying assumption is usually made that the excipient itself, rather than a metabolite of the excipient is the active agent. When it is a metabolite that is of interest, additional data describing the production of the metabolite following absorption of the parent will be required.

2. Product Usage Data

To effectively characterize product-specific usage patterns, it is necessary to estimate the size of the target population that could be exposed, the characteristics of the individuals in the exposure population (e.g., age and gender), and quantitative aspects of that exposure (e.g., amount used per day). Two parameters, the *contact rate* (CR) and the *exposure duration and frequency* (EDF), vary with individual usage patterns. The CR is the amount of excipient contacted per unit time or event (e.g., milliliters cough syrup taken per treatment, or grams of shampoo used per shampooing event). The EDF describes how long and how often the exposure occurs. It is calculated using the *exposure frequency* (EF), which is the number of days per year or times per day, that a medication is used, and the *exposure duration* (ED), which is the number of days, weeks, or years that a product is used. For prescription drugs, the usage patterns can be assumed to be those recommended. For example, a patient may be instructed to take a certain medication twice daily for 2 weeks. In this example, CR is twice per day, EF is 2 weeks per treatment, and ED is one treatment. For a maintenance prescription drug, an OTC drug, or a consumer product, CR and EF may be known or assumed to be those recommended in the labeling, but ED, that is, the number of months or years of use, may have to be inferred from other sources, such as those described in the following. The *average time* (AT) over which the exposure event occurs is used to determine the *average daily dose* (ADD), which is the amount of exposure averaged over the period of time over which the actual exposure occurs. For example, if a drug is administered daily for 30 days, then the value for both EDF and AT is 30. However, if the drug is administered 2 days per week for 4 weeks, the EDF would be 8 days (2 days/week × 4 weeks) and the AT would be 28 days (7 days/week × 4 weeks).

3. General Default Values

Many of the parameters required for exposure assessment calculations are rarely derived through actual measurement. The size of the target populations typically of interest for drug exposure assessments makes this impractical, if not impossible. For example, it is not realistic to consider obtaining the body weights of individuals, or even representative samples of individuals, from populations of patients who are potential recipients of any particular commercial drug product. In situations such as this, the use of so-called default values is an acceptable practice. Default values are based on scientifically documented experience, and serve as generic substitutes for values that might be derived from situation-specific data. Some common default values are provided in Table 7.

Sources of default values include the EPA *Exposure Factors Handbook* (11), and the National Health and Nutrition Examination Surveys (NHANES; 12). In addition to age, gender, and ethnic exposure information, the NHANES databases include drug utilization data that, in some cases, is product-specific.

C. Exposed Populations and Usage Patterns

Within exposed populations, one should attempt to identify any subpopulations having potentially enhanced susceptibility to a particular excipient. These subpopulations might include low birth weight infants, the elderly, patients with large surface treatment areas (e.g., burns), and patients with a history of disease that might influence susceptibility (e.g., asthma or contact dermatitis).

An estimate of the size of the particular target population should be derived. This could be important for a variety of reasons, which would include the quantitative evaluation of the incidence of any adverse reactions that might be observed.

Drug products are designed and labeled to be used in a relatively specific way, but the actual use of a product may differ significantly from the prescribed or labeled directions. It can be quite a challenge to predict the behavior of medical prescribers and their patients and, thus, the limits of deviations from specified exposure levels. One must, if possible, take this range of potential exposure scenarios into consideration when designing safe dosage forms. To account for these deviations, some type of exaggerated exposure assessment that encompasses the range of possibilities is necessary. The degree of exaggeration is subjective and dependent on the degree of conservatism that both the product developers and

Table 7 Default Values, Useful Conversion, and Common Measures

Default values (11)		
Body weight (kg)		
Adult male		70
Adult female		58
Adolescent (11–18 yr)		40
Child (2–10 yr)		20
Infant (0–2 yr)		10
Lifetime (yr)		70
Surface areas (m^2)		
	Man	Woman
Head	0.118	0.110
Trunk	0.569	0.542
Arms	0.228	0.210
Hands	0.084	0.075
Thighs	0.198	0.258
Lower legs	0.207	0.194
Feet	0.112	0.098
Whole body	1.94	1.69
Inhalation rates (light activity, m^3/h)		
Adult man	1.2	
Adult woman	1.14	
Child (10 yr)	0.78	
Infant (1 yr)	0.25	
Newborn	0.09	
Common measures		
Teaspoon	5 mL	
Tablespoon	15 mL	
Drop	20/mL	
Fluid ounce	30 mL	
Conversion factors		
1 kg = 2.2 lb		
mg/kg = ppm, µg/g (liquid)		
µg/kg = ppb, ng/g (liquid)		
1 g = 0.03527 oz		
1 oz = 28.35 g		
1 lb = 0.4536 kg		

the regulators deem appropriate for any given situation. Clearly, the more information available on how people may actually use a product, the better an assessment can be. One sometimes can obtain exposure data by conducting actual use studies with the product or by studying marketing information for similar products.

The setting within which a drug product is administered can be an important factor for assessing potential exposure because this may markedly influence the degree of control exerted over exposures (e.g., hospital vs. home or prescription vs. OTC).

D. Information Sources

1. Case Studies

Recent regulatory evaluations of new excipients could be consulted for approaches to exposure assessments that might be required to satisfy regulatory requirements. These would include cyclodextrins (13), HFA-134a aerosol propellant (14), chitosan (15), Azone (16) and dimethyl sufoxide (DMSO; 16). A review of the literature on any of these excipients could provide creative ideas and approaches for exposure assessments, as well as insight into current regulatory thinking. The approval process for these agents illustrates the tortuous regulatory pathways and hurdles new excipients may face during development.

2. Marketing Databases and Specialized Consumer Studies

To estimate the size of a target population, the characteristics of the individuals in the exposed population (age, disease state, and such) and quantitative aspects of exposure (amount used per dose, per day, per lifetime, and such), marketing information can be very useful. Indeed, it is often the only basis for estimating the magnitude of potential population exposure. One can look at actual user data from the marketed product that will contain a new excipient, or if not marketed, data from a similar product that is marketed. There are many well-known marketing databases available for both prescription and OTC drugs, such as IMS America (Plymouth Meeting, Pennsylvania) and Nielsen Marketing Research (Northbrook, Illinois). Information on these databases can be found in comprehensive compilations of marketing research organizations (17,18). In addition to these types of data that are collected on a regular basis for various commercial products, there are companies that periodically do specialized reports on specific market segments that contain detailed information on use and users, such as Frost & Sullivan (Mountain View, California), and FIND/SVP (New York, New York). If useful data cannot be obtained using these resources, there are many market research companies that will design and undertake survey-type studies to generate data on specific products (17,18).

E. Uncertainty and Assumptions

1. Uncertainty: Data Gaps and Variability

Any exposure assessment that uses a single value to characterize an important parameter, (e.g., the frequency of product use or body weight), will be inherently uncertain. Uncertainty takes two forms: data gaps and variability. In carrying out exposure assessments, one can expect to encounter gaps in the data and knowledge required for accurate calculations. In these instances, it is necessary to make assumptions. One should be mindful that assumptions, such as assuming that an excipient is 100% bioavailable, might tend to exaggerate exposures so that decisions based on assumptions may be overly conservative. In any event, all assumptions should be clearly identified in assessment documentation, together with their bases.

Variability is a function of the distribution of values for a known data point (e.g., the range of body weights in an identified population). The choice of the point in the distribution to use (i.e., the mean value or the 95% upper confidence limit [UCL] of the mean) will have an influence on exposure estimates and may contribute to the conservatism of these estimates.

It is common to perform sensitivity analyses to determine the magnitude of the influence of any change in the value of a parameter or assumption might have on estimated exposure. When the assumption used to fill a data gap provides an unrealistically high estimate of exposure, then it may be advantageous to generate additional data to fill that data gap and refine the uncertainty. Quantitative methods, termed stochastic or probabilistic (e.g., Monte Carlo) analyses, to evaluate uncertainty have been applied in exposure assessments (19).

F. Impurities

In certain circumstances it may be necessary to take into account the exposure to impurities contained within excipients. The presence of such impurities may alter the absorption characteristics of the excipient, or may contribute to localized irritation or systemic toxicity. Heavy metals, for example, could be an important safety consideration in an exposure assessment. A current example one might encounter is the presence of lead in calcium salts (20). Questions have also come up in the recent past on the possible presence of dioxins in cellulose derivatives, such as carboxymethylcellulose (21). Impurities may also result from predictable degradation processes following formulation and storage. This is typified by the formation of nitrosamines from nitrogen-containing compounds in certain formulations (22). Impurities that are more toxic (or less safe) than the "carrier" excipient often require specialized exposure assessments.

III. ESTIMATES OF INTAKE BY ROUTE OF EXPOSURE

The potential sites employed for drug administration include all the orifices of the body, body surfaces, and tissues and organs accessible by injection. The number of potential dosage forms and drug delivery systems designed for administering drugs is great and increasing. Here only the major routes of administration are discussed. As stated earlier, the key pieces of information to conduct an exposure assessment are the product-specific usage data and the product-specific chemical data. The key chemical-specific data are those that characterize the bioavailability of the excipient. Typically, absorption across the initial biological barrier into the systemic circulation is assumed to be a representative surrogate for the relevant tissue dose. This section focuses on those issues relevant to the absorption of an excipient by each of the major routes of exposure. This section also provides sample calculations for estimating exposure by these routes that will serve as a template for sample calculations and provide the framework for more complex analyses. Numerous comprehensive references are readily available that discuss almost every conceivable facet of drug administration. As a starting point for more specific research, Ansel and his colleagues provide a very good comprehensive overview of the entire area (23). Exhaustive discussions of route-specific exposure assessments are also available (11,24,25).

A. Oral Route of Exposure

Exposure to numerous drugs, and consequently, excipients contained in those products is by the oral route. The amount of the excipient delivered to the gastrointestinal tract (i.e., the administered dose), is the important dose-metric when the potential site of action for the excipient is localized irritation. For systemic effects, the most important considerations when assessing exposure by this route are the amount of absorption of the excipient from the gastrointestinal tract and distribution of the absorbed excipient into the systemic circulation. An understanding of those factors that influence absorption by this route will help identify the data needed for the exposure assessment and allow consideration of the uncertainties in the estimates of dose.

1. *Issues with Oral Exposure*

Following dissociation of the excipient from the ingested dosage form, free excipient is presented to the intestinal mucosa where it may be absorbed and pass into the blood. In some cases, not all of the excipient present in the dosage form will be released into the lumen of the gastrointestinal tract. Also, in some cases not all of the released excipient will cross the gastrointestinal membranes and reach the circulatory system. Occasionally, none of the excipient is absorbed either because of the physicochemical properties of the excipient, or it is converted to nonabsorbable forms by metabolic processes in the gut.

The availability of excipients for absorption at various sites throughout the gastrointestinal tract is largely determined by the physicochemical properties of the excipient (e.g., lipid solubility, pK_a), the pH of the sections of the gut, and very importantly, the particular type of oral dosage form in which it is contained and the oral dosing regimen. The absorption of some substances can vary considerably in different portions of the gastrointestinal tract (i.e., stomach, upper small intestine, proximal intestine, colon, or rectum). The absorption of an excipient from an enteric-coated tablet, which passes through the stomach intact before undergoing dissolution in the intestine, could be much different from a conventionally coated tablet. Similarly, excipients in sustained- or controlled-release tablets or capsules deliver excipients to potential sites for absorption in patterns not produced by immediate-release dosage forms. Absorption of small, divided doses may result in greater uptake than from a single bolus dose of equivalent amount. On occasion, it may be important to determine the approximate regional locations where excipients become available for absorption, and noninvasive methodology for this type of investigation is available (26,27).

Clearly, the simplest approach is to assume that 100% of an orally administered dose of an excipient is absorbed and reaches the systemic circulation. In that case, one can then ignore the influence of particular dosage forms or dosing regimens, the excipient physicochemical properties, or other potential influencing factors. However, when this assumption is unrealistic and absorption is an important characteristic, additional data may have to be generated to characterize the potential for absorption and subsequent delivery in the systemic circulation.

Although a number of the factors that influence absorption cannot be evaluated quantitatively, the potential influence on absorption should be considered qualitatively to assess the potential uncertainty in the assumptions made. For example, an absorption coefficient for an excipient may be available from data using fasted animals. If the directions for actual usage instruct an individual to take the drug containing the excipient only with food, then the potential absorption in that individual will reflect not only species differences but differences owing to the status of the gastrointestinal tract. Numerous other examples could be generated to illustrate the need to evaluate, even qualitatively, the uncertainty in any data used or assumptions made relative to bioavailability by the oral route.

2. Estimates of Dose by the Oral Route

Estimates of the dose of an excipient by the oral route can be calculated using the general equation given in Table 6. An examination of the quantitative composition of an oral dosage form, and the anticipated maximum single and daily dose of an oral drug product directly provides the quantitative exposure information for any of the product's excipients. The key data needed to assess exposure usually includes the following:

1. Concentration (C) of the excipient in the oral dosage form (i.e., milligrams excipient per tablet or milligrams excipient per milliliter of cough syrup)
2. The dosing regimen (CR; i.e., the number of tablets per day or milliliters per day)
3. The intended duration and frequency (EDF) of therapy or product usage
4. The bioavailability (absorption; AB) of the excipient under these specified conditions (i.e., single dose vs. multiple doses, with or without food)
5. Qualitative description of factors that may influence absorption (e.g., use of an animal model for absorption, use of pure excipient rather than the dosage form of interest, biological factors that may be important in sensitive individuals or individuals with relevant disease states)

A sample calculation is given in Table 8.

B. Dermal Route of Exposure

Drug products can be applied to the skin or mucous membranes for local as well as systemic effects. For dermal applications, a variety of dosage forms are avail-

---

Table 8 Oral Exposure Example: Treatment for Seasonal Allergy

Objective:
Estimate the average daily dose (ADD) of specific excipient (E) from seasonal treatment for hay fever
Assumptions:
The formulation matrix does not effect the absorption.
Information required:
C   = Concentration of E in tablet (10 mg/tablet)
CR   = Number of tablets/day (2 tablets/day)
EF   = Number of days/month (14)
ED   = Number of months/year
AB   = Absorption coefficient (50%)
BW   = Adult female (60 kg)
AT   = Total days over which exposure occurred (3 × 30 days)
Calculation:

$$\text{ADD} = \frac{10 \text{ mg/tablet} \times 2 \text{ tablets/day} \times 14 \text{ days/mo} \times 3 \text{ mo} \times 0.5}{60 \text{ kg} \times 90 \text{ days}}$$

ADD = 0.08 mg/kg/day

able and include semisolid products (ointments, creams), solids (powders) and liquids (aerosols, lotions). Transdermal delivery systems (patches) are also in widespread use. All topically applied products expose patients to excipients used in the drug formulation, and in transdermal delivery systems, patients are additionally exposed to the excipients that make up the platform of the dosage form (e.g., adhesives). When doing exposure assessments for dosage forms applied topically, one should be aware of the potential for local irritation owing to excessive concentrations of excipients at sites of application. Because of irritation potential, there is frequently a maximal tolerated concentration that cannot be exceeded.

1. Issues with Dermal Exposure

Penetration through the stratum corneum is the major rate-limiting step for percutaneous absorption of any excipient applied topically. Once it has passed this initial layer, excipients may accumulate in epidermal skin compartments, thereby providing a depot effect or a slow release to the blood and lymph capillaries in the underlying dermis. Skin can also act as an active metabolizing compartment that may further reduce the bioavailability of an excipient.

It is well to keep in mind that absorption will differ, depending on the part of the body exposed, as different areas of the skin surface absorb agents more readily than others. The general health and condition of the skin are important determinants of absorption and are influenced by age, hydration of the skin, circulation to the skin, and so forth. A review of the principles of dermal exposure assessment, along with a description of the kinetic models to estimate dermal absorption, is available (28,29).

2. Estimates of Dose by the Dermal Route

To estimate the exposure (systemic availability) of an excipient in a topical dosage form, one usually requires the following information:

Concentration of excipient in the product formulation
Amount of the formulation used in each application
The surface area over which the formulation is applied
The length of time each application remains in contact with the site of application
The absorption coefficient

The absorption coefficient can be estimated from data on the fraction of the material absorbed over a stated period of time (typically a 24-h cycle), or by use of a specific permeability constant (cm/h), which is the amount of excipient absorbed per unit of time per specified unit of surface area. Excellent references that describe methods for determining absorption coefficients are available (24,28,29).

Table 9 Dermal Exposure Example: Shampoo

Objective:

Estimate the average daily dose (ADD) of specific excipient (E) during a single shampooing event.

Assumptions:

The formulation matrix does not effect the absorption of E.

Neither percutaneous metabolism of E nor binding of E within the skin occurs.

Information required:

C = Concentration of E in shampoo (0.0475 mg/g shampoo)

CR = Amount of shampoo per application × number of applications per shampooing event (3.4 g × 2)

EF = Number of days (events) per week shampooed (3 days/week)

ED = Number of weeks/year product used (16 weeks/year)

AB = Absorption coefficient [(24%/24 hours) × hours per shampooing event (0.017 or 60 s)]

BW = Adult male (70 kg)

AT = Total days over which exposure occurred (112 days)

Calculation:

$$ADD = \frac{\begin{array}{c} 0.0475 \text{ mg/g shampoo} \times 3.4 \text{ g shampoo/application} \times 2 \text{ applications/day} \\ \times 3 \text{ days/wk} \times 16 \text{ wks} \times 0.24 \times 0.017 \text{ h/24 h} \end{array}}{70 \text{ kg} \times 112 \text{ days}}$$

$ADD = 3.4 \times 10^{-7}$ mg/kg/day

An example of a topical exposure assessment for a medicated shampoo product is provided in Table 9.

C. Inhalation Route of Exposure

1. Issues with Inhalation Exposure

With the exception of general anesthetic gases, most drugs administered by the inhalation route are formulated in metered-dose inhalers, which are pressurized systems that deliver measured amounts of particles, either solids or liquids dispersed in a gaseous medium. The size range of the particles has a major influence on the areas of lung surface exposed, the relative amount of the dose that is expired, and the amount that reaches the esophagus and subsequently absorbed in the gastrointestinal tract.

Despite the sophisticated delivery devices and influential exposure factors, exposure assessments for excipients in metered-dose inhalers can be carried out without a great deal of difficulty if a number of simplifying assumptions are made. With metered-dose inhalers this is usually satisfactory because the absolute amount of excipients contained in a delivered dose is extremely small.

2. Estimates of Exposure by the Inhalation Route

To estimate the amount of an excipient that reaches the systemic circulation when administered by a metered-dose inhaler, one usually requires the following information:

> The concentration of the excipient in product (μg/L)
> Amount of product encountered per event (L/puff)
> Number of events per unit of time (puffs/day)
> Absorption coefficient.

An example of a calculation is provided in Table 10.

D. Parenteral Routes of Exposure

1. Issues with Parenteral Routes of Exposure

Although drug formulations may be injected into almost any organ or area of the body, most commonly injections are made into veins, muscles, or into or

Table 10 Inhalation Exposure Example: Metered-Dose Inhaler

Objective:
Estimate the average daily dose (ADD) of a specific excipient (E) during a single
  dosing event.
Assumptions:
  Systemic absorption of E is 100%.
  100% of the administered formulation reaches the mucosal absorption surface.
  The formulation matrix does not influence the absorption of E.
Information required:
C = Concentration of E in formulation (9.8 μg/L)
CR = Volume of one puff (50 μl) $\times$ number of puffs/event (day) (2)
EF = Number of events (days) per week used (3)
ED = Number of weeks/year used (52 weeks/year)
AB = Assumed to be 100%
BW = Adolescent (11–18 yr) (40 kg)
AT = Total days over which exposure occurred
Calculation:

$$ADD = \frac{\begin{array}{c} 9.8 \ \mu g/L \times 50 \ \mu L/puff \times 2 \ puffs/day \times 3 \ days/wk \\ \times \ 52 \ wk/yr \times 1 \times 1 \times 10^{-6} \ \mu l/L \end{array}}{40 \ kg \times 7 \ days/wk \times 52 \ wk/yr}$$

ADD = 6.0×10^{-6} mg/kg/day

---

under the skin. Sites differ primarily in their potential for controlling the rate of absorption of administered medications; the extent of absorption from these sites is almost always complete. Accordingly, to evaluate exposure, one usually assumes that bioavailability is 100%.

2. Estimates of Exposure by the Parenteral Route

To estimate the exposure for an excipient in an injectable dosage form, one requires the following information:

> The concentration of the excipient in the dosage form
> The volume of the dosage form injected

IV. SPECIAL CONSIDERATIONS

The exposure assessment approaches described in this chapter are applicable to relatively simplistic, uncomplicated exposure assessments. Several factors may influence estimates of exposure that typically are not considered in an exposure assessment. Also, whereas an estimated dose by one route of administration or dosing regimen has been deemed safe, the equivalent total dose by another route of exposure or dosing regimen or in another target population (e.g., children or the elderly) may not provide the same level of safety. Some factors that could affect exposure assessments are listed in Table 11.

An exposure assessment is conducted to provide quantitative estimates of dose for the given scenario under consideration. The feasibility that one or more factors may influence estimates of dose resulting in a change in the estimated

Table 11   Potential Factors That May Impact Excipient Exposure Assessments

Potential for excipient to accumulate in tissues or organs following repeated or continuous exposures

Potential for excipients with long half-lives to reach high blood levels when administered frequently at short time intervals

Potential effect of simultaneous exposure to an excipient from sources (e.g., food) other than the pharmaceutical product under consideration

Potential differences in exposure when administered by other than the designated route of exposure

Potential influence of nutrition factors (e.g., fasting, composition of diet)

Potential influence of compromised hepatic or renal function

Potential interaction with other components within the pharmaceutical product that influences exposure dose (i.e., greater or smaller) is considered as part of the overall evaluation of uncertainty. In the absence of actual data on the quantitative effect of some of these considerations, estimates of dose may be adjusted upward or downward to accommodate the influence of these factors. For example, if it is assumed that persons with impaired renal function would excrete less of the excipient, then the dose to the individual would be higher than estimated for the general population. If the magnitude of that uncertainty elevates the estimated dose above the level that has been identified through toxicity testing to be safe, additional information may be required to quantify the effect of these other factors and to ''fine-tune'' estimates of exposure. Some factors could be evaluated experimentally, such as the influence of nutritional status or hepatic function. The influence of other factors, such as accumulation with multiple dosing or with contributions from other sources, could be evaluated using pharmacokinetic models (8–10).

V. FUTURE OUTLOOK

Relative to the active ingredients delivered to the body by the wide variety of pharmaceutical products commercially available, the vast majority of excipients at levels in common use can be considered almost biologically inert for most formulation applications. In fact, this feature is the primary consideration for their initial selection and continued use. However, these traditional excipients are now being supplemented with agents with increasing levels of biological activity and attendant safety concerns. Examples of these newer excipients would include the vast array of penetration enhancers being developed for transdermal drug delivery products, agents included in oral products that predictably alter gastric emptying, mucosal adhesives for selective targeting of drug delivery systems, and the use of newer antimicrobial agents in preservative systems.

With the increasing use of excipients with enhanced potential for pharmacological and toxicological activity, exposure assessments will need to be more sophisticated than is now usually appropriate. Simply calculating the amount of excipients delivered to the body when a recommended or likely dosage regimen is followed will no longer be sufficient for safety evaluations. More use will need to be made of the assessment tools now widely employed for therapeutic agents and for environmental toxicants. Meaningful exposure assessments for biologically active excipients require more actual clinical measurements of pharmacokinetic parameters, and the application of physiologically based mathematical models for predicting parameters in situations that may limit actual measurements (e.g., neonates or pregnant patients). In tandem with the application of mathematical modeling, the predictability of early provisional assessments can be enhanced by applying the rapidly expanding knowledge bases of chemical structure–activity-relations.

REFERENCES

1. JM Smith, TRP Dodd. Adverse reactions to pharmaceutical excipients. Adv Drug React Acute Poisoning Rev 1:93–142, 1982.
2. LK Golightly, SS Smolinske, ML Bennett, EW Sutherland III, BH Rumac. Pharmaceutical excipients: adverse effects associated with 'inactive' ingredients in drug products (part I). Med Toxicol 3:128–165, 1988.
3. LK Golightly, SS Smolinske, ML Bennett, EW Sutherland III, BH Rumac. Pharmaceutical excipients: adverse effects associated with "inactive" ingredients in drug products (part II). Med Toxicol 3:209–240, 1988.
4. American Academy of Pediatrics, Committee on Drugs. "Inactive" ingredients in pharmaceutical products. Pediatrics 76:635–643, 1985.
5. H Boxenbaum, C DiLea. First-time-in-human dose selection: allometric thoughts and perspectives. J Clin Pharmacol 35:957–966, 1995.
6. A Mouro, D Mehta. Are single-dose toxicology studies in animals adequate to support single-doses of a new drug in humans? Clin Pharmacol Ther 59:258–264, 1996.
7. SE Hrudey, W Chen, CG Rousseaux. Bioavailability in Environmental Risk Assessment. Boca Raton, FL: Lewis Publishers—CRC Press, 1996, pp 7–73.
8. SB Charnick, R Kawai, JR Nedelman, M Lemaire, W Niederberger, H Sato. Perspectives in pharmacokinetics: physiologically based pharmacokinetic modeling as a tool for drug development. J Pharmacokinet Biopharm 23:217–229, 1995.
9. KB Bishoff, RL Dedrick, DS Zaharko, JA Longstreth. Methotrexate pharmacokinetics. J Pharm Sci 60:1128–1133, 1971.
10. H-W Leung. Use of physiologically based pharmacokinetic models to establish biological exposure indexes. Am Ind Hyg Assoc J 53:376–374, 1992.
11. U.S. Environmental Protection Agency (EPA). Exposure Factors Handbook. Washington, DC: Office of Health and Environmental Assessment, USEPA/600-1/8-89/043, 1997.
12. National Health and Nutrition Examination Survey, 1988–1991 (NHANES III), Version 1. Hyattsville, MD: National Center for Health Statistics, 1995.
13. VJ Stella, RA Rajewski. Cyclodextrins: their future in drug formulation and delivery. Pharm Res 14:556–567, 1997.
14. ME Boulder. HFC 134a: the latest in propellant alternatives. Spray Technol Market August 32–35, 1994.
15. L Illum. Chitosan and its use as a pharmaceutical excipient. Pharm Res 15:1326–1331, 1998.
16. TK Ghosh, AK Banga. Methods of enhancement of transdermal drug delivery: part IIA, Chemical permeation enhancers. Pharmacol Technol April: 62–89, 1993.
17. HC Barksdale Jr, JL Goldstucker. Marketing Information: A Professional Reference Guide. 3rd ed. Atlanta, GA: Georgia State University Business Press, 1995.
18. PMD: Pharmaceutical Marketers Directory. Boca Raton, FL: CPS Communications, 1998.
19. U.S. Environmental Protection Agency. Guiding principles for Monte Carlo analysis. EPA/630/R-97/001. Washington, DC: Risk Assessment Forum, 1998.

20. BP Bourgoin, DR Evans, JR Cornett, SM Lingard, AJ Quattrone. Lead content in 70 brands of dietary calcium supplements. Am J Public Health 83:1155–1160, 1993.

21. U.S. EPA. The national dioxin study report to Congress. September 24, 1987.

22. DC Harvey, HJ Chou. N-Nitrosamines in cosmetic products: an overview. Cosmet Toil 109:53–62, 1994.

23. HC Ansel, NG Popovich, LV Allen Jr. Pharmaceutical Dosage Forms and Drug Delivery Systems, 6th ed. Malvern, PA: Lea & Febiger, 1995.

24. U.S. EPA. Dermal exposure assessment: principles and applications. EPA/600/8-91/011B. Washington, DC: Office of Health and Environmental Assessment, 1992.

25. U.S. EPA. Methods for derivation of inhalation reference concentrations and application of inhalation dosimetry. EPA/600/8-90/066F. Washington, DC: Office of Research and Development, 1994.

26. D Gardner, R Casper, F Leith, I Wilding. Noninvasive methodology for assessing regional drug absorption from the gastrointestinal tract. Pharmacol Technol Oct:82–89, 1997.

27. A Burch, WH Barr. Absorption of propranolol in humans following oral, jejunal, and ileal administration. Pharm Res 15:953–957, 1998.

28. RH Guy, AH Guy, HI Moibach, VP Shah. The bioavailability of dermatological and other topically administered drugs. Pharm Res 3:253:262, 1986.

29. MS Roberts, KA Walters, eds. Dermal Absorption and Toxicity Assessment. New York: Marcel Dekker, 1998.

30. Physicians Desk Reference for Nonprescription Drugs, Advil Monograph. Montvale, NJ: Medical Economics, 1998, p. 428.

31. PC Rock, International Pharmaceutical Aerosol Consortium for Toxicity Testing Secretariat, Washington, DC, personal communication, 1998.

13
Risk Assessment and Risk Communication

Anthony D. Dayan
*St. Bartholomew's and The Royal London School of Medicine and
Dentistry, London, United Kingdom*

I. INTRODUCTION

The goal of this chapter is to discuss the general principles of risk assessment
and risk communication as they relate to the use of excipients in drug products.
A vital need of the user of any product is to understand its "safety," to know
when and under what conditions its employment will not be *harmful*, and corre-
spondingly when it may carry some risk. *Safety*, in this sense, is the complement
of *toxicity*, and it represents the best judgment of the circumstances when any
hazard of the product will not be realized as a *risk*.

  The particular meanings of the term involved in risk assessment and communication must be clear to understand the nature, supporting information, and
process of risk assessment. The following definitions serve to clarify the use of
these terms:

> *Harm*: change in the function or structure (pathological) of tissues or or-
> gans, or dysfunction of the person, which exceeds a normal adaptive
> response.
> *Hazard*: the intrinsic property of a substance that makes it harmful (e.g.,
> high pH causing irritancy, hepatotoxicity, and such). As a concept it is
> divorced from considerations of dose or exposure.
> *Risk*: conventionally defined as the likelihood (probability) that a substance
> will cause harm under given circumstances) (i.e., dose, duration of expo-
> sure, and any special susceptibility of those exposed). It is helpful to

extend this beyond the probability to include the nature and the severity of the harm, because what matters to those at risk is not only how likely they are to be affected, but just as strongly in what way they may be damaged and how severely; for example, a high risk of mild local irritation may be as unacceptable as even a low risk of sensitization, or a very low probability of a treatable cancer.

Safety: in the present context is the judgment that a particular exposure to a substance carries either no foreseeable risk of harm, or only such a negligible risk that it is of no practical importance. In rare instances, safety may be demonstrated by studies in humans showing lack of harm under appropriate circumstances of exposure and on investigation by relevant techniques.

Toxicity: the occurrence and nature of the harm due to exposure to a substance. It may be caused directly (e.g., local irritation at the site of application or systemic target organ damage following absorption), or it may arise indirectly, as when the toxicant reduces (or enhances) the absorption or availability of a medicine or an essential nutrient. Toxicity also carries the dimension of time. A toxic response may be acute—immediate (e.g., local inflammation); delayed—that is, appearing some time after the causal exposure (e.g., peripheral neuropathy caused by certain organophosphorus pesticides, or the development of cancer after many years); or latent—not apparent until further challenge by a toxicant, such as reexposure to the appropriate antigen after immunological sensitization. It may also be apparent in the next generation, such as a teratogenic effect on the developing fetus; in fertility, owing to an action on the gonads; or a transmissible induced mutation in germ cells.

As toxicity may also affect function (physiological, biochemical, or other) as well as structure (e.g., intestinal ulceration or renal tubular necrosis), it is really a universal term for any chemically induced harm (ionizing radiation and other physical causes of injury can be excluded for the present purpose). Delineation of its nature, time relations and consequences are necessary in considering the totality of the risk that a given exposure to a hazard may cause.

These concepts are more fully discussed in standard sources on toxicology (1–3).

The same concepts require equal consideration from a different viewpoint; namely, the philosophical and sociological ideas of ''risk,'' and the processes that underlie our approaches to its detection and assessment as essential steps before deciding whether a particular toxic risk is acceptable or unacceptable. In turn, that is linked to how we manage circumstances so that potentially harmful substances are made available and are used with what is considered as ''sufficient'' assurance of safety. This is the area of risk analysis and risk management, which goes from toxicity testing to demonstrate a hazard, through a formal ordered structure for predicting the risk to consumers and for deciding if the risk is acceptable. Following that comes the difficult but vital area of *risk communication*, concerned with how people are best informed about risks so that they can decide rationally whether to accept or reject them, and what precautions may be needed to avoid or minimize risks. In the extreme, all substances are harmful, but in the real world various means can be employed to limit the risk to a low and acceptable level—risk management.

There are very readable, formal accounts of risk and risk analysis in several monographs that deal with chemical and the better-studied engineering and natural hazards of modern life, and their position in modern society [see, e.g., the works by National Research Council (NRC, 4,5), Lowrance (6), Royal Society (7), and British Medical Association (BMA; 8)]. Popular perceptions and misperceptions of risk are discussed, for example, by BMA (8) and Slovic (9,10).

To be considered acceptable in modern society, governmental, regulatory, legal, industrial, and other public requirements about the nature and magnitude of risks must be adequately met, usually by a combination of expert judgment of experimental and human evidence and other information, plus the societal perception of how much risk of what sort is acceptable. That is closely joined to concepts of the responsibilities of the inventor, the manufacturer, the supplier, and the user of the product that creates the risk, as well as to more legalistic notions of product liability, legal negligence, labeling, consumer information, etc.

It will be apparent that *risk*, as the term is commonly employed, is not an absolute or fixed property. The probability of its occurrence will depend on exposure and individual susceptibility, and its acceptability is influenced both by perceptions and the reality of its severity and magnitude, and by the essential trade-off between risk (of injury) and benefit (of use). In conventional usage, the term *risk* covers all these quite different aspects, from laboratory detection of hazard, to a societal, official, industrial, or individual judgment of acceptability in a given circumstance (4,7–10).

II. RISK ANALYSIS

A. Steps in Risk Analysis

The formal structure of the processes of risk detection, assessment, and management has been gathered into the overall discipline of risk analysis, which comprises the sequence:

1. *Hazard identification*: consideration of physical and chemical nature of a substance and of the results of toxicity testing.

2. *Exposure assessment*: to what concentration and for how long will persons (or some other target system) be exposed.
3. *Risk identification*: nature and numbers of the exposed population and circumstances (dose or concentration and duration) of their exposure.
4. *Risk estimation*: what is the probability that harm will be produced in those individuals, and what will be its nature, including type, duration, magnitude of damage, and the possibility of recovery.
5. *Risk characterization*: matching the risk of exposure against anticipated benefit (expressed in the same units).
6. *Risk management*: if the risk is deemed acceptable, because of the linked benefit, how the risk is controlled to ensure that it is minimized.
7. *Risk communication*: how to explain the probability and nature of the harm to those likely to be affected, so that they understand the value and danger of the exposure, and behave in such a way that the risk is minimized.

This is a complex matrix of experimental, medical, societal, philosophical, financial, and ultimately politically based questions and decisions, as shown in Fig. 1. It must be appreciated that today much of the questioning and most of the decision-making is made by expert groups on behalf of the consuming public. To retain public confidence, decisions should be biased toward a considerable level of safety by minimizing any likelihood of a risk.

For excipients that, by virtue of their nature and function, are intended to be inactive, albeit technically vital in producing medicines and other products that are accepted and properly used by the public, there is a particular need for precautions to avoid substances or uses that might carry a toxic risk. It would be a disaster if use of a medicine by patients who needed the active principle were to be prevented by concern about "inactive" excipients. For this reason, the risk assessment of excipients will be biased toward a very considerable margin of safety, and even a limited suggestion or hint of a toxic risk will be evaluated with great care. Many excipients, too, are either used in the food industry, or are allied to or derived from food ingredients and additives. Safety of foods is no less important, so many of these excipients have probably already been evaluated to demonstrate a very considerable degree of safety, and anything more than a remote laboratory suggestion of toxicity, which can be strongly denied by good experimental or other results, will almost certainly result in termination of the development or use of that excipient.

B. Factors Involved in the Risk Assessment of Any Substance

Many aspects need to be considered when evaluating the risk of a particular excipient. Knowledge of the chemical and physical nature of the substance and

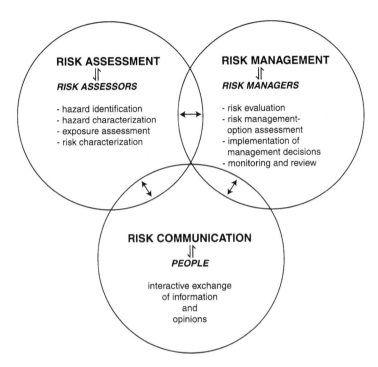

Figure 1 Framework for risk analysis.

its biological and toxicological properties help in understanding the inherent hazard of the material. The circumstances, magnitude and duration of exposure of persons are critical factors in understanding human exposure and relating it to the effects noted in animal studies at high doses. The cautious extrapolation of laboratory and clinical findings to the nature of the patients who will be exposed to the substance (e.g., during treatment if it is a component of a medicine) is the final outcome of the risk assessment process.

Other chapters in this book present critical accounts of the types of data required and the means to obtain them, so it is only necessary to briefly recall the principal points to be considered:

1. Nature of the Substance

The chemical and physicochemical nature and any known biological activities together, will indicate the chemical and biological class of the substance, and so will suggest possible toxic properties. They include consideration of purity and stability, and any likely effect on the active ingredient of the formulation of the product (e.g., adsorption or other means of affecting bioavailability), as well as the possibility of an interaction between different excipients (e.g., a chemical reaction between, say, a preservative and a solution aid, or a tablet disintegrant).

2. Dose

What is the quantity of excipient in a unit dose? What is the recommended range of doses in different subjects? What are the likely duration and frequency of courses of treatment? In other words, what is the rate and total exposure of someone to the excipient, and by what route?

3. Recipient

Is there any special feature about the formulated product and the intended recipient that may create a particular risk? For example, might a tablet to be swallowed become lodged in the esophagus because of the properties of its constituents, and could that cause local damage? Does an inhaled preparation carry any risk of immunological sensitization and thus of causing a reaction in the airways? Might the total quantity of a nonabsorbed bulking agent in an oral formulation be so high that it could act as a bulk laxative? These types of risks can be recognized for subsequent formal evaluation by a comprehensive, sequential consideration of the constituents of the product, its route of administration, and any special characteristics of persons who will consume it.

4. Toxicity

The next step is to evaluate information about the toxicity and pharmacokinetics of the excipient, as that, in addition to the dose and exposure, determine whatever risk it may pose. The consideration here must comprise any hazard of local as well as systemic exposure.

The information may come from experimentation designed to explore specific properties or a particular route of exposure to the excipient, or from the results of tests done to comply with regulatory requirements or expectations. The data must cover toxicity, pharmacokinetics, and metabolism. As many excipients have multiple industrial uses (e.g., bulking, flavoring and coloring agents may be used in confectionery, other manufactured food products, other medicines, and sometimes in cosmetics), there may already be extensive human and regulatory experience, perhaps in published form (e.g., a monograph by JECFA or the EU's Scientific Committee on Food, a pharmacopeial monograph or a U.S. "Generally Regarded As Safe" [GRAS] listing), or as a confidential Drug Master File already lodged with a regulatory agency. The availability and usefulness of published information will depend on its completeness and the age and adequacy of the studies on which it is based compared with current regulatory requirements or expectations.

A comprehensive search through the medical literature and adverse reaction databases is necessary to ensure that all possible risks have been evaluated. Whenever feasible, it is also helpful to evaluate records of occupational exposure and of the health of workers manufacturing the excipient. Conditions at work should prevent harmful effects, but if any have occurred, they would directly indicate the risk to humans, just as, conversely, the absence of reported harm is at least suggestive of safety.

C. Extrapolation and Risk Prediction

This is the most interesting and also the most difficult step, as it requires critical analysis of the findings in all the available experiments, including formal GLP-assured toxicity tests or pragmatic and perhaps more academic inquiries; information from humans, which may range from a formal clinical study or anecdotal, single-case reports; and any relevant reports from the veterinary literature.

All those results, and the "null" aspects of potential toxic actions not investigated, must then be related to the likely exposure of humans, both the recommended (or average) dose and the maximum likely dose. Given that, the risk to humans can be predicted by extrapolation from the experimental results, combined whenever possible with human experience.

The use of experimental toxicity information to predict toxic risks in humans typically involves the combination of an understanding of the dose–response relation and knowledge of exposure of humans as the target species. This accurate but simplistic statement must be heavily qualified in a number of ways:

1. Does information about the dose–response relation come from comprehensive and properly conducted experiments?
2. Were the experiments done in relevant species that are likely to respond in a way similar to humans? Findings in animals, or in in vitro tests, that come from bodily processes and cellular responses unique to a given species may not accurately indicate a relevant toxic response in humans.
3. Was the dose and the resultant exposure at a level at which the normal physiological mechanisms of the body were still operative, or did it exceed, say, the absorptive or detoxifying metabolic capacity; or did it disturb the normal nutritional state of test animals? These considerations are important because excipients, largely being toxicologically inert, can easily be fed in such quantities that they cause toxicologically irrelevant diarrhea, cecal enlargement, osmotic diuresis, or shortage of essential nutrients owing to physical displacement from the diet.
4. In addition, what is understood of the causal mechanism underlying

the toxic response? Is it known or at least likely that that process also operates in humans, or is it specific to the test species or in vitro system? For example, local irritancy is common to all species, subject to sufficient local exposure, whereas many endocrine glands in the rat respond differently from those in humans. The liver in rats and mice is susceptible to enzyme induction followed by tumor development owing to chemicals in a way different from humans; whereas experimentally induced cutaneous sensitization in the guinea pig can be a reliable indicator of that potential risk in persons. Thus, a full understanding of the test species or test system is mandatory in accurately interpreting the relevance of the findings to humans.

An essential observation about the dose–response relation in toxicity experiments, or in any available clinical reports, is the level of exposure at which no toxic or other harmful effect can be observed (i.e., the no-observed adverse effect level; NOAEL), or the more stringent level at which no response at all is seen (the no-observed effect level; NOEL). It is usually on that information, rather than on more sophisticated mathematical models, that successful risk and safety predictions are based. There are differences in the ways in which official agencies in the United States, Japan, and Europe may approach the modeling of a risk posed by a genotoxic carcinogen (1–3), but that is unlikely to be relevant to an excipient.

D. Risk Estimation

The elements required for risk estimation, based on experimental observations, and its extrapolation to humans, as the target species, comprise the following:

1. Appropriate and Valid Experiments Demonstrating Toxic Effects

Standard animal experiments are necessary to demonstrate toxic effects and the dose range over which those effects are manifest (i.e., a dose–response). Any actions considered to be physiological or pharmacological responses that are not harmful, and the relevant doses, or better plasma concentrations as an index of absorption and systemic exposure, unless the key harmful action is a topical one at the site of application, should be studied.

2. Demonstration of the NOAEL or NOEL

For most excipients, which are commonly selected for their relative biological inertness, it is quite likely that a NOEL will have been discovered, although nonspecific effects (e.g., caused by the bulk of an ingredient in the diet) may cause physiological disturbances that result in attribution of a NOAEL. However, either type of value is appropriate and acceptable.

In considering what effects have been detected in experiments, whether in vivo or in vitro, their biological importance must be considered before the NOAEL can be determined (i.e., interanimal variation and chance may also result in spurious effects and chance actions). The risk analyst must decide whether any, and if so which effects form part of the toxicity of the substance, and thus should be considered in the analysis, and which effects can be excluded as irrelevant.

3. Relevance to Humans

A further part of that analysis is the decision whether effects demonstrated are relevant to humans, or whether they should be considered as specific responses of the test species or system. This important phase is commonly labeled *weight of evidence* assessment. It combines expert knowledge and common sense.

E. Weight of the Evidence Assessment

The assessment of the various studies that make up the database on a given substance often involves assessing both positive and negative data from multiple studies and determining the relevance of each study to the overall risk. Key features include:

- Is the action known to be specific to the test species or system and not to occur in humans (e.g., the ready production of hepatic tumors in rodents exposed to substances that cause enzyme or peroxisome proliferation in the liver in certain ways; the considerable sensitivity of the dog to substances detoxified by acetylation, as it has a low capacity for this reaction; the deliberate excess of activating oxidative metabolism in S9 hepatic microsomes and the virtual exclusion of most other phase I and phase II xenobiotic metabolizing enzymes.
- Is the action related to the dose and duration of treatment and does it occur in a statistically significant proportion of animals or in vitro experiments?
- A low incidence of positive findings not related, say, to dose or duration of exposure is more likely to be a chance finding than a true toxic action.
- Is the effect internally consistent and confirmed?

In many toxicity experiments there are overlapping measures of many organ functions; hence, an abnormality in one should also be apparent in the others. Examples include the fact that a rise in one plasma transaminase level would usually be associated with an increase in other indicators of liver (or muscle) damage and corresponding histopathological changes; and changes in organ weights are commonly reflected in histological abnormalities. Related to this is the important distinction between "biological" and statistical "significance." It is common now to make so many estimations, usually with procedures of great precision, that statistically significant deviations ($p < 0.05$) are quite common findings. However, it is essential to ask whether the change is of such a magnitude that it has real importance as an indicator of an action that would harm the animal or subject (e.g., the precision of many hematological and biochemical estimations is such that differences of 1–2% are calculated to be "significant," ($p < 0.05$). In practice, they are very unlikely to be clinically meaningful, because changes of this or greater magnitude can readily be caused by minor changes in physiological state, such as variation in the time of feeding or drinking, period awake, and others.

Even if the experiments conform to GLP, it is necessary to confirm that the appropriate substance was correctly administered (purity and stability); that the formulation or vehicle alone were not responsible for the effect, and that it was not due to spontaneous degeneration or incidental disease in the test animals. In choosing a vehicle for studies on pharmaceutical excipients, it is important not to choose an existing excipient (e.g., methylcellulose, corn oil). The choice of a vehicle can be particularly challenging, for many excipients are often vehicles.

There are other key factors that may be relevant to the nature of the particular experiments done; for example, anticipated response to any positive control treatment, and no more than the usual random effects in vehicle-only controls, and lack of evidence of inadequate nutrition of animals, or of substrate exhaustion in in vitro systems.

An analogous process of evaluation to demonstrate validity should be applied to clinical reports, as findings in humans are no less subject to artifact and misinterpretation.

At the end of this stage, which may best be done formally and that certainly should be fully recorded if apparent toxic actions are to be excluded from the risk analysis, there should be a high level of confidence in the key toxic effects and their relation to treatment.

F.   What is a "Safe" Dose or Exposure for a Human?

This question requires both scientific and politicosocietal considerations, because it involves the decision whether any risk, or a risk of a particular magnitude, is to be accepted in a given circumstance. Because the decision is often made by small groups of technical and regulatory experts on behalf of exposed consumers, who are ordinary members of the public, the nature of the decision and how it is made should be as open and transparent as possible.

That statement is generally applicable to most circumstances in which toxicity may occur. For pharmaceutical excipients, which may be technically necessary, but may not directly contribute to the desired effect of the preparation, the expectation would be that they effectively carry no risk of causing toxicity in humans.

Risks will be considered to be "acceptable," or at least "tolerable," depending on the circumstances under which they occur. Thus, the acceptable risk of toxicity during treatment of a grave disease may be much higher than that during relief of a minor ailment; contrast, for example, the risks we may willingly accept in treating a headache and a heart attack. With an excipient, because it will not itself carry any direct benefit, the acceptable risk would be very low indeed, and effectively close to no realistic expectation that it will cause any harm. It is not easy to put a numerical value on this sort of risk, and it might differ depending on the nature of the risk: contrast a brief feeling of malaise or nausea and life-long immunological sensitization. In approximate terms, an acceptable risk for this type of material might be put at "1 in a million in a lifetime of exposure," which effectively means that there is no likelihood of any serious harm, and only an extremely small possibility of even a minor toxic effect (4,7–8).

G. Methods of Extrapolation to Predict Safe Exposure of Humans

Because excipients are usually biologically inert, or almost so, and because almost no risk from their use is the goal, then direct mathematical extrapolation from the dose–response curve is not a reasonable procedure. It would require careful definition of the low-dose threshold at the NOAEL and, thereby, impossibly large experiments for statistical reasons, and considerable unvalidatable assumptions about the shape of response curve and its downward extrapolation over several orders of magnitude.

Based on extensive use and validated by experience, uncertainty factors ('safety factors') are commonly employed, as is almost universally done for foodstuffs and other substances that may be present in the diet.

This method of predicting safe exposure in humans from experimental results is based on the idea that, if the exposure of persons is much lower than the NOAEL in appropriate toxicity tests in animals, there will be no appreciable risk to the human. It is usual for a factor of 100 to be applied in case of conventional forms of toxicity, and for it to be increased to 500 if there is toxicity to the fetus, or if the only experimental data is for a low effect level (LOAEL or LOEL), rather than a NOAEL (11,12). This means that the value of the experimentally determined NOAEL is divided by 100 or 500 to give the dose or exposure that is considered as not carrying any risk to humans on lifetime exposure. The strength of the evidence supporting this assumption is discussed in standard monographs on toxicology. It has been supported by a detailed review by Renwick (13), who has also confirmed the original understanding that the uncertainty factor of 100 was the multiple of separate factors of 10 to cover interspecies differences and a further 10 to account for intraspecies variability. It is suggested, too, that it may be possible to subdivide each of the values of 10 into two factors of about 3 that represent pharmacodynamic and pharmacokinetic differences.

The validity of applying this approach to the range of individuals found in the community (e.g., to children as well as to adults), has also been demonstrated (14). The principal exception that might occur would be if there were immunological sensitization to the excipient, as elicitation of an allergic response in a presensitized subject often requires a very small dose. In practice, an allergen would not be employed as an excipient, excluding that problem. It may be suggested that the very elderly, or those suffering certain diseases, might also be more susceptible to an excipient because of impairment of its metabolism or clearance by age or the illness. In practice that is almost unknown, probably because of the selection of excipients for their pharmacological and toxicological inertness or lack of activity, and the limited disturbance that, by those means, is still compatible with life.

Thus, although pragmatically based, use of the appropriate uncertainty or safety factor with good quality toxicological data has proved to be an effective and realistic means of predicting toxicological safety.

III. RISK MANAGEMENT

Once the data are available to determine the risk of a potential excipient, the process of risk management can be undertaken. Initially it will be done by the manufacturer, whose decision will result in development or abandonment of the substance as an excipient. Subsequently, the data will have to form part of an application for approval of the substance by regulatory agencies in countries where it is intended to use it. Official approval, as by the FDA in the United States or a European Commission agency or committee in the European Union, is based on formal evaluation of the manufacturing, pharmaceutical, and biological data, plus, in each case, additional consideration of the circumstances of each proposed use, including assessment of the need for the excipient in each medicinal formulation, and any influence it may have on stability and bioavailability of the formulated medicine.

Similar to the active moiety in a medicinal preparation, any indication of risk will be evaluated against the benefit (for an excipient likely to be concerned with efficiency of manufacture, release characteristics, or improved patient acceptability of a pharmaceutical) of the new substance. The value of its use will have to far outweigh any concern about potential toxicity.

Once licensed, use of the excipient and reports of adverse effects in patients, or of technical problems in manufacture and storage, will be regularly declared to the regulatory authorities, and any problems will be monitored. In principle, there is no difference between the monitoring of an excipient for adverse effects and that of the active principle of the medicine. The likelihood of a harmful effect of an excipient may be much less than that of a drug, but the need to survey and consider the possibility is no less important.

Risk management of what is almost always a bland, relatively inactive substance, such as an excipient, therefore, follows the same principles and pathway as any other substance, even though the likelihood of finding any harmful response should virtually have been eliminated by the process of development and industrial and official review. Part of this process includes whatever national legal requirements are placed on an excipient as part of the official licensing of its use; for example covering manufacture, storage, analytical specification, and formulation, as well as labeling. The latter covers both warning procedures appropriate to maintain health and safety at work, and any information that should be given to the consumer, either directly or by the health professional who prescribes the medicine containing the excipient—risk communication.

IV. RISK COMMUNICATION

There are at least two distinct aspects to risk communication. One is the need to alert the consumer to any known risk, so that he or she can take precautions, information to be given either directly or by the professionals who have prescribed and dispensed a medicinal formulation carrying the risk. Another is the general need to explain to everyone how the official process of risk evaluation and management works, so that decision making by distant groups of experts is understood and supported. A third, and quite different component is recognizing the imminence of an emergency, or worse a crisis, and how to deal with it. The latter should be too remote from any consideration of excipients to require attention here. The second point, although increasingly important in public affairs, is also not specific to excipients and does not deserve more attention here.

The importance of telling consumers what they are taking and what its possible effects may be has increased greatly in the past few years, partly because of the conscious move to greater personal choice, which comes from our growing acceptance of consumerist ideology, and in part from realization that in a democratic society the onus is on revealing everything unless a case can be made for its limited concealment.

Inspection of food packages labels and leaflets on containers of medicines will show increasingly detailed lists of ingredients, sometimes including quantitative information, or at least an indication of relative amount from the position in the list. The exact detail will depend on the country where the information is being given, for national requirements still differ.

That sort of information will at least show the consumer what is present; thus, it will permit the individual with an unusual susceptibility to take precautions not needed by the general population (e.g., the sufferer from celiac disease can avoid gluten-containing preparations). The implications are that the manufacturer must provide the information in a form understandable by members of the public, and the user of the product must make the effort to read the label and consider the information supplied. When we consider medicines in general, the practice has developed over the years, and is now a formal requirement, of providing a leaflet for the patient, which sets out in some detail the composition of the product, its anticipated beneficial and potential harmful actions, instructions on how to take it, and advice on avoiding or minimizing possible adverse actions. Although most of this information will correctly be focused on the active principle, there is certainly the possibility of also discussing any relevant effects of the excipients. It is very unusual however, to find messages about the latter type of substance, doubtless because they are selected for lack of toxicity. Understandably, the approach to foodstuffs is different, but labelling to indicate ingredients is increasingly important.

In providing information about risk and benefit, it is essential not to frighten the reader nor to blind him or her with unintelligible technical terms, while being frank about the actions and risks involved. This is difficult, because it demands clear explanations of highly technical matters in terms acceptable to a lay person. It will also be specific to each language and each country. Helpful information about various approaches is available (15–17).

V. CONCLUSIONS

Their necessary lack of pharmacological and often, also, of physiological activity means that excipients are often regarded as analogous to foodstuffs and food additives. For that reason the general processes and standards of risk assessment applied to them are much the same as those applied to foodstuffs. In fact, many of them are derived from closely comparable products used in food manufacture; for example, types of starch and cellulose derivatives, simple powder lubricants, and more or less inert wetting agents. However, as for any substance to be used as part of a medicine, scientific caution, professional standards, and regulatory demand, together force an evaluation to be made of the composition and potential toxicity of each excipient, and of its possible effects on the complete pharmaceutical preparation to exclude any direct or indirect mechanism for causing harm. The assessment of the chemical, biological, and toxicological information, taken separately and then combined into a comprehensive analysis, means that consideration of a medicinal excipient is as rigorous as that of the active drug itself. The process of characterization of the risk and its extrapolation to a safe level of human exposure is very similar in most instances to that successfully applied to foodstuffs over many years.

Communication of any likely risk to the consuming patient is likely to resemble the messages and approaches adopted for medicines as a whole, as labeling of food with warning and other messages is understandably less well developed to cope with the specific needs of individual consumers. The need to inform without frightening and to educate without confusing makes great demands on the writer of the information, and assumes that the patient will read and comprehend the messages conveyed. Given the low potential for harm by excipients, that may usually be correct, but experience of the response of patients to leaflets and labels on medicines as a whole shows that misunderstanding is easy. The risk assessor and communicator for an excipient may not have a much easier time than their colleagues dealing with a pharmacologically active drug principle.

REFERENCES

1. B Ballantyne, TC Marrs, P Turner, eds. General and Applied Toxicology, vols 1 and 2. London: Macmillan Press, 1992.
2. AW Hayes, ed. Principles and Methods of Toxicology. 3rd ed. New York: Raven Press, 1994.
3. CD Klaassen, MO Amdur, J Doull, eds. Casarett and Doull's Toxicology: The Basic Science of Poisons. 5th ed. New York: McGraw Hill, 1996.
4. NRC (National Research Council). Risk Assessment in the Federal Government: Managing the Process. Washington, DC: National Academy Press, 1983.
5. NRC. Science and Judgement in Risk Assessment. Washington, DC: National Academy Press, 1994.
6. WW Lowrance. Of Acceptable Risk: Science and the Determination of Safety. Los Altos, CA: Kaufman, 1976.
7. The Royal Society. Risk: Analysis, Perception and Management. London: The Royal Society, 1992.
8. BMA Book of Risk. Penguin: London, 1990.
9. P Slovic. Informing and educating the public about risk. Risk Anal 6:403–415, 1986.
10. P Slovic. Perception of risk. Science 236:280–285, 1987.
11. International Programme on Chemical Safety. Environmental Health Criteria 70. Principles for the Safety Assessment of Food Additives and Contaminants in Food. Geneva: World Health Organization, 1987.
12. International Programme on Chemical Safety. Environmental Health Criteria 170. Assessing Human Health Risks of Chemicals: Derivation of Guidance Values for Health-Based Exposure Limits. Geneva: World Health Organization, 1994.

13. AG Renwick. Data-derived safety factors for the evaluation of food additives and environmental contaminants. Food Addit Contam 10:275–305, 1993.
14. ILSI Europe. Applicability of the Acceptable Daily Intake (ADI) to Infants and Children. ILSI Report Series. Brussels: ILSI Europe, 1998.
15. B Fischoff. Risk perception and communication unplugged: twenty years of process. Risk Anal 15:137–145, 1995.
16. Department of Health. Communicating About Risks to Public Health. London: Department of Health, 1997.
17. VT Covello. Risk communication. In: P Calow, ed. Handbook of Environmental Risk Assessment and Management. Oxford: Blackwell Science, 1997, pp 520–541.

14

Harmonization of Excipient Standards

Zak T. Chowhan
Pharmaceutical Development Consultant, Cockeysville, Maryland

I. INTRODUCTION

International harmonization is a complex process as indicated by the failure of previous attempts (before 1989) at world harmonization of pharmacopeial standards. The pharmacopeial text, test methods, purity specifications, and limits for the impurities are not only scientific and technical, but also are legally binding. Pharmacopeias differ in procedures and policies that make harmonization efforts more difficult. Harmonization of excipients is complicated because large numbers of excipients are not a single chemical entity but are rather complex mixtures of similar chemical compounds. Many are derived from natural sources and some are synthetic polymers. They are produced by many suppliers, and many manufacturing methods are used in their production. These substances are also used in the food industry, chemical industry, cosmetic industry, and agriculture. The pharmaceutical use of many excipients is only a small part of the total business.

Although the major component of interpharmacopeial harmonization is "retrospective" (already published and official text in pharmacopeias), harmonization "prospective" (new monographs and test methods) cannot be ignored. Some of the harmonization efforts on prospective harmonization are focused on biotechnology products. Retrospective harmonization is more difficult because the standards and test methods are legally binding and the marketed drug products in the respective countries must comply with these standards.

In spite of these difficulties, the secretariats of the three pharmacopeias, the *United States Pharmacopeia (USP)*, the *European Pharmacopoeia* (EP), and the *Japanese Pharmacopoeia* (*JP*) agreed at an informal meeting in 1989 that closer cooperation be established among different pharmacopeial authorities to achieve greater harmonization among the different pharmacopeial standards. Excipients were chosen to begin the harmonization process because it was generally agreed that the harmonized standards would facilitate registration, manufacture, and shipping of the drug products around the world. The three pharmacopeias formed a voluntary alliance, the Pharmacopoeial Discussion Group (PDG), in September 1989, to work on harmonization of excipient standards and test methods.

To begin the harmonization process, a survey of excipient suppliers, users, and regulators was conducted by the *European*, *Japanese*, and *United States Pharmacopeias* in May 1990, and a second survey conducted in May 1992. As a result of the survey, the PDG prepared a rank-ordered list of the ten most important excipients for harmonization. The PDG then ranked the 25 top excipients (42 monographs: 14 cellulose derivatives, 5 starches, and so on) and tests and assays for harmonization and established a lead pharmacopeia system to make assignments among the pharmacopeias (to represent the actual function, it was changed to coordinating pharmacopeia). PDG also elaborated a seven-step harmonization procedure in October 1993, based on experiences gained in the development of harmonized monographs for lactose and magnesium stearate, which was later modified to suit the harmonization needs.

In view of the complex issues, PDG recognized at the onset of their discussions that full harmonization is a worthy goal that is not always attainable and noted that pharmacopeial harmonization does not mean unification or identical requirements. Harmonization is an evolutionary process, and there are degrees of harmonization, advancing from minimally acceptable level of harmonization to the level of complete harmony. The main emphasis of harmonization is on nondivergence of methods and specifications. Disharmony occurs if different methods are required to analyze the same characteristic or a different pass–fail criteria is required that results in different conclusions on actual sample, using the same analytical method.

Pharmacopeial Forum (*PF*), published bimonthly by USP, provides a public forum for an open revision process. Standards development goes through several distinct stages and periods for public comments. These are stimuli to the revision process, pharmacopeial previews, and in-process revision. Publication and republication in *PF* are central to the *USP* revision process.

Since early 1990, the USP elevated the priority on harmonization of excipient monographs. A harmonization procedure that would meet the needs of the three pharmacopeias had to be developed. The process of the selection of monographs for harmonization, setting priorities, assigning monographs to the pharmacopeias, the lead (coordinating) pharmacopeia system, the type of information needed in the preparation of the draft monographs, publication of the drafts for public comments, and the action needed when an impasse is reached, had to be established. Since the harmonization effort began, publication of the *Japanese Pharmacopoeial Forum* (*JPF*) by the *Japanese Pharmacopoeia* was initiated. Together with *PF* and *Pharmeuropa* (*PE*), these publications provide a vehicle for public notice and comments on harmonization proposals. Another related harmonization accomplishment is the decision by the *JP* to begin publication of *JP Supplements* in October 1993, thus providing a new mechanism for adoption of harmonized standards on a frequent basis by the *JP* to match the revision publication process of the *USP* and the *EP*.

The overall approach to harmonization embraces two phases: diagnostic and prescriptive. The diagnostic phase (see following items a and b) includes a thorough evaluation of the intent of the standard, assay method, acceptance criteria, and a comparison of the differences among the three pharmacopeias. The prospective phase (the following items c–e) involves proposing a common standard, assay method, and acceptance criteria that would meet the objective(s) and be acceptable to each pharmacopeia. The approach includes (a) objective comparison of the monograph requirements in the three compendia, *USP*, *EP*, and *JP*; (b) starting from ground zero and trying to find a scientific rationale for the standards and the test methods; (c) the coordinating pharmacopeia making a proposal and explaining the scientific basis of the proposal; (d) continuously communicating with the other compendia committees, and each compendia, in turn, publishing the proposals at different stages in their respective forums for public comments; and (e) discussing the differences in the standards and the test methods with other compendial committees and reaching a consensus.

As a result of the experience gained, the PDG developed a stepwise harmonization procedure (Table 1) that was implemented. Stage 5 was further elaborated into 5A, Provisional Harmonized Text and 5B, Consensus. Complete harmonization is not a single-stage process; it involves three different and difficult steps:

1. Dissection and scientific evaluation of the standards and test methods based on current knowledge and technology.
2. Discussion of the pharmacopeial differences in test limits and test methods leading to a consensus (or lack of it).
3. Differences in general policies of each pharmacopeia, which could lead to inclusion of a requirement in one pharmacopeia and not in the other.

When the pharmacopeials are working toward harmonization of a compendial article that may exist in different forms (different hydrates, salts, and such) and that may have more than one monograph (e.g., lactose), the pharmacopeias will harmonize on those forms on which agreement can be reached and later publish harmonized monographs for the other forms. Harmonization is not to be delayed until agreement is achieved on all forms.

The progress in harmonization continues to improve as experiences gained

Table 1 The Harmonization Process

Stage 1: Identification
  Need for PDG effort
    Priorities are assigned for monographs and general chapters
    The lead pharmacopeia is identified
Stage 2: Investigation
  Input from users, producers, and industry groups
  Standards and test methods are dissected and evaluated
  First draft prepared in the style of the lead pharmacopeia
  The lead pharmacopeia publishes the draft
Stage 3: Proposal
  The lead pharmacopeia revises the draft in view of the comments received
  The second draft is sent to the other PDG members
  PDG members inform the expert bodies and begin local inquiry of producers and
    user on all issues
Stage 4: Official inquiry
  All three PDG members publish the second draft in their periodicals
  Expert bodies review the second draft and complete local inquiries and communicate
    to the lead pharmacopeia
  The lead pharmacopeia revises the second draft and sends the text to other PDG
    members
Stage 5A: Provisional harmonized text
  The third draft is sent to the experts by PDG for review
  Remaining issues are discussed to minimize divergence
  The tentative official date is identified
  Each pharmacopeia prepares the "harmonized text" in its style and shares the
    statements of divergences
Stage 5B: Consensus
Not published in *PF*: End of Harmonization
Stage 6: Adoption
  Formal adoption process of the consensus draft begins
Stage 7: Implementation
  The "harmonized text" is published and becomes official

Source: Ref. 1.

in the past are used in establishing a complete and practical system that fits the requirements of excipient suppliers, users, and regulators. The total commitment to harmonization by PDG, industry, and the regulatory agencies is the key element to success. Several hurdles still need to be worked out, and the process continues to evolve as new experiences and problems present themselves. With the three pharmacopeias sharing the work responsibility and the assistance provided by the trade association that was formed in 1991, International Pharmaceutical Excipients Council (IPEC America, IPEC Europe, and JPEC), the remaining hurdles are expected to be overcome.

Significant challenges have been overcome and concrete progress made in "harmonizing" the most difficult and most important monographs. As new knowledge and data become available, a procedure to revise and improve the monographs that have gone through the harmonization process was established by the PDG. It was recognized that there must be full communication on revision proposals of the monographs that have completed the harmonization process. All revision proposals must go through the coordinating pharmacopeia, and this policy must be fully established. Otherwise, the monographs will become deharmonized very quickly.

II. THE COMPONENTS OF HARMONIZATION

Figure 1 outlines the components of harmonization and its complexities. Harmonization has to start from the origin of the materials. Local sources for the materials can vary for natural and synthetic materials. The origin of excipients is varied. Excipients derived from the mining industry vary based on the geographical loca-

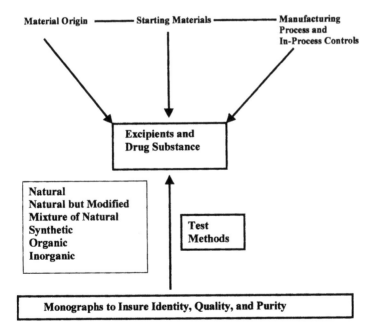

Figure 1 The components of harmonization. (From *Ref.* 1.)

tion. Differences in local conditions may dictate the need for differences in standards and analytical methods. For example, local sources of sucrose could be based on cane sugar or beet sugar. Sucrose derived from beet but not sucrose derived from cane, uses sulfitation as a bleaching process. Thus, a harmonized limit on sulfite content for the sucrose monograph may not be justified for sucrose derived from cane. Similarly, a limit on dextrin may be appropriate for cane sugar only if a similar limit is proposed for nonstarch polysaccharides for beet-derived sucrose. Another example of the local source differences is sorbitol derived from cane sugar versus sorbitol derived from beet sugar.

In addition, the manufacturing process and the in-process controls vary from manufacturer to manufacturer. For example, in the final neutralization step in hydroxyethylcellulose synthesis, one of the major manufacturers in Europe uses hydrochloric acid as a neutralizing acid and the major producer of hydroxyethylcellulose in the United States uses nitric acid. As a result, the residual levels of chloride and nitrate are different for the excipient available in Europe and in the United States. The residual limits of inorganic ions set by the *USP* and *EP* were the level of residue on ignition or sulfated ash, which is a nonspecific test. However, *EP* had an official limit of 0.1% on nitrate. The excipient produced and used in drug products in the United States had a much higher level of nitrate. These variations obviously affect the composition and the type and quantity of the residual impurities.

The other equally important aspect is the test methods that vary from qualitative, semiquantitative, to quantitative. Without harmonized test methods, harmonization of the monograph standards have very little meaning. For example, the methods for the determination of heavy metals in the *USP* and the *EP* are such that 10 ppm limit in the *USP* is equivalent to 20 ppm limit in the *EP*.

III. INTERESTED PARTIES IN HARMONIZATION

Because most pharmaceutical excipients represent only a small fraction of the total use of the article in commerce, only a fraction of the excipient producers are members of IPEC. In addition to the producers and users of excipients, the other partners in the harmonization of excipient standards and test methods are the three pharmacopeias, and regulatory agencies in the three political regions; United States, Europe Union (EU), and Japan. Figure 2 shows that communication between all interested parties is essential for the process to become successful.

IV. THE GOALS OF HARMONIZATION

The goal of harmonization is to bring the policies, monograph specifications, analytical methods, and acceptance criteria of the pharmacopeias into agreement.

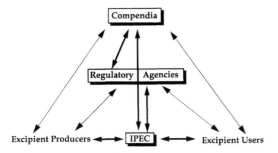

Figure 2 The interested parties in harmonization.

The policy recognizes the value of unity (i.e., a single, common set of tests and specifications, policies, and general methods), but recognizes that unity may not always be achievable. Where unity cannot be achieved, harmonization means agreement based on objective comparability and a clear statement of any differences. The goal, therefore, is harmony, not unison.

Harmonization of analytical methods is centered on the principles of analytical validation, availability of reagents, and equipment. It is essential that the methods can be applied worldwide in the same manner with appropriate supplies, equipment, and training.

The ideal situation is a single method that satisfies the criteria for validation of all pharmacopeias. The PDG recognizes, however, that it is not always possible to harmonize on a single, ideal method. International differences in analytical supplies (e.g., diatomaceous earth, column packings, reagents, and so forth) or in equipment and prospects for training may imply different analytical procedures, but in such cases results obtained must be objectively comparable. When different lots or methods yield the same comparable results, provision is made by the three pharmacopeias to allow alternative methods (subject to validation against the standard analytical procedure). If the "ideal situation" is not reached, the pharmacopeial committees inform their users of the remaining differences (e.g., legal requirements, reagents not available, and such) in their forum and continue to work toward unification.

V. CRITERIA OF IDENTITY, PURITY, QUALITY, AND STRENGTH

The excipient monographs in the compendia followed the same basic criteria as used for the monographs on drug substances. The criteria of identity, purity, quality, and strength are important for the characterization of excipients. However, many excipients are derived from botanical, animal, and mineral sources and represent variable complex mixtures of similar chemical compounds. It is a challenge to chemically characterize some of these excipients and set purity standards. Some of these excipients with variable composition have a long history of use and full chemical characterization with modern analytical techniques becomes technically difficult. With many excipients, full chemical characterization just for the sake of quality control and setting specifications with wide ranges does not provide any useful information because there are no safety issues.

VI. HARMONIZATION OF LIMIT TESTS AND STANDARDS ON IMPURITIES

The limit tests and standards for impurities in compendial monographs is intended to ensure a high level of purity of drug substance or excipient and to establish maximum content levels of specific toxic contaminants. If the contaminant can be defined precisely, the limits are expressed numerically. Otherwise, the tolerance of the limit test is given based on endpoints (e.g., a visual observation of a physical change, such as color, odor, turbidity, effervescence, or other). The limit test standards apply only under conditions in which the drug substance or excipient is customarily used. *USP/National Formulary (NF)* chapter ⟨1086⟩ Impurities in Official Articles points out the complexity of the process in setting limits for impurities in bulk drug substance and discusses eight factors that should be considered: toxicity, route of administration, daily dose, target population, pharmacology, source, duration of therapy, and cost. The precise basis of the establishment of limits in the compendial monographs is not clearly defined. In the absence of reliable data and good rationale, the result is false sense of security or unnecessary purity requirements that increases the cost of the product. The ultimate goal is to produce and supply to the patient the final drug product that is safe, efficacious, and cost-effective. Excipients are often used in drug products that are administered in different regimens, which makes the complex process of establishing limits even more complex.

Confirmation that absolute assurance of purity cannot be guaranteed is made clear in General Notices section of *USP/NF*, which states that "it is manifestly impossible to include in each monograph a test for every impurity, contaminant, or adulterant that might be present, including microbial contamination" (2). The source of these impurities, such as a change in the source of material or in the process or introduction from extraneous sources, are discussed. The suggestion is made that suitable additional tests be used if any of these changes occurs. Excipient manufacturers and users are becoming increasingly aware of the importance and consequences of any changes in the materials and processes and their implications in complying with current Good Manufacturing Practices (CGMPs).

The excipient monographs in the compendia cover gases, liquids, and solids, representing a wide variety of natural products, synthetic materials, and polymers. *USP/NF* contains approximately 300 excipient monographs that cover innumerable tests for impurities. It is not within the scope of this chapter to analyze the limit tests and standards and examine the complex problem of finding a rationale for all tests and standards in the compendia. The complex situation was examined earlier (3) and presented at the National Industrial Conference in 1978.

Three different standards: Organic Volatile Impurities (OVIs), Microbial Contamination, and Nitrate/Glyoxal Limits and their harmonization status between the *USP, EP,* and *JP* are discussed in the following sections.

A. Harmonization of Organic Volatile Impurities

1. Background

The USP Subcommittee on Chemical Purity was concerned with solvent residues and volatile contaminants resulting from synthesis, processing, and transfer among containers. In 1985, the USP Convention received a letter from the U.S. Food and Drug Administration (FDA) expressing concern that certain pharmaceuticals occasionally become contaminated with residual cleaning solvents that were not purged before refilling reusable containers.

The *USP* policy on OVIs was formulated as a result of extensive discussions at several open conferences and *PF* proposals since 1985. The following is the historical background:

At a USP Open Conference held in October 1986, a proposal was made for a new chapter on Organic Volatile Impurities ⟨467⟩. The requirements for testing OVIs would be proposed for monographs on articles that are generally administered for the systemic treatment of chronic conditions. Chronic was defined as 30 days or longer. The initial focus of the OVI testing was on seven widely used organic volatile liquids that are known to be toxic and their limits were expressed as maximum daily exposure per patient per day, as microgram of each OVI, based on gram of formulation(s). The solvents and their limits were ethylene oxide (1 ppm), benzene (10 ppm), chloroform (10 ppm), 1,4-dioxane (50 ppm), methylene chloride (100 ppm), tetrahydrofuran (100 ppm), and trichloroethylene (100 ppm).

The original subcommittee proposal and USP Open Conference (April 1988) discussion circled around a test for a dosage unit to which the patient would be exposed. It was proposed to limit the total number of micrograms of each of six residues in the finished dose unit from whatever source. The pharmaceutical industry, however, pointed out that it would be much better to control this aspect at the level of the drug substance and excipients. This would vastly reduce the total amount of testing and holding up of valuable product owing to increase in production time. The subcommittee was informed that suppliers preferred a single set of limits to individual customer-oriented limits. Thus, the subcommittee proposed the limits for the drug substance and excipients.

In 1988 (4), it was proposed to consider solvents that cause irreversible toxic effects, such as carcinogenicity, teratogenicity, and mutagenicity. New limits were based on tumor–dose bioassay studies and the proposed safety factor was 10^5. On this basis, tetrahydrofuran was removed from the list and the limits of approximately 10, 50, 100, 150, 500, and 600 ppm for ethylene oxide, chloroform, benzene, 1,4-dioxane, methylene chloride, and trichloroethylene, respectively, were proposed. The ppm limits were chosen as the equivalent of micrograms per day for 1 g of drug substance ingested.

A list of 757 monographs that would require OVI test based on the definition of chronic as longer than 30 days of consecutive therapy was published (5) in the *PF* in 1989. Only those drug substances that had chronic systemic indication received the requirement. However, there was no way of knowing which excipients were used in one or another therapeutic pattern. The general principle was that the public should be assured that unwanted OVIs having irreversible toxic effects were not present in bulk pharmaceuticals.

Three new solvents and their limits were proposed (6) in 1993. The solvents and their limits were acetonitrile (50 ppm), 1,2-dichloroethane (100 ppm), and pyridine (100 ppm). These solvents were of concern to the EP and Japanese Pharmaceutical Manufacturers Association and were consistent with the harmonization initiative.

The list of OVIs and their limits from the *USP* are given in Table 2. The *USP* lists only five solvents. As the *USP* policy on OVIs was being applied to excipient monographs, the International Pharmaceutical Excipient Council Americas (IPEC Americas) criticized the *USP* policy and considered the OVI requirements as inappropriate except when the manufacturing process for the excipients might reasonably be expected to involve organic solvents. The main issue was the requirement of OVIs for excipients that do not involve solvents in their manufacturing process, but could absorb OVIs owing to improper packaging, shipping, and handling.

In view of further discussions and to avoid unnecessary testing, the subcom-

Table 2 Official Limits of OVI

Organic volatile impurity	Limit (ppm)
Benzene	100
Chloroform	50
1,4-Dioxane	100
Methylene chloride	500
Trichloroethylene	100

Source: Ref. 7.

mittee revised chapter ⟨467⟩ Organic Volatile Impurities, which was published in the sixth supplement to the *USP/NF* and reads:

> Unnecessary testing may be avoided where a manufacturer has assurance, based on knowledge of the manufacturing process and controlled handling, shipping, and storage of an article, that there is no potential for specific toxic solvents to be present and the material, if tested will comply with the established standards. In particular, items shipped in nontight containers, within the regulations that apply to food additives [49 CFR 177.841 (e) (1), (3) (1995)], can be considered not to have acquired toxic solvents during transportation (8).

The following are the excipient monographs for which the proposed OVIs requirement in *PF* were canceled through mutual GEN/EX1 (General Chapters/ Excipient 1) Subcommittees action: Alginic Acid, Almond Oil, Butyl Alcohol, Carrageenan, Chlorocresol, Crospovidone, Ethyl Oleate, Hexylene Glycol, Light Mineral Oil, Propylene Glycol Alginate, Rose Oil, Sodium Alginate, Sodium Metabisulfite, and Squalene. The following are the excipient monographs for which deletion of the OVI requirement has been proposed by EX1 Subcommittee as part of the international harmonization effort: Carboxymethylcellulose Calcium, Lactose Monohydrate, Anhydrous Lactose, Sodium Chloride, Corn Starch, and Sodium Starch Glycolate.

The International Conference on Harmonization (ICH) draft Consensus Guideline (9), *Impurities: Guideline for Residual Solvents* were released for consultation at step 2 of the ICH Process on 7 November 1996. *Residual solvents in pharmaceuticals* were defined as organic volatile chemicals that are used or produced in the synthesis of drug substances or excipients, or in the preparation of drug products. They are not completely removed by practical manufacturing techniques. The guidelines emphasize that all residual solvents should be removed to the extent possible to meet product specifications, good manufacturing practices, or other quality-based requirements. The requirement of the removal of residual solvents is on drug products. The product should contain no higher levels of residual solvents than can be supported by the safety data.

The residual solvents are classified into four categories:

- Class 1 solvents that are known to cause unacceptable toxicities. The class 1 solvents that should not be employed are benzene, carbon tetrachloride, 1,2-dichloroethane, 1,1-dichloroethane, and 1,1,1-trichloroethane.
- Class 2 solvents (26 chemicals) that should be limited to protect the patients from potential adverse effects.
- Class 3 solvents (28 chemicals) should be used when practical.
- Additional solvents (10 chemicals) for which no toxicological data was found.

2. Harmonization Issues

Because the limits in the ICH guidelines are different for the five residual solvents in the *USP*, the first action that was approved by the USP Drugs Standards Development Executive Committee of the Committee of Revision is to change the *USP* OVI limits so that they are in line with the ICH guidelines. The *USP* limits apply to the drug substances and excipients. The ICH limits, as published in their tables, apply to formulations, although they allow another option that involves testing of the drug substance and excipients to be used in the formulation. The other issue that USP has to resolve is the number of residual solvents. ICH lists some 30 class 1 and class 2 solvents. *USP* lists only 5 of these. How many residual solvents *USP* could include is not yet clear. There are doubts about whether any of the current *USP* methods would be capable of determining benzene at 2 ppm (the ICH limit).

B. Harmonization of Microbial Requirements

1. Background

One of the major sources of microbiological contamination in the nonsterile pharmaceutical dosage forms is the bioburden of excipients. Therefore, control of the bioburden of excipients used in nonsterile pharmaceutical dosage forms is important in assessing the microbiological quality of the drug product. Excipients that are obtained by chemical synthesis have the least microbial contamination level because they do not possess necessary conditions to allow the growth of microorganisms. However, microbial contamination has occurred in the past by contamination from the package and during shipment. Excipients derived from minerals are less likely to be contaminated with microorganisms. There are, however, exceptions (e.g., talc, aluminum and magnesium salts, bentonites, kaolin, and phosphates) that can be contaminated with pathogenic microorganisms such as *Pseudomonas aeruginosa, Staphylococcus aureus, Salmonella* spp. and *Clostridium perfringens* (10–12).

Excipients obtained from animal and botanical origins, present a higher risk of contamination owing to their characteristics and to the recollection, extraction, manipulation, and storage process that they must undergo (13). Gram-negative bacteria (*Pseudomonas*); gram-positive bacteria (*Lactobacillus, Bacillus* and *Streptococcus*), and molds (*Penicillium* and *Aspergillus*) can be found in raw materials of vegetable origin, although they are not usually pathogenic agents for humans. The excipients derived from raw materials of animal origin present a very high risk of contamination, mainly by nonpathogenic enterobacteria, although pathogenic strains of *Salmonella* and *Shigella* have also been found (14).

In 1995, Rosa et al. (15) reported the analysis of a total of 115 samples of excipients: 36 lactose, 27 talc, 19 corn starch, 18 arabic gum, 8 gelatin, 3 pregelatinized starch, 3 cellulose, and 1 tragacanth gum. Fewer than 10^2 bacteria per gram were found in 69.6% of the samples and 95.2% of the samples showed fewer than 10^2 fungi per gram. Arabic and tragacanth gum were the most-contaminated products by bacteria and fungi, respectively. Pregelatinized starch, cellulose, and lactose were the least-contaminated excipients because of the manufacturing process. In none of the samples were *Escherichia coli* or *Salmonella–Shigella* spp. detected; however, strains of *Enterobactor*, *Seratia*, and *Proteus* were isolated from ten samples of five different excipients. Only five samples did not comply with the microbiological standards established by the *USP* and *EP*. A sample of corn starch did not comply because the number of fungi was greater than the established limits; and another sample of corn starch and three samples of lactose failed to comply because of the high number of aerobic bacteria.

It was noted (15) that the present microbiological requirements for excipients, established by the *USP/NF*, are insufficient and heterogeneous, which is the reason for confusion. It was also proposed that each article must comply with a different microbiological pattern relative to the total number of microorganisms as well as the absence of the specified pathogens. Furthermore, the fungi limits must be extended to every excipient, because this type of contamination can produce serious problems: the organoleptic properties can be changed and mycotoxin injuries to human health can be produced (16,17). Without good microbial quality, contamination of the finished product as well as manufacturing plant could cause problems that may be difficult to eliminate. Therefore, it is essential that the standards in the compendia be adequate.

The proposed revision to the *USP* information chapter ⟨1111⟩ Microbiological Attributes of Non-Sterile Pharmaceutical Products states that

> the drug product chemical components can be a primary source of microbial contamination. The nature and extent of microbiological testing should be based upon the knowledge of the material, its origin, how it is to be used, and historical data and experience. Materials of animal or botanical origin, for instance should receive special attention (18).

An overall index of relative level of contamination and an indication of the quality of the manufacturing process was adopted into the information chapter by establishing target values for the assessment of microbial levels for drug substance, excipients, and components. The target value for total aerobic microbial count is 1000 colony-forming units (CFU)/g (or mL), and the target value for total yeast and mold count is 100 CFU/g (mL). The target values for microbial contamination of nonsterile pharmaceutical dosage forms are based primarily on the type of dosage form (inhalant, topical, oral liquid, oral solid, or other), water activity, and its route of administration.

Table 3 Comparison of Microbial Requirements After the Harmonization Process

Excipient	USP/NF microbial limits			EP microbial limits			JP microbial limits		
	Total aerobic counts (NMT/g)	Combined yeast and molds (NMT/g)	Absence of	Total aerobic counts	Combined yeast and molds (NMT/g)	Absence of	Total aerobic counts (NMT/g)	Combined yeast and molds (NMT/g)	Absence of
Lactose monohydrate, anhydrous lactose	100	50	C	100	None	C	100	50	C, D
Magnesium stearate	1000	500	C and D	None	None	None	1000	500	C, D
Microcrystalline cellulose, powdered cellulose	1000	100	A, B, C, D	1000	100	A, B, C, D	1000	100	A, B, C, D
Sucrose	None	None	None	None	None	None	None	None	None
Povidone	None	None	None	None	None	None	None	None	None
Talc[a]	500	None	None	For topical: NMT 100 aerobic bacteria and fungi For oral: 1000	100		None	None	None
Corn starch[a]	1000	100	C, D For absorbable dusting powder: A, B	1000	100	C	None	None	None
Wheat and potato starch[a]	1000	100	C	1000	100	C	None	None	None
Rice starch[a]	1000	100	C, D	None	None	None	None	None	None
Sodium starch glycolate**	None	100	B, C, D	None	None	C, D	None	None	None
Croscarmellose sodium***	1000	100	C, D	None	None	None	None	None	None

[a] The proposals are at different stages of the harmonization process under the lead pharmacopeia.
A, *Staphylococcus aureus*; B, *Pseudomonas aeruginosa*; C, *Escherichia coli*; D, *Salmonella* species; *in JPE and not in JP, **not in JP
Source: Ref. 19.

2. Harmonization Issues

The adoption of the foregoing recommendation in the general information chapter for excipients is illustrated in Table 3. Under corn starch, for absorbable dusting powders, additional requirement for the absence of *Staphylococcus aereus* and *Pseudomonas aeruginosa* are proposed. Similarly, under talc, different requirements for the aerobic bacteria and fungi are proposed for topical and oral route of administration.

During the process of harmonization of excipient standards and test methods among *USP*, *JP*, and *EP*, the microbial requirements for each excipient are reviewed and discussed in an effort to reach a consensus on a scientific basis. As a result, there is an improvement in the microbial standards for those monographs.

Table 4 gives a summary of the microbial requirements of some excipients in the *USP* before and after the beginning of the harmonization process. For the monographs that have gone through the harmonization process, the requirements are official and for others that are at different stages of harmonization, the proposals are not yet finalized. Table 3 shows that the new specifications and proposals have considered the origin of the excipient, the route of administration, dosage form, the manufacturing process, and so on.

An important issue in the harmonization of microbial quality is the harmonization of the test methods for microbial contamination. The harmonization effort between *USP*, *JP*, and *EP* of this area is underway. From a harmonization viewpoint, this is a major concern and perhaps a high priority issue for the PDG.

C. Glyoxal and Nitrate Limits in Hydroxyethyl Cellulose Monograph

1. Background

The coordinating pharmacopeia for the international harmonization of compendial standards of hydroxyethyl cellulose (HEC) is *EP*. HEC can contain glyoxal (ethanedial), which is added to improve the dispersion of the polymer in water, and residual nitrate from the nitric acid neutralization at the conclusion of the reaction step in the synthetic process. Other acids used in the neutralization step include hydrochloric acid and acetic acid. The *USP* monograph does not limit or specify the anticaking agent and the residual nitrate and chloride are limited by the residue on ignition limit of 5.0%. During the harmonization process, *EP* proposed a limit for glyoxal to differentiate the pharmaceutical grade from the technical grade. The nitrate limit was proposed because it is in the official monograph of *EP*.

To establish the limits of glyoxal and nitrate, IPEC Americas reviewed the contents of glyoxal and nitrate in lots produced by a major producer in the United States. These lots were used by the pharmaceutical industry in manufacturing

Table 4 Comparison of the *USP/NF* Microbial Requirements Before and After the Harmonization Process

Excipient	*USP* microbial limits before harmonization		*USP/NF* microbial limits after harmonization process and *PF* proposals		
	Total aerobic counts (NMT)	Absence of	Total aerobic counts (NMT)	Combined yeast and molds (NMT)	Absence of
Lactose monyhydrate anhydrous lactose	Lactose, none	Lactose: *Salmonella* and *E. coli*	100/g	50/g	*E. coli*
Magnesium stearate	1000/g	*E. coli*	100/g	50/g	*E. coli*
Microcrystalline cellulose	None	None	1000/g	500/g	*Salmonella* spp., *E. coli.* *Staphylococcus aureus, Pseudomonas aeruginosa*
Starch corn[a]	None	*Salmonella* spp. and *E. coli*	1000/g	100/g	*Salmonella* spp., *E. coli* For absorbable dusting powder: *S. aereus, P. aeruginosa*
Wheat and potato starch[a]			1000/g	1000/g	Absence of *E. coli*
Rice starch[a]			1000/g	100/g	*Salmonella* spp., *E. coli*
Sodium starch glycolate[a]	None	*Salmonella* spp., *E. coli*	1000/g	1000/g	*Salmonella* spp., *E. coli, P. aeruginosa*
Croscarmellose sodium[a]	None	None	1000/g	100/g	*Salmonella* spp., *E. coli*
Talc[a]	Total bacterial count does not exceed 500	None	For topical administration total viable aerobic count NMT 100 aerobic bacteria and fungi/per gram For oral administration: total viable aerobic bacteria NMT 1000/g and NMT fungi 100/g	1000/g	*Salmonella* spp., *E. coli, S. aereus, P. aeruginosa*
Sucrose	None	None	None	None	None
Povidone	None	None	None	None	None

[a] Indicates that these excipients have not completed the harmonization process
Source: Ref. 19.

Table 5 Glyoxal and Nitrate Contents of Different Lots
of Hydroxyethyl Cellulose *NF*

Lot no.	Nitrate (%)	Glyoxal (ppm)
High viscosity grade		
A	4.2	218
B	3.6	120
C	4.1	131
D	3.7	186
E	3.0	148
F	4.0	105
G	3.6	128
H	3.4	126
I	3.6	122
J	3.7	116
Low viscosity grade		
A	3.1	56
B	3.7	56
C	4.3	194
D	4.7	252
E	4.4	391
F	4.2	373
G	3.6	419

drug products. The results are given in Table 5. IPEC Americas also conducted a toxicology critical review and evaluation of glyoxal and nitrate. Given the results of the toxicology review and regulatory guidelines for nitrate and sodium nitrate, acceptable levels of these chemicals in HEC intended for use in oral pharmaceuticals were estimated. In the estimation, it was assumed that the daily intake of HEC in oral pharmaceuticals would be 450 mg (150 mg/tablet intended to be given three times a day) for a prolonged time period.

2. Harmonization of Glyoxal Limit Based on Toxicity and Safety Evaluation

Because no information on the effects of orally ingested glyoxal by humans was available, the most appropriate animal studies were used to develop a no-observed adverse effect level (NOAEL) for glyoxal present in HEC. It was concluded that the most appropriate laboratory animal studies were those in which glyoxal was mixed with the animal feed and orally ingested. There are reports of a 90-day rat-feeding study with a NOAEL of 125 mg/kg per day glyoxal and a 90-day dog-feeding study in which the highest dose tested was 115 mg/kg per day glyoxal, which caused no adverse effects (20).

By applying a safety factor of 10^5 (10 for the use of animal data, 10 for individual variations in humans, 10 for the use of data from subchronic animal studies, 100 additional safety factor) to the NOAEL from the 90-day dog study, the human dose of 0.00115 mg/kg per day, or a daily dose of 0.081 mg for a 70-kg man is obtained. To ingest 0.081 mg glyoxal in 450 mg of HEC, the limit of glyoxal would be 180 ppm.

From the commercial batch analysis and the toxicology data analysis, IPEC proposed a glyoxal limit of not more than 200 ppm in HEC intended for use in oral pharmaceuticals. *EP* accepted the limit and is now harmonized with the *USP* and *JP*. The stage 5A draft recommendation of 200 ppm glyoxal limit is reportedly based on application of an extraordinarily large safety factor (100,000) which far exceeds that traditionally employed by regulatory authorities in the United States and other countries. In many countries, a safety factor of 100 or 1000 applied to a NOAEL from a subchronic animal toxicity study is considered more than adequate to derive an acceptable daily intake for humans. Thus, a glyoxal level in the range of 2,000–20,000 ppm is supportable, and is more consistent with the risk assessment practice in the United States and other countries.

After a review of the IPEC report and related information on glyoxal, Dr. Sharon Northrup, Chairperson of the USP DSD Toxicity, Biocompatibility, and Cell Culture Subcommittee, commented that glyoxal is among the numerous Maillard reaction products that are produced whenever foods are cooked (S. Northrup, personal communication, 1998). Pharmaceutical products that are subjected to heat sterilization have also been analyzed for glyoxal. Wieslander et al. (21) reported dialysis solutions, having 3.86 and 4.25% glucose, contained glyoxal and other aldehydes from Maillard reaction. In addition to human exposure from cooked foodstuffs, glyoxal is formed endogenously from the oxidation of arachidonic acid (22) and carbohydrate metabolism (23). Glyoxal is very reactive and readily undergoes reduction by the glyoxalase enzymes. The issue of chronic toxicity of glyoxal was indirectly evaluated in the IARC evaluation of the carcinogenic potential of coffee (24). The IARC conclusion led Dr. Northrup to believe that safety factors of 100,000 may be more conservative than necessary to establish a limit of glyoxal for excipients (S Northrup, personal communication, 1998). That is, the human exposure and epidemiology data likely contradict positive rodent bioassay data on methylglyoxal, a congener of glyoxal.

A major U.S. supplier of HEC manufactures a family of HEC products, with varying contents of added glyoxal. The glyoxal in these products is not a synthetic by-product or HEC degradation product. Rather, glyoxal is purposely added to these HEC products as a dispersing aid. The HEC supplier indicated that these HEC products with added glyoxal have been used in pharmaceuticals in the United States for over 10 years. Some of the dosage forms in which HEC is used in marketed products in the United States are ophthalmic solution and suspension, prompt-release and sustained-release oral tablets, otic solution, topical solution and topical sponge, and controlled-release transdermal film. The HEC supplier has suggested a glyoxal limit of 0.75% to the USP, and production data in support of this limit were provided. The USP subcommittee learned from IPEC that the member who originally suggested the 200 ppm glyoxal limit based on the 100,000 safety factor now believes that a much lower safety factor is acceptable, and specifically that a glyoxal level of 7500 ppm in HEC is safe.

3. Harmonization of Nitrate Limit Based on Safety and Toxicity Evaluation

Because one of the major producers of HEC in the United States uses nitric acid for the neutralization step, HEC containing higher levels of nitrate has been used in oral pharmaceuticals for many years. The amount of nitrate and other inorganic ions was limited in the *NF* by the total inorganic salt content in residue on ignition (not more than 5.0%). On the other hand, *EP* had a limit on sulfated ash of 4% and a limit on nitrate of not more than 0.1% based on the commercial material available in Europe.

To harmonize the limit on nitrate content of HEC, IPEC Americas carried out a critical review and evaluation of the available toxicology information for nitrate. Several guidelines for permissible levels of nitrate in drinking water and sodium nitrate in food products have been established. Among these are the Joint FAO/WHO Expert Committee on Food Additives (JECFA) acceptable daily intake (ADI) of 0–5 mg/kg per day sodium nitrate, which is equivalent to 3.7 mg/kg per day nitrate; a World Health Organization (WHO) drinking water quality guideline of 10 mg nitrate, measured as nitrogen per liter and a U.S. Environmental Protection Agency (EPA) safe drinking water criteria of 10 mg nitrate, measured as nitrate per liter; and the U.S. FDA limit of 200 ppm sodium nitrate in smoked and cured fish and meat products.

Both the WHO and U.S. EPA safe drinking water guidelines were established based on information that 10 mg nitrate per liter would be protective of the most sensitive populations (i.e., human infants), who are most susceptible to methemoglobinemia that could result from the conversion of nitrate to nitrite in the intestinal tract. The safe drinking water criteria of 10 mg nitrate per liter would result in the intake of 15–20 mg nitrate per day by adult humans weighing 70 kg who consume 1.5–2.0 L of water daily. This level of consumption is equivalent to 0.214–0.286 mg/kg per day nitrate. From these levels of permissible nitrate intake, the presence of 3.3–4.4% nitrate in HEC consumed at a rate of 450 mg/day would not be expected to cause adverse effects in humans.

There was complete agreement among IPEC Americas, IPEC Europe, and JPEC on the proposed limit of not more than (NMT) 4.4% nitrate in HEC. IPEC

Europe proposed a nitrate limit of NMT 4.4% in HEC to the lead pharmacopeia. As a result, *EP* widened the limit of nitrate from 0.1 to 0.2% in the Official Inquiry stage monograph. IPEC and USP are working with the *EP* commission, and it appears that the nitrate limit may be widened to NMT 3.0% for the low-viscosity grades of HEC.

VII. MONOGRAPH HARMONIZATION CASE STUDIES

The process of harmonization is complex, slow, difficult, and at times frustrating. However, progress has been made, and excipients that have gone through the harmonization process continue to be revised and harmonized. Other excipients that were selected for harmonization are at different stages of harmonization. Two case studies presented in this section illustrate and answer several questions, especially the following:

> Why does harmonization take so long?
> Why is harmonization so difficult?
> Why is the harmonization process so frustrating?

A. Case Study of Lactose and Magnesium Stearate

Two of the most widely used excipients, lactose and magnesium stearate, were the first selected for harmonization. The *USP* was selected as the lead pharmacopeia for harmonization of these monographs. In retrospect, these may have been the most difficult excipients with which to begin harmonization. The stepwise harmonization procedure that evolved after experiences with lactose and magnesium stearate did not exist. Also, both excipients are widely used worldwide, have long histories of use and standards development, and are produced by many manufacturers. Magnesium stearate is a variable mixture of magnesium salts derived from solid organic acids that are obtained from fats, and consists chiefly of variable proportions of magnesium stearate and magnesium palmitate. Lactose on the other hand, is a single chemical compound that exists in two isomeric forms: α and β. Spatial orientation of the hydroxyl group on the glucose C-1 carbon atom is designated as α, for axial orientation, or β for equatorial orientation. In solid state, at least four types are available. These are α-lactose monohydrate, α-lactose anhydrous stable, β-lactose, and spray-dried lactose.

The variable composition of magnesium stearate leads to variable physical properties that have been reported in the literature. The differences in physical properties from lot-to-lot and supplier-to-supplier are also caused by the type of method used in the manufacture of magnesium stearate. The manufacture of magnesium stearate may use one of two methods: (a) the melting of the starting materials, or (b) the precipitation of aqueous suspension of fatty acids and magnesium salt; flat or needle-shaped crystals are obtained, depending on the pH and other precipitation conditions.

The control of important physical properties, such as particle size, particle shape, specific surface area, and bulk density, of an excipient with variable composition is a difficult challenge. A major producer of magnesium stearate in the United States claims that physical and functional properties of magnesium stearate are influenced by the method of manufacturing, rather than by the purity of magnesium stearate or the fatty acid composition. It is also claimed that their material is predominantly monohydrate and the crystal habit is mostly plates or flakes. This material is widely used as a lubricant in the United States and has excellent lubrication properties. A recent report (25) used differential scanning calorimetry (DSC) to investigate the lubricating properties of commercial samples of magnesium stearate. The report claimed that milling, drying, and storage decreased the lubricating properties. The shear face is the long lattice of crystal. Milling breaks the crystal structure to give an amorphous form, which resulted in a decrease in the lubricating properties owing to loss of the shear face at which water or gas molecules can act. The mechanism of lubrication seemed to involve water or gas molecules, or both, entering the spaces of the crystal lattice, causing a decrease in the interactive forces of the crystal lattice which, in turn, led to easier shearing of the lubricant powder particles. According to this report, lubricating properties depended on the moisture content and the total enthalpy, as seen from the DSC peaks corresponding to desorption of water, with the latter having a greater effect. Another report (26) classified magnesium stearate into five types: type A, amorphous, no crystal water, waxy fragments; type B, crystalline, dihydrate, mostly platelets; type C, crystalline, mixture of mono- and trihydrate, often needles; type D, crystalline, mixture of mono- and trihydrate, mixture of platelets and needles; type E, crystalline, mixture of all hydrates, mixture of all shapes. The conclusion of the study was that the batches of mainly magnesium stearate dihydrate with specific area less than 5 m^2/g (air permeability) gave the best results for lubrication, crushing strength, disintegration time, and dissolution of the active ingredient.

The November–December 1988 *PF* Stimuli article (27) discussed the difficulties in setting meaningful standards for excipients, using magnesium stearate as an example. Setting meaningful standards related to quality, identity, purity, and strength of excipients derived from natural sources is not an easy task. They are either variable in composition or limited chemical composition information is available because of their proprietary nature. Excipients are used in dosage forms to perform certain functions, and functionality depends on chemical purity as well as physical properties, such as particle size, particle shape, and surface area. The determination of chemical purity of excipients that are variable mixtures is difficult. Similarly, very little attention is paid to physical tests that affect the functionality of these components of the dosage forms.

Two Stimuli to the Revision Process articles containing suggested revisions for lactose (28) and magnesium stearate (29) for harmonizing these monographs with *EP* and *JP* were published in *PF* in 1990. These articles served as focal points for discussions at a miniconference held in October 1990 at USP headquarters at which the excipient subcommittee members and American and European suppliers and users of excipients participated. Further discussions were held during the January 1991 Joint Pharmacopeial Open Conference on International Harmonization of Excipient Standards (30).

The major issues on lactose were discussed at the 1990 and 1991 meetings and served as a guide in drafting a revised Stimuli article by the subcommittee. The two approaches, individual monographs for the four available types (α-lactose monohydrate, α-lactose anhydrous stable, β-lactose, and spray-dried lactose) versus family monographs covering the most important types were discussed. A consensus was reached to delete the term α, to propose individual monographs and not to include β-lactose because of its infrequent use.

The major issues on magnesium stearate were also discussed at the 1990 and 1991 meetings and served as a guide to the Subcommittee in drafting a revised Stimuli article. Other comments on the 1990 Stimuli articles were received and reviewed by the subcommittee. In the revised Stimuli article (31), commentary on all proposals was included after each section of the monograph. Basically, every standard was dissected, justification or lack of justification was discussed, and PDG members were asked to find the rationale for the standards. Copies of the Stimuli article were sent to other PDG members for their comments. The major issues were considered in the lactose monograph proposals published in the September–October 1991 *PF* as a Stimuli article (31).

Comments to the 1990 Stimuli article on magnesium stearate were received and reviewed by the subcommittee. The major issues were considered in preparing the Magnesium Stearate monograph proposal published in the September–October 1991 *PF* as a revised Stimuli article. The *Pharmacopeial Preview* proposal on lactose were based on the comments received in response to the September–October, 1991 Stimuli article on lactose. The major revisions on magnesium stearate were based on the comments received in response to the September 1991 Stimuli article. They were used in preparing the *Pharmacopeial Preview* proposal (32) for magnesium stearate.

Two new monographs, one for lactose monohydrate and one for anhydrous lactose were published in the In-Process Revision Section of *PF* (33) as proposed harmonized replacements for the current Lactose monograph. The comments received after the publication of the Preview draft were considered in preparing the In Process Revision draft. The Lactose Monohydrate and Anhydrous Lactose monographs were published again in the In-Process Revision section of *PF* (34) for adoption in the ninth supplement to *USP 12–NF17* and became official on January 1, 1994. The Lactose monohydrate monograph also became official in *EP* in January 1994. Harmonization was completed in both monographs for monograph sections, including Title, Rubric Definition, Packaging and Storage, Clarity and Color of Solution, Protein and Light Absorbing Impurities, Acidity or Alkalinity, Specific Rotation, Heavy Metals, Water, and Residue on Ignition. A technical aspect in progress at the time for harmonization was adoption of an official compendial procedure for the determination of particle size distribution. Possible differences among the pharmacopeias involved the presence or absence of tests for loss on drying, microbial limits, and organic volatile impurities.

Comments received on the Preview (32) article on magnesium stearate were considered by the subcommittee in preparing the major revisions to the In-Process Revision (35) monograph of Magnesium Stearate. Based on the correspondence and proposals (35,36), the following harmonization issues were identified and actions proposed to the other PDG members:

- Labeling (specific surface area): This section must be deferred until a general test chapter providing a method for determining the specific surface area of magnesium stearate is also ready for adoption; thus, the consensus draft should not show a Labeling section.
- Microbial Limits: This test is included in the revision proposal because the final step in the production of magnesium stearate may involve aqueous precipitation, which could be the source of microbial contamination. The proposed requirements are consistent with the microbial count limits for excipients from natural origin (37), as contained in the proposed general information chapter ⟨1111⟩ Microbial Attributes of Pharmaceutical Raw Ingredients, Excipients, Drug Substances, and Nonsterile Dosage Forms..
- Limit for Chloride and Sulfate: IPEC Europe reported in November 1992 that the conclusion of discussions with suppliers was to retain the present *EP*. December 1992 limits of 250 ppm for chloride and 0.5% for sulfate. However, *PE*'s Revised Magnesium Stearate draft specified limits of 0.1% for chloride and 0.3% for sulfate. These *PE* limits were proposed in the July–August 1993 *PF* In-Process Revision draft (35). The proposed sulfate limit was revised to 1.0% in the January–February 1994 issue of *PF* (36) based on data that the previously proposed limit of 0.3% is not characteristic and reasonable for material of commerce prepared by certain precipitation processes. The other PDG members were notified of this change.

A February 1994 response from the EP commission indicated that "The pharmacopeia limits that are currently in force (250 ppm for *Chloride* and 0.5% for *Sulfate*) have not given rise to any criticism so far." The PE commission also stated in this response that "The limits for *Chloride* and *Sulfate* could be set at 0.5% at most for each of these." Having considered the foregoing adverse comment from EP regarding the chloride and sulfate limits proposed in *PF* (35), the USP Subcommittee concluded that both the Limits of Chloride and Limits of Sulfate tests, harmonized for methods, can be established. Sulfate and chloride levels of 1.0% and 0.1%, respectively, are unobnoxious in terms of daily intake, especially when the dosage forms contain only 0.1% to a maximum of 2% magnesium stearate. Thus, inclusion of these tests in the monograph may serve only as tests for usual manufacturing process cleanup control.

As a part of the subsequent harmonization efforts, the EX1 Subcommittee again reviewed the sulfate limit topic. Both the current limit of 0.5% based on the corresponding limit of *EP* monograph and the 0.3% limit suggested by the EP commission were considered. The subcommittee is not aware of any safety issues involving sulfate content in magnesium stearate and considers the sulfate limit of 0.3% suggested by EP to be a potential lock-out specification. The sulfate test and limit in the monograph seems to serve only as a test for usual manufacturing process cleanup control.

On the basis of this information, the excipient subcommittee approved a sulfate limit of 0.5% for the Magnesium Stearate Stage 5B draft monograph. This limit is based on the precipitation manufacturing process capability necessary to produce magnesium stearate that is consistent with its long history of application as the most effective lubricant in tablet and capsule manufacturing and has no safety issues.

1. Limits for Cadmium, Lead, and Nickel or Lead or Heavy Metals

A major U.S. supplier of magnesium stearate submitted comments to the USP on testing magnesium stearate for cadmium (Cd), lead (Pb), and nickel (Ni); testing should be done on the magnesium source, not on the stearate salt. This was discussed (38) with EP and *JP* members at the Second Joint Pharmacopeial Open Conference on International Harmonization of Excipient Standards. A salt is considered to be an entity and not two separate ions by the PDG (i.e., the prescribed tests do not relate to the cation on one hand and the anion on the other, but to the whole molecule). Cd, Pb, and Ni cations may come from various sources. Nickel is used as a catalyst during the hydrogenation of fatty acids and Cd may be a contaminant. Thus it is the position of the PDG that, as a matter of principle, magnesium stearate, not magnesium, should be tested for Cd, Pb, and Ni.

The flameless atomic absorption (AA) spectrophotometric procedure first appeared in *PF* in September–October 1991 Stimuli (31) article and then in a July–August 1992 *PF* Previews (32) draft. A more detailed AA procedure, based on similar procedures in other *USP* monographs, was published in the July–August 1993 *PF* In-Process Revision draft (35). Three different laboratories, including the USP laboratory evaluated this procedure. In the past 7 years, during the harmonization of magnesium stearate monograph, USP has gone around in a circle by proposing to replace the Pb test by the test on Pb, Cd, and Ni and after public comments and discussions replacing these tests by the Pb test. The Drugs Standards Development Executive Committee after considering the public comments decided to adopt the Pb test and limit.

These recommendations for the introduction of test limits for Cd, Pb, and Ni were based on test results showing high levels of Cd and Ni in some technological-grade magnesium stearate samples from a worldwide source. Limits of 3 ppm Cd, 10 ppm Pb, and 5 ppm Ni were suggested. These suggestions were subsequently proposed as a draft under *Pharmacopeial Previews* (32) and as a proposal under In-Process Revision (35) in *Pharmacopeial Forum* for harmonized magnesium stearate monograph.

Comments received in response to the proposed limit tests for Cd, Pb, and Ni focused on the significant capital costs of acquiring graphite furnace AA spectrophotometers, the added number of analyst hours required to perform these tests, the costs of having the test performed by outside laboratories, and the need for these tests from a toxicity standpoint. One correspondent inquired as to whether any specific toxicity problems have been directly linked to magnesium stearate contaminated with these metals. Another correspondent wondered what problems would be resolved if these new tests were implemented, presuming that commercially available material could meet the proposed limits based on comments received, the subcommittee decided to review this issue from a toxicity–safety viewpoint.

2. Limits Based on Toxicity–Safety Evaluation

The review plan was to estimate (39) the maximum daily intake of Cd, Pb, and Ni in a worst-case setting (i.e., under conditions of maximum daily dosing of three currently marketed pharmaceutical products formulated with above-average levels of magnesium stearate containing maximum proposed levels of these metals). From a comparison of these maximum daily intake values with documented toxicity values for Cd, Pb, and Ni, a decision can then be made about the need, from a safety–toxicity viewpoint, for including these tests in the magnesium stearate monograph.

Literature toxicity values, provisional tolerable total intake levels (PTTILs), and no-observed adverse effect levels (NOAELs) obtained from the *Federal Reg-*

Table 6 Worst-Case Daily Intake of Magnesium Stearate

Capsule	Fill (mg)	Magnesium stearate level (mg) 4%	5%	Maximum daily dose + 25% no. capsules	Magnesium stearate intake (mg)
A	232.9	9.32	—	7.5	69.9
B	170.5	6.82	—	5.0	34.1
C	—	—	13.5	5.0	67.5

Source: Ref. 39.

ister and the EPA Integrated Risk Information System (IRIS) database are as follows:

Pb: 75 µg/day (25 µg/day for pregnant women)
Cd: 10 µg/kg per day (400 µg/day)
Ni: 20 µg/kg per day (800 µg/day)

(Note: The Pb value is based on the PTTIL proposed by FDA in the 4 February 1994 *Federal Register*. The Cd and Ni values in parentheses are values for a 40-kg "adult," to simulate a worst-case situation.) The data are summarized in Tables 6 and 7. These results indicate that the worst-case Cd, Pb, and Ni daily intake levels (capsule A containing 4% magnesium stearate) are far below the reported PTTILs or NOAELs. The Pb PTTIL for adults is more than 100 times higher than the maximum daily Pb intake level from capsule A and the lead PTTIL for pregnant women is more than 35 times greater than the maximum daily Pb intake level from capsule A (see Table 7). Likewise, the Cd NOAEL is more than 1900 times higher than the maximum daily Cd intake level, and the Ni NOAEL is almost 2300 times higher than the maximum daily Ni intake level.

Table 7 Worst-Case Daily Intake of Lead, Cadmium, and Nickel

Capsule	Lead (µg) 10 ppm	Cadmium (µg) 3 ppm	108 ppm	Nickel (µg) 5 ppm	215 ppm
A	0.70	0.21	7.55	0.35	15.03
B	0.34	0.10	—	0.17	—
C	0.68	0.20	7.29	0.34	14.51

Source: Ref. 39.

The California Environmental Protection Agency, through the Safe Drinking Water and Toxic Enforcement Act of 1986 (Proposition 65), has adopted an acceptable daily intake level of 0.5 μg for Pb. This level represents the NOAEL for Pb divided by 1000. Based on a 69.9-mg–maximum daily magnesium stearate intake with capsule A (see Table 6), the Proposition 65 daily Pb limit of 0.5 μg corresponds to 7.15 ppm (0.5 μg/0.0699 g) for Pb in magnesium stearate. A proposal to tighten the Pb limit in magnesium stearate monograph from the current 10 ppm value to 7 ppm would, therefore, seem consistent with the Proposition 65 limit. It might also be appropriate to consider a more conservative limit of 5 ppm. In either case (7 ppm or 5 ppm), the current "wet chemistry" procedure under the *USP/NF* general test chapter ⟨251⟩ would be sufficiently sensitive to be applicable to magnesium stearate. Five *USP/NF* monographs have lead limit of 5 ppm, and one *USP* monograph—Calcium Carbonate—has a lead limit of 3 ppm, all determined by the procedure under chapter ⟨251⟩ Lead.

A report of 108 ppm Cd or less and 215 ppm Ni in some technical-grade magnesium stearate samples were submitted to USP. The Cd content is reportedly a result of cross-contamination, and Ni reportedly may be present because of its use as a catalyst in hydrogenation process. Even at these Cd and Ni levels, the Cd NOAEL is more than 50 times greater than the 7.55 μg–maximum daily Cd intake level (400/7.55), and Ni NOAEL is also more than 50 times greater than the 15.03-μg–maximum daily nickel intake level (800/15.03; see Table 7).

Given the analysis, it was concluded (39,40) that from a safety–toxicity perspective, the data do not support the need for inclusion of AA tests for Cd, Pb, and Ni in the magnesium stearate monograph. This analysis is based on the presumption that the trace metals are ingested only from magnesium stearate. However, most ingested trace metals that contribute to the daily intake could come from other sources.

To find a rationale approach to establishing safety- and toxicity-based test limits, USP initiated a survey of all official USP correspondents to obtain maximum daily intake data for 28 frequently used excipients. Based on the survey results, a hypothetical product containing four ingredients in a formulation was examined by the individual component and composite component approaches (41).

Table 8 gives the daily intake of Pb from the daily intake of the formulation based on the individual component approach. The contribution of Pb from the four ingredients is calculated from the current limits specified under ⟨251⟩ Lead in the *NF* monographs for the four ingredients. These individual ingredients do not provide Pb levels that exceed the 75 μg/day PTTIL for Pb, as provided in the *Federal Register*. However, the total daily intake of Pb from this formulation, based only on the four ingredients, is 108.5 μg in the worst case, which exceeds the 75 μg limit by 33.5 μg.

Table 8 Daily Lead Intake from the Survey Products

Ingredient	Maximum daily intake of ingredient (mg)	Compendial heavy metals/ lead limit (%)	Maximum daily intake lead (μg)
Ingredient A	7803	0.0003	23.4
Ingredient B	6624	0.001	66.2
Ingredient C	3132	0.0005	15.7
Ingredient D	315	0.001	3.2
			Total = 108.5 μg

Source: Ref. 41.

Table 9 gives the PTTIL-based lead limits based on the composite component approach. This approach requires that the total Pb contributed by the four components does not exceed the PTTIL limit of 75 μg Pb. These limits are computed, for example, for ingredient A, the contribution of 23.4 μg Pb is revised by multiplying with 75:108.5 ratio. The limits based on the California Proposition 65 requirement of 0.5 μg/day of Pb per product are much lower. Based on this limit, the resulting Pb limit for ingredient A is 0.000001% (10 ppb) (Table 10). This value is at or close to the lowest reasonably achievable detection limit with graphite furnace AA instruments. To readily achieve readings in this 10 ppb region, it is reportedly necessary to use an inductively coupled plasma spectrophotometric procedure.

Comments to the foregoing approaches of establishing safety- and toxicity-based test limits, were published in *PF* (42). It was pointed out that toxicity cannot be the sole criterion for testing and limiting impurities in excipients. One

Table 9 PTTIL-Based Limits

Ingredient	Maximum daily intake lead based on PTTIL (μg)	Maximum limit (%)
Ingredient A	16.2	0.0002
Ingredient B	45.8	0.0007
Ingredient C	10.8	0.0003
Ingredient D	2.2	0.0007
	Total = 75.0 μg	

Source: Ref. 41.

Table 10 Proposition 65-Based Lead Limits

Ingredient	Maximum daily intake lead based on Calif. proposition 65 (μg)	Maximum limit (%)
Ingredient A	0.11	0.000001
Ingredient B	0.30	0.000004
Ingredient C	0.07	0.000002
Ingredient D	0.02	0.000006
	Total = 0.5 μg	

Source: Ref. 41.

concern related to the presence of metallic impurities in excipients is the potential effect of these impurities, notably iron, on stability of the formulation. This physicochemical incompatibility concern is product formulation-specific. If there are no toxicity concerns associated with the presence of metallic impurities in a given article, should the control of the presence of these metallic impurities for physicochemical incompatibility reasons, perhaps by graphite furnace AA procedure, be within the purview of the compendia, or should they be addressed through individual chemical supplier–pharmaceutical manufacturer purchase specifications. If there are no toxicological concerns to reduce the level of impurities, it was emphasized (42) that the limits should be based on the current state of technology, but with a view to ensuring adequate margin of safety to compensate for production-related fluctuations.

The "harmonized" monograph of magnesium stearate adopted by the *USP/NF* requires a limit of not more than 10 ppm Pb. The excipient subcommittee was not convinced that tight limits on Ni and Cd were warranted. Although there was a harmony among the three pharmacopeias, the unresolved point is that *EP* in their stage 5A draft requires a 20 ppm limit for heavy metals. For the sake of harmonization, the EX1 Subcommittee concentrated efforts on a proposal to delete the lead test from the *NF* monograph and add the *JP* heavy metal test with modification as per *PF* Blake Stimuli article (43) with a 20-ppm heavy metals limit to the Magnesium Stearate Consensus Stage 5B draft monograph. It was thought that the heavy metals test is broader-based in its coverage than the current lead test. Substances that typically respond to the *USP* heavy metals test include lead, mercury, bismuth, arsenic, antimony, tin, cadmium, silver, copper, ruthenium, and molybdenum. A major U.S. supplier of magnesium stearate reported their adoption of the heavy metals test for testing of their material. They expressed concerns about the safety and stability of thioacetamide and cited the fact that the *JP* test method uses "self-generated" hydrogen sulfide, rather than requiring use of hydrogen sulfide compressed gas cylinder. Further, the JP committee indicated in their 21 March 1994 letter that they preferred adoption of the heavy metals test, rather than the lead test, in the magnesium stearate monograph.

As USP was proposing to harmonize the heavy metals test, it was found that the EP commission proposed in *Pharmeuropa* 10.1 the replacement of heavy metals test in the monograph with AA test for Cd, Pb, and Ni. USP has asked the Toxicity, Biocompatibility, and Cell Culture Subcommittee and the EX1 Subcommittee to review this dilemma and propose a scientifically sound approach.

The USP adopted the monograph on 15 August 1994 in the 10th Supplement of *NF 17*. JP adopted the "harmonized monograph" on 15 December 1994 in 2nd Supplement of *JP12*. The monograph was reverted to stage 5A of the harmonization process. The following are the unresolved issues of harmonization between the *USP* monograph and the stage 5A draft.

> *Definition* section: *NF* "edible source"
> *ID* test A in *EP*: not in *NF*
> *ID* test B in *EP*: not in *NF*
> *Sulfate limit*: 1% in *NF* versus 0.3% *EP*
> *Heavy metals* test in *EP* (10 ppm) versus lead test in *NF* (10 ppm)
> *Mold and yeast* limits and *Salmonella* specifications in *NF* and not in *EP*
> *Packaging*: NF "tight" versus "well-closed" in *EP*
> *Fatty acid composition* conditions differ in *EP* and *NF*
> *Specific surface area NF* method not in *EP*

Several revisions to the "harmonized" lactose monographs were made in further efforts to harmonize them. The monograph was reverted back to the consensus stage and the drafts (stage 5B) were submitted by the USP to EP and JP in March 1998. Noteworthy points are:

- Microbial Limits: The requirement for the absence of *Salmonella* was deleted. However, the requirements for the count of molds and yeast was retained because a total aerobic count will not detect many of these molds and yeasts. The deletion of this requirement would require the development of a single-step method and medium that detects both bacteria and molds and yeast effectively.
- Loss on Drying and Water: Both tests were retained for the Lactose Monohydrate monograph. The water in lactose originates from water of crystallization and free adsorbed water. Karl Fischer titration is used for the water of crystallization and loss on drying is used for free adsorbed water. The *JP* condition of 80°C and 2 h are adequate to determine free and adsorbed water.
- Content of α- and β-Lactose Anomers: The silylation procedure and derivatization procedure specified in the *NF* monograph are proposed

in the stage 5B consensus draft. Changes proposed for the silylation reagent and derivatization procedure section of this test are intended to facilitate the dissolution of the anhydrous lactose sample by separate use of dimethyl sulfoxide as the dissolution solvent. Most commercial anhydrous lactose samples contain a high percentage of β-anomer and require prolonged and vigorous mixing to dissolve if the current silylation reagent mixture is used as the dissolution solvent.

B. Case Study of Microcrystalline Cellulose Harmonization

The *USP* is the coordinating pharmacopeia for the international harmonization of compendial standards for the Microcrystalline Cellulose (MCC) monograph, as part of the process of international harmonization of monographs and general analytical methods of the *USP*, *JP* and *EP*. The harmonization efforts on this monograph began in 1991 with a mini-conference of microcrystalline cellulose suppliers–users at USP headquarters.

A MCC revision draft based on 1991 meeting discussions was published as a Stimuli article (44). A revised draft, based on comments received in response to the Stimuli article was published in the Preview section of the July–August 1993 issue of *PF* (45). EP published the preview draft in September 1993 issue of *Pharmeuropa* (46). The *PF* Preview draft was also published in the October 1993 issue of *JP Forum* (47). A subsequent draft was published in the July–August 1994 issue of *PF* (48) and in the July 1995 issue of *JP Forum* (49). USP notified EP and JP in November 1994 of its intent to adopt this "harmonized" monograph in the *Second Supplement to USP 23* and to *NF 18*. A chronological summary of MCC revision activities was submitted to the EP commission and JP in February 1995. The monograph became official in the *NF* on May 15, 1995. The July–August 1994 *PF* (48) harmonized monograph draft was submitted to EP in June 1994, but EP did not publish the draft in *Pharmeuropa*. Instead, EP published (50) a different draft, one adopted by the European Pharmacopoeia Commission, in the March 1996 issue of *Pharmeuropa*, and adopted this draft on July 1996.

In April 1996 a major producer of MCC submitted their concerns about the *Pharmeuropa* 1996 draft to the USP and to EP committees. The primary concern related to the absence of the requirements for labeled parameters as well as the overall degree of harmonization. The labeled parameters were considered important because they address the need of varying requirements for physical characteristics.

The EP response underlined the philosophy of the European Pharmacopoeia Commission: to provide quality specifications for chemical and microbiological quality of excipients, but not to include functionality-related testing in the legally binding parts of the monograph, because technological quality of a raw material is not a question of public safety. The specifications related to the safeguard of the patient are purity and quality of the excipient. According to EP, the functionality or better characterization of the physical properties of the excipients is not a matter of public health and should be covered by an agreement between the manufacturer and the user of an excipient (pharmaceutical manufacturer).

In principle, the thought process about dealing with the functionality issue between the USP and EP commissions is the same. Both compendia are against specifications for functionality-type tests and both compendia agree that the functionality-type issues should be discussed between the excipient manufacturer and the excipient user. In practice, this is already done. The USP decided to develop standardized methods so that the excipient manufacturer and the excipient user could interpret each others data. The labeling requirements of the *USP* apply only when there is an agreement between the excipient producer and the excipient user to meet certain agreed on test criteria.

On other differences in the "harmonized" monographs, the EP commission disagreed that there are more differences in the monographs after the harmonization process than there were before the initiation of harmonization. The monograph adopted by EP in July 1996 differs significantly from the *NF* monograph. In fact, the *USP* and *EP* "harmonized" monographs are more different than they were before the harmonization process started. This is unfortunate, especially because the USP excipient subcommittee worked with the worldwide producers and users of the excipient and made considerable improvements in the monograph. These improvements were made to ensure identity, quality, purity, and consistent physical properties that control functionality of the excipient.

Because of the differences in the "harmonized" monograph, and because the formal harmonization procedure was not developed when the harmonization effort started on lactose, magnesium stearate, and microcrystalline cellulose, PDG at the December 1996 meeting agreed that microcrystalline cellulose and powdered cellulose monographs were at stage 5. USP has submitted a revised stage 5A draft of these monographs to the *JP* and *EP* commissions.

VIII. CONCLUSION

The progress in harmonization will improve as previous experiences are used in evolving a complete and practical system that fits the requirements of the excipient suppliers, users and regulators. The ambitious and complex nature of the work the *USP* staff and the Excipients Subcommittee have undertaken, and the difficulties and obstacles that had to be overcome have been illustrated in case studies in this chapter.

REFERENCES

1. ZT Chowhan. Pharm Technol 21:76, 1997.
2. The United States Pharmacopeial (USP) Convention, Inc. United States Pharmacopeia 23/National Formulary 18. Rockville, MD: The USP Convention, 1997.
3. ZT Chowhan. Regulatory controls of pharmaceutical excipients. Proceedings of the 18th Annual Conference on Pharmaceutical Analysis. Madison, WI: University of Wisconsin Extension Services, Pharmacy and Health Science Unit, 1978.
4. Pharmacopeial Forum 14:4570, 1988.
5. Pharmacopeial Forum 15:5256, 1989.
6. Pharmacopeial Forum 19:6306, 1993.
7. ⟨467⟩ Organic Volatile Impurities. In: United States Pharmacopeia 23/National Formulary 18. Rockville, MD: The USP Convention, 1997, p 1747.
8. ⟨467⟩ Organic Volatile Impurities. In: Sixth Supplement to the United States Pharmacopeia 23/National Formulary 18. Rockville, MD: The USP Convention, 1977, p 3766.
9. Impurities: Guidelines for Residual Solvents. Released for Consultation at Step 2 of the ICH Process on 7 November 1996 by the ICH Steering Committee.
10. X Buhlmann, M Gay, HU Gubler, H Hess, A Kabay, F. Krusel, W Sackman, I Schiller, S Urban. Microbiological quality of pharmaceutical preparations. Am J Pharmacol 144:165–185, 1972.
11. D Kruger. Ein Beitrag zum Thema der mikrobiellen Kontamination von Wirk- und Hilfsstoffen. Pharm Ind 35:569–577, 1973.
12. G Sykes. The control of microbiological contamination in pharmaceutical products for oral and topical use. J Mond Pharm 14:78–81, 1971.
13. ML Garcia Arribas, MA Mosso, MC De La Rossa, de Iriarte Gaston. Relacion entre la contaminacion microbiana de medicamentosde administration oral y la naturaleza de sus componentes. Ciencia Ind Farm 2:376–378, 1983.
14. E Underwood. Ecology of microorganisms as it affects the pharmaceutical industry. In: WB Hugo, AD Russel, eds. Pharmaceutical Microbiology. Oxford: Blackwell 1992, pp 353–368.
15. MC de la Rosa, MR Medina, C Viver. Microbiological quality of pharmaceutical raw materials. Pharm Acta Helv 70:227–232, 1995.
16. H Hiticoto, S Morozumi, T Wauke, S Sakai, and H Kurata. Fundamental contamination and mycotoxin detection of powdered herbal drugs. Appl Environ Microbiol 36:252–256, 1978.
17. G Suarez Fernandez, M Ylla-Catala. The formation of aflatoxines in different types of starches for pharmaceutical uses. Pharm Acta Helv 54:78–81, 1979.
18. Pharmacopeial Forum 22:3098, 1996.
19. ZT Chowhan. Pharm Technol 21(12):60–64, 1997.
20. OECD, SIDS Profile for High Production Volume (HPV) Chemicals. Glyoxal, CAS 107-22-2, summary of responses to the OECD Request for Available Data on HPV Chemicals. May 1992
21. A Wieslander, G Fozsback, E Svenson, T Linden. Cytotoxicity, pH, and glucose degradation products in four different brands of PD fluid. Adv Peritoneal Dial 12: 57–60, 1996.

22. A Mlakar, G Spiteller. Previously unkown aldehydic lipic peroxidation compounds of arachidonic acid. Chem Phys Lipids 79:47–53, 1996
23. KJ Wells-Knecht, E Brinkmann, MC Wells-Knecht, JE Litchfield, MU Armed, S. Reddy, DV Zyzak, SR Thorpe, JW Baynes. Nephrol Dial Transplant 11(suppl 5): 41–47, 1996
24. International Agency for Research on Cancer (IARC). Monographs on the evaluation of carcinogenic risks to humans. IARC Monogr 51:41–206, 1991.
25. Y Wada, T. Matsubara. Powder Technol 78:109, 1994.
26. KJ Steffens, and J. Koglin. Manuf Chem pp 16, December 1993
27. ZT Chowhan. Pharmacopeial Forum 14:4621, 1988.
28. ZT Chowhan. Pharmacopeial Forum 16:1281, 1990.
29. ZT Chowhan. Pharmacopeial Forum 16:999, 1990.
30. Joint Pharmacopeial Open Conference on International Harmonization of Excipient Standards, 1991, p 48.
31. ZT Chowhan. Pharmacopeial Forum 17(5):2419, 1991
32. Pharmacopeial Previews. Anhydrous lactose, lactose monohydrate, lactose monohydrate modified, and magnesium stearate. Pharmacopeial Forum 18(4):3591, 1992
33. In-Process Revision. Anhydrous lactose and lactose monohydrate. Pharmacopeial Previ 19(3):5325, 1993
34. In-Process Revision. Anhydrous lactose and lactose monohydrate. Pharmacopeial Forum 19(4):5751, 1993.
35. In-Process Revision. Magnesium stearate. Pharmacopeial Forum 19(4):5754, 1993.
36. Pharmacopeial Forum 20(1):6889, 1994.
37. Pharmacopeial Forum 18(4):3596, 1992.
38. Second Joint Pharmacopeial Open Conference on International Harmonization of Excipient Standards. St. Petersburg Beach, FL. January 30–February 2, 1994.
39. ZT Chowhan, WL Paul, LT Grady. Magnesium stearate—proposed limits for cadmium, lead, and nickel. Pharmacopeial Forum 21(1):157, 1995.
40. ZT Chowhan. Pharm Technol 19(8):43, 1995.
41. WL Paul. Excipient intake and heavy metal limits. Pharmacopeial Forum 21(6): 1629, 1995.
42. D Jakel, H Ludwig, M Rock, M Thevenin. Impurity tests and impurity limits for pharmaceutical excipients. Pharmacopeial Forum 21(6):1641, 1995.
43. KB Blake. Harmonization of the *USP, EP,* and *JP* Heavy Metals Testing Procedures. Pharmacopeial Forum 21(6):1632, 1995
44. ZT Chowhan. Revision of microcrystalline cellulose monograph and powdered cellulose monograph. Pharmacopeial Forum 18(2):3194, 1992.
45. In-Process Revision. Microcrystalline cellulose and powdered cellulose. Pharmacopeial Forum 19(4):5596, 1993.
46. Microcrystalline cellulose and powdered cellulose. Pharmeuropa 5(3):232, 1993.
47. Microcrystalline cellulose and powdered cellulose. JP Forum 2(4):11, 1993.
48. In-Process Revision. Microcrystalline cellulose and powdered cellulose. Pharmacopeial Forum 20(4):7767, 1994.
49. Microcrystalline cellulose and powdered cellulose. JP Forum 4(3):380, 1995.
50. Cellulose, microcrystalline. Pharmeuropa 8(1):122, 1996.

Index

Abbreviated new drug applications
    (ANDA), 80
Absorption, 269–271
    excipients, 288–289
Absorption bases, 13
Acceptable risk, 315
Acute dermal toxicity assay, 158–
    159
    protocol design, 158
    skin scoring, 158–159
Acute inhalation studies, 189–191
Acute systemic toxicity
    dermal excipient safety program,
    156–167
Acute toxicity data
    new excipients, 109–113
Acute toxicity intranasal studies, 196
Acute toxicity parenteral tests, 215,
    217–218
Acute toxicity studies
    data interpretation, 277–278
    eye irritation, 112
    ophthalmic excipients, 233–234,
    237–238
    parenteral testing, 215, 217–218
    vaginal excipient studies, 252–254

ADD, 289–290
Additives, food
    pharmaceutical excipients, 30
ADME-PK studies, 114–116
    parenteral tests, 216, 219–220
ADME studies
    dermal excipients, 149–155
        alternative studies, 150–153
        multiple dose, 154–155
        single dose, 153–154
        in vitro human skin transport, 150–
        153
    ophthalmic excipients, 234
    oral/parenteral, 155
Adulterated
    defined, 75
Aerosol studies
    intranasal studies, 201
Afebril, 64
Agar overlay cytotoxicity test
    ophthalmic excipients, 236–237
Age, toxicokinetics, 269
Aivcel PH MCC, 5
Albino rabbits, ophthalmic excipient
    studies, 234–236, 238
Alcohol, asthma products, 66

Althea (marshmallow root), 61
Ames test, 113–114
ANDA, 80, 82–83
Andersen impactor, 195
Anhydrous lactose, 6
Anhydrous lanolin, 13
Animal models
    dermal excipient safety program,
        147–148
    intranasal studies, 199–202
    ophthalmic excipient studies, 234–
        236
    oral pharmaceutical excipient tests,
        130–131
    parenteral tests, 226
    rectal excipient studies, 260
    safety evaluation guidelines, 105
    28-day toxicity, 167
    vaginal excipient studies, 248–
        252
Annex of Directive 91/507/EEC, 96
Antiadherents, 11
Antimicrobial agents, 15–16
Antioxidants, 15
Aqueous nasal solutions, 16–17
Aqueous vehicles, 15
Aspartame toxicity, 69
Assays
    acute dermal toxicity, 158–159
    bacterial reverse mutation, 197
    cytotoxicity
        TSS excipients, 166
    mammalian in vitro, 197–198
    mouse local lymph node
        delayed-type contact hypersensitiv-
            ity, 198
    MTT cell viability
        TSS excipients, 166
    USP elution
        TSS excipients, 166
Asthma products
    alcohol content, 66
AT, 289–290
Augmentin, 69
Average daily dose (ADD), 289–290
Average time (AT), 289–290

Bacterial reverse mutation assay, 197
Base set tests, 108–117, 111t
    new excipients, 109–114
    parenteral studies, 215–216
    tiered-testing strategy, 108–109
    toxicokinetics, 114–117
Benzalkonium chloride
    contact allergy, 69
Benzyl alcohol toxicity, 69
Beta-emitters, in radiolabeled excipient
    deposition studies, 191
Bioavailability, defined, 268, 287
Bioburden, excipients, 332–333
Blood compatibility, parenteral tests,
    226–227
Blood volume, intravenous parenteral
    tests, 220
Body weight, oral pharmaceutical excipi-
    ent tests, 135
BPCs, 87
Bronchospastic activity, assessment,
    190
Buccal excipients, 262
Buffers, 15
Bulk pharmaceutical chemicals (BPCs),
    87
Bulk stability, test material, 210

Cadmium, work-case daily intake, 346t
Cadmium limits, 344–351
    safety evaluation, 345–352
Calcium stearate, 10
Calcium Sulfate NF, 7
Capsules, 12
Carbon 14, in radiolabeled excipient de-
    position studies, 191
Carcinogenicity studies
    dermal excipients, 173
    intranasal excipients, 197
    ophthalmic excipients, 236
    vaginal excipients, 255
Cardiac sensitization
    defined, 190–191
    studies, 190–191
Cascade impactors, 195
Cascade impactor tests, 187

Cationic emulsifiers, 14
Cecal enlargement, oral pharmaceutical
  excipient tests, 135
Cellulose, 4t, 5
21 CFR 201.20, 85
21 CFR 312.31, 80
21 CFR 312.33, 80
21 CFR 314.127, 81–82
21 CFR 314.503, 80
21 CFR 312.23(a)(7), 79
21 CFR 201.100(b)5, 84–85
21 CFR 210.3(b), 75
21 CFR 211.180(b), 83
21 CFR 207.10(e), 88
21 CFR 330.1(e), 78
21 CFR Part 207, 83–84
21 CFR Part 328, 85
cGMPs, 87
Chloride limits, magnesium stearate,
  343
Chlorobutanol, hypersensitivity, 69
Chromosomal damage in vivo test,
  198
Chronic toxicity studies
  in two species, 120
  vaginal excipient studies, 254–256
Cocoa butter, 18
Coloring agents, 11–12
*Compliance Program Guidance Man-
  ual*, 84
Compressed pills, 63
Compressible Sugar NF, 6
Concentration, excipients, 288–289
Consumer studies, exposure assessment,
  292
Contact allergy
  benzalkonium chloride, 69
  lanolin, 69
Contact rate (CR), 289
Cornstarch, 9
Council Directive 65/65/EEC(31),
  89
Council Directive 75/318/EEC, 89
Council Directive 75/319/EEC(32),
  89
Council Directive 91/356/EEC, 89

Council Directive 92/27/EEC, 89–
  90
CR, 289
Creams, 13
Cynomolgus monkeys
  ophthalmic excipient studies, 235–
    236, 238
  vaginal excipient studies, 252
Cytotoxicity assays
  TSS excipients, 166

Default values, dosage estimation, 290
Delayed-type contact hypersensitivity,
  MLLN, 198
Delivery systems, intranasal studies,
  199
Dermal (*see* Topical *and* Transdermal)
Dermal excipient safety program
  acute systemic toxicity, 156–167
    acute dermal toxicity assay, 158–
      159
    oral route, 163–164
    parenteral route, 164
    skin sensitization, 159–163
  ADME, 149–155
  animal models, 147–148
  application site exposure, 155
  carcinogenicity, 173
  design, 145–146, 156–173
  developmental toxicity, 171–173
    alternative dosing, 172–173
  literature search, 146–147
  oral/parenteral ADME studies,
    155
  pharmacokinetic/toxicokinetic de-
    sign, 148–155
  repeated dose toxicity, 167–170
    90-day toxicity, 169–170
    six-month toxicity, 170
    twelve-month toxicity, 170
    28-day toxicity, 167–169
  reproductive toxicity, 171–173
  special studies, 173–175
    photocarcinogenicity, 174–175
    phototoxicity, 174
  TSS excipients, 165–167

Dermal exposure routes
  dose estimate, 297–298
  exposure assessment, 296–298
  pharmaceutical excipients, 141–176
    literature search, 146–147
    pharmacokinetic/toxicokinetic
      study design, 148–155
    safety study design, 156–173
    special studies, 173–175
    species selection, 147–148
    vs. oral route, 270
Dermal toxicity assay
  acute, 158–159
    protocol design, 158
    skin scoring, 158–159
Developmental studies (*see* Teratology
  studies)
Developmental toxicity
  alternative dosing, 172–173
  dermal excipient safety program,
    171–173
Dextrates, 6
Dextrose excipient NF, 6
Diarrhea
  sorbitol, 67–68
  sugar alcohols, 67–68
  valproic acid syrup, 67–68
*Dibasic Calcium Phosphate Dihydrate
  USP*, 7
Dicalcium Phosphate Dihydrate,
  7
Diethylene glycol toxicity, 64, 66
Dilution potential, 5
Direct compression, 3, 4–7
Disintegrants, 9–10
  formulation levels, 10t
Dispersible suppository bases, 18
Distribution, 271
Di-Tab, 7
Dogs
  drug delivery studies, 201
  injection site irritation, 218–219
  intranasal toxicology studies, 201
  ophthalmic excipient studies, 236,
    239

Dose, risk analysis, 310
Dose selection
  oral pharmaceutical excipient tests,
    131–133
  safety evaluation guidelines, 105–108
  28-day toxicity, 167–168
Draize system
  ocular irritation, 241–242, 246–247
Draize test, 235, 238
  international issues, 247
Drug concentrate, 17
Drug, defined, 73, 75
Drug delivery studies
  dog, 201
Drug for injectable suspension, 208
Drug for injection, 208
Drug injectable emulsion, 208
Drug injectable suspension, 208
Drug injection, 208
Drug product application requirements
  FDA, 81–84
  ophthalmic drugs, 80–81
  parenteral drugs, 80
  pharmaceutical excipient regulation,
    79–84
  topical drugs, 81–84
Dry granulation, 3–4
Dual-brush generators, 187

EC No 541/95, 93
EC No 542/95, 93
EDF, 289
EF, 289
E-Ferol toxicity, 69
Effervescent tablets, 12
Elcema, 5
Electroosmosis, 145
Electrotransport system, 145
Emulsion bases, 13
Emulsions, 16
Endpoints, oral pharmaceutical excipient
  tests, 136–137
Environmental impact studies, 117
Epinephrine, hypersensitivity, 68–69

Ethanol, 29, 31–35, 66–67, 186
  impurities, 34t
  monograph
    tests, 35t
  pharmacopeial monograph comparison, 26t–27t
Ethyl alcohol, overdose, 66
European Pharmacopoeia (EP), 90, 95–96
  excipient standardization, 25
European Union, 76t–77t
European Union for prescription medicinal products
  excipient data requirements, 90–94
  excipient labeling and nomenclature requirements, 94–95
  excipient regulatory status, 89–90
  medicinal product regulation, 88–89
  pharmaceutical excipient regulation, 88–96
Evidence assessment, 313–314
Excipient data requirements, European Union for prescription medicinal products, 90–94
Excipient deposition studies, radiolabeled, 191–192
Excipient exposure
  assessment equation, 289t
  checklist, 288t
Excipient Guidance, 92
Excipients, 1–18
  absorption, 288–289
  bioburden, 332–333
  buccal, 262
  classifications, 3
  concentration, 288–289
  defined, 1–2, 59–60, 75, 89, 185–186
  exposure routes guidelines, 110t
    humans, 189t
  function, 3
  gas, 187–188
  history, 59–61
  impurities, 36–56
    inorganic, 36–41
    organic, 41–56
  inhaled route, 185–195
  intranasal route, 195–202

[Excipients]
  labeling and nomenclature requirements
    European Union for prescription medicinal products, 94–95
  liquid, 186–187
  mucosal, 262
  new
    base set tests, 109–114
    genotoxicity studies, 113–114
    worker exposure, 109–113
  ophthalmic, 231–247
  in oral medicines, 2–12
  origin and production, 22
  parenteral route, 207–227
  penile, 262
  pharmaceutical (*see* Pharmaceutical excipients)
  pharmaceutical grade, 22–25
  purity, 21–35
  ratio to active ingredient
    pharmaceutical preparations, 31t
  rectal, 258–261
  regulation, 60–61
    European Union for prescription medicinal products, 89–90
  safety, 59–70
  solid, 186–187
  standardization
    criteria, 29–30
    legal aspects, 25–29
    scientific and technical aspects, 29–35
  standards, harmonization (*see* Harmonization)
  sublingual, 262
  technical grade, 22–25
  toxicity, 59–70
    historical events, 62t
    hypersensitivity, 68–69
    intentional overdose, 66–67
    sugar alcohols
      diarrhea, 67–68
    vs. safety, 64
  use, 61–64
  vaginal, 248–258
  various routes, 12–18

Excretion, 273–275
Exposed populations, usage patterns,
    290–292
Exposure assessment, 283–301, 308
    assumptions, 293
    dose, 287–290
    exposed populations, 290–292
    future, 301
    impurities, 293
    information sources, 292
    intake estimates, 294–300
        dermal route, 296–298
        inhalation route, 298–299
        oral route, 294–296
        parenteral route, 299–300
    usage patterns, 290–292
Exposure duration, 117–118
Exposure duration and frequency
    (EDF), 289
Exposure frequency (EF), 289
Exposure routes, guidelines, 105, 110t,
    189t
Extrapolation, risk prediction, 311–312
Eye irritation, acute toxicity studies, 112
Eye irritation studies, primary, 237–238

FDA, 78–79
    drug product application require-
        ments, 81–84
    manufacturing and quality require-
        ments, 87
FD&C Act, 75, 78–79
Flameless atomic absorption spectropho-
        tometric procedure, 345
Flavoring agents, 12
Flow-past nose-only chambers, 194
Fluidized bed generators, 187
Food additives
    oral pharmaceutical excipient tests,
        137
    pharmaceutical excipients, 30
Food and Drug Modernization Act of
        1997, 85
Formulation stability
    test material, 210

Gamma-emitter
    in radiolabeled excipient deposition
        studies, 191
Gases, generation, 187–188
Gas excipients, 187–188
Gelatin, safety, 69
Gelatin NF, 9
Generally recognized as safe (GRAS),
    64
Generators, 187
Genotoxicity studies, 119
    data interpretation, 279
    intranasal studies, 197–198
    new excipients, 113–114
Genotoxic substances
    defined, 113
Glidants, 11
GLP regulations, test material, 209
Glycerinated gelatin, 18
Glycerin from tallow, production routes,
    23f
Glyoxal limits
    hydroxyethyl cellulose monograph,
        335–340
    NOAEL, 337–338
    safety evaluation, 337–338
Good Laboratory Practice (GLP) regula-
        tions
    test material, 209
Good Manufacturing Practices
        (cGMPs), 87
    excipient production, 22–24
GRAS, 64–65
Guidance on Development Pharmaceu-
        tics, 91
Guinea pigs
    hairless
        dermal excipients, 148
        intranasal studies, 199
        septal window, 201

Hairless guinea pigs, dermal excipient
        safety program, 148
*Handbook of Pharmaceutical Excipi-
        ents*, excipient definition, 1
Harm, defined, 305

Harmonization, 321–352
  components, 325
  excipient standards, 321–352
  goals, 326–327
  identity criteria, 326–327
  impurities standards and litmus tests,
      328–352
    HEC, 335–340
    microbial requirements, 332–335
    monograph case studies, 340–352
    OVIs, 329–332
  interested parties, 326
  limit tests, 328–340
  MCC monograph, 351–352
  procedure, 323–324
  purity criteria, 326–327
  quality criteria, 326–327
  strength criteria, 326–327
Harmonized monograph
  ethanol, 35t
  lactose, 350–351
  magnesium stearate, 349–350
Hazard
  defined, 267, 305
Hazard identification, 267–280, 307
    (*see also* Toxicokinetics)
Heavy metals
  excipient inorganic impurities, 36–39
  test materials, 213
HEC harmonization, 335–340
Hemolysis, 211
HFA-227, 187
HFA-134a, 187, 192
Hopper-type generators, 187
Hydrocarbon bases, 13
Hydrophilic Petrolatum USP, 13
Hydrous lactose, 6
Hydroxyethyl cellulose (HEC)
  harmonization, 335–340
Hydroxyethyl cellulose monograph
  glyoxal limits, 335–340
  nitrate limits, 335–340
Hypersensitivity, 68–69, 198
  chlorobutanol, 69
  epinephrine, 68–69
  tartrazine, 69

Hypersensitivity responses
  intranasal studies, 198

Ibuprofen, excipients, 284
ICH guidelines, solvent categorization, 213
Impurities
  ethanol, 34t
  excipients, 36–56
  exposure assessment, 293
  spirits, 33t
  standards
    harmonization (*see* Harmonization,
        impurities standards and litmus
        test)
Inactive ingredient, defined, 75
Inactive Ingredient Guide, 83
Indwelling cannula, intravenous paren-
    teral tests, 223
Infusion pumps, intravenous parenteral
    tests, 222
Inhalational exposure (*see* Inhalation
    studies)
Inhalation delivery systems, 16–17
Inhalation exposure systems
  larger animals, 194–195
  small animals, 194
Inhalation route, exposure assessment,
    298–299
Inhalation studies, 188–191
  acute toxicity, 189–191
  conduct, 188
  design, 188–191
  dose selection, 192–193
  inhalation exposure systems, 193–195
  MDI, 299t
  pharmacology-toxicology, 188–189
  physiological parameters, 192
  pulmonary parameters
    species comparison, 192t
  radiolabeled excipient deposition stud-
      ies, 191–192
  safety evaluation program, 188
  test atmosphere monitoring, 195
Injection sites
  intravenous parenteral tests, 220–222
  irritation, parenteral tests, 218–219

Inorganic impurities, excipients, 36–41
Inorganic salts, 7
International Conference on harmoniza-
    tion (ICH) guidelines, for sol-
    vent categorization, 213
International harmonization (*see* Harmo-
    nization)
International Harmonization of Excipi-
    ent Quality Standards, 28
International issues, ophthalmic excipi-
    ent studies, 247
International Pharmaceutical Excipients
    Council (IPEC)
    excipient definition, 1–2
    guidelines, 215
    manufacturing and quality require-
        ments, 87–88, 96
    pharmaceutical excipient definition, 2
Intramuscular parenteral tests, 224–225
Intranasal delivery systems, 16–17
Intranasal studies, 195–202
    acute toxicity, 196
    aerosol studies, 201
    animal models, 199–202
    carcinogenic risk, 197
    delivery systems, 199
    dog, 201
    dose frequency in, 202
    dose level selection in, 202
    exposure duration in, 202
    genetic toxicity studies, 197–198
    hypersensitivity responses, 198
    intranasal delivery, 198–199
    monkeys, 201–202
    primates, 201–202
    repeat-dose, 196–197, 201
    reproductive outcome, 197
    teratology studies, 197, 201
Intraperitoneal parenteral tests, 223–224
Intravenous parenteral tests, 220–223
    blood volume, 220
    bolus injection vs. infusion, 221
    indwelling cannula, 223
    infusion pumps, 222
    injection sites, 220–222
    osmotic pumps, 223

Iontophoretic transdermal formulation,
    145
IPEC
    guidelines, 215
    manufacturing and quality require-
        ments, 87–88, 96
Isotonic vehicles, 15

Japanese Pharmacopoeia, 323
*Japanese Pharmacopoeia/Japanese
    Pharmaceutical Excipients* (JP/
    JPE), 25
Japanese Pharmacopoeial Forum,
    323

Labeling
    magnesium stearate, 343
    pharmaceutical excipient regulation,
        84–85
Lactose, 5–6
    monograph harmonization case stud-
        ies, 340–344
Laskin-type nebulizers, 187
Laxative effects, oral pharmaceutical ex-
    cipient tests, 135
LD_{50}, 112
Lead
    daily intake, 348t
    maximum daily intake, 349
    work-case daily intake, 346t
Lead limits
    magnesium stearate, 344–351
    safety evaluation, 345–352
Lecithin, 186
Lethal dose 50 (LD_{50}), 112
Light-scattering particle size devices,
    195
Limit doses, toxicological studies,
    106t
Limit tests, harmonization (*see* Harmoni-
    zation, impurities standards and
    litmus tests)
Limulus test, 213–214
Liquid aerosols, vapor generation,
    186
Liquid excipients, 186–187

Literature search
  dermal excipient safety program,
    146–147
  pharmaceutical excipients, 126
  safety evaluation guidelines, 102–103
Liver weight, oral pharmaceutical excipient tests, 135
Lubricants, 10

Mad cow disease, 69–70
Magnesium stearate, 10
  chloride limits, 343
  harmonized monograph, 349–350
  labeling, 343
  lead limits, 344–351
  microbial limits, 343
  monograph harmonization case studies, 340–344
  nickel limits, 344–351
  safety evaluation, 345–352
  sulfate limits, 343
  work-case daily intake, 346t
Magnetic resonance imaging (MRI), in radiolabeled excipient deposition studies, 191
Mammalian in vitro assay, 197–198
Mannitol, 7
Manufacturing and quality requirements
  EU, 96
  FDA, 87
Marketing Authorization for Medicinal Products, 90
Marketing databases, exposure assessment, 292
Marshmallow root, 61
Masses, 63
Material Safety Data Sheet (MSDS), 125
Maximum feasible concentration, inhalation studies, 193
Maximum injectable volumes, parenteral tests, 226
Maximum tolerated dose (MTD), inhalation studies, 193
MCC, 5, 8
MCC monograph harmonization, 351–352

McDonald systems, ocular irritation, 241–242, 246–247
MDIs, 186, 187
  inhalation exposure, 299t
Medicinal products
  defined, 89
  regulation
    European Union for prescription medicinal products, 88–89
Metabolism, 271–273
Metals, excipient inorganic impurities, 36–39
Metered-dose inhalers (MDIs), 186, 187
  inhalation exposure, 299t
Microbial limits, magnesium stearate, 343
Microbial requirements, harmonization, 332–335
Microcrystalline cellulose (MCC), 5, 8
Microcrystalline cellulose (MCC) monograph harmonization, 351–352
Microscopic examination, vaginal tissue, 257–258
Mineral oil, 13
MLLN, delayed-type contact hypersensitivity, 198
Monkeys
  cynomolgus
    ophthalmic excipient studies, 235–236, 238
    vaginal excipient studies, 252
  intranasal studies, 201–202
  rhesus, ophthalmic excipient studies, 235–236
Monograph harmonization case studies, 340–352
  lactose, 340–344
  magnesium stearate, 340–344
Mouse local lymph node assay (MLLN)
  delayed-type contact hypersensitivity, 198
Mouse micronucleus test, 114
MRI, in radiolabeled excipient deposition studies, 191
MSDS, 125

MTD, inhalation studies, 193
MTT cell viability assay, TSS excipients, 166
Mucosal excipients, 262
  guidelines, 232t
Mucosal preparations, 17–18
Mutagenicity studies, parenteral testing, 215, 216

Nasal epithelium, species differences in, 200
Nasal sprays, formulation aids, 186–187
Nasal volumes, species comparison, 200
National Formulary (NF), 86
  excipient definition, 1
Natural sweeteners, 60
New Drug Application (NDA), safety, 78–84
New excipients
  base set tests, 109–114
  genotoxicity studies, 113–114
  worker exposure, 109–113
New pharmaceutical excipients
  chemical and physical properties, 125–126
  exposure assessment, 126–127
  toxicology tests
    oral exposure route, 123–124
New Zealand rabbits (*see* Albino rabbits)
NF, 86
Nickel, work-case daily intake, 346t
Nickel limits
  magnesium stearate, 344–351
  safety evaluation, 345–352
90-day toxicity dermal excipients, 169–170
90-day toxicity studies, 118–119
Nitrate limits
  hydroxyethyl cellulose monograph, 335–340
  safety evaluation, 339–400
NOAEL
  glyoxal limits, 337–338
  oral pharmaceutical excipient tests, 134–135
  risk estimation, 312–313
NOEL, risk estimation, 312–313

Nomenclature requirements, pharmaceutical excipient regulation, 84–85
Nonanimal models, ophthalmic excipient studies, 235–236
Nonhuman primates, ophthalmic excipient studies, 235–236
Nonrodents, oral pharmaceutical excipient tests, 131
Nose-only chambers, 194
Note for Guidance: Chemistry of Active Ingredients, 92–93
Note for Guidance: Specifications and Control Tests on the Finished Product, 91

Ocular irritation, evaluation, 241–246
OECD guidelines, ocular irritation testing, 247
Ointment bases, classification and properties, 13–15, 14t
Oleaginous bases, 13
Oleic acid, 186
One-generation reproduction studies, 120, 136
Open cylinder nose-only chambers, 194
Ophthalmic drugs, drug product application requirements, 80–81
Ophthalmic excipient studies, 231–247
  acute toxicity studies, 233–234, 237–238
  ADME studies, 234
  administration, 240–241
  agar overlay cytotoxicity test, 236–237
  albino rabbits, 234–236, 238
  animal models, 234–236
  carcinogenicity studies, 236, 239
  chronic toxicity, 238–239
  cynomolgus monkeys, 235–236, 238
  dogs, 236, 239
  dose selection, 239–240
  evaluation criteria, 241–246
  guidelines, 232t
  international issues, 247
  interpretation, 246–247
  nonanimal models, 235–236
  nonhuman primates, 235–236

[Ophthalmic excipient studies]
  ph, 233
  pharmacokinetic studies, 234
  pigmented rabbits, 235
  rats, 236
  reproduction studies, 236
  rhesus monkeys, 235–236
  safety evaluation studies, 233–246
  study design, 236–239
  teratology studies, 236
  tonicity, 233
  in vitro tests, 236–238
Oral bioavailability, pharmaceutical ex-
    cipients, 287–288
Oral exposure routes
  pharmaceutical excipients, 123–137
    background information assess-
        ment, 127
    exposure assessment, 126–127
    literature search, 126
    study design, 127–134
    toxicology tests, 123–124
Oral/parenteral ADME studies, dermal
    excipient safety program, 155
Oral pharmaceutical excipient tests
  analytical considerations, 133–134
  animal models, 130–131
  dose selection, 131–133
    dose group number, 132
  endpoints, 136–137
  NOAEL determination, 134–135
  oral administration method
    human exposure pattern, 128–129
  test material, 129–130
    physical state, 129
    physiochemical properties, 130
Oral route
  acute systemic toxicity
    dermal excipient safety program,
        163–164
  dose estimate, 295–296
  exposure assessment, 294–296
Organic volatile impurities (OVIs)
  harmonization, 329–332
Osmotic pumps, intravenous parenteral
    tests, 223

Osmotonicity, 211
OVIs
  harmonization, 329–332
  official limits, 330

Parenteral, defined, 207
Parenteral drugs, drug product applica-
    tion requirements, 80
Parenteral preparations, types, 208
Parenteral route
  acute systemic toxicity
    dermal excipient safety program,
        164
  advantages, 207–208
  disadvantages, 208
  exposure assessment, 299–300
  oral studies, 270–271
  route comparison, 209t
Parenteral studies, 207–227
  ADME-PK studies, 216
  base set testing, 215–216
  mutagenicity studies, 215, 216
  repeat-dosing studies, 215
  subchronic studies, 216
  teratology studies, 216
  test material, 208–215
    physical and chemical properties,
        208–211
    physiochemical properties, 209–211
    purity and stability, 212–214
Parenteral systems, 15–16
Parenteral tests, 217–220
  acute toxicity, 215, 217–218
  ADME-PK studies, 219–220
  dose route selection, 226–227
    blood compatibility, 226–227
    maximum injectable volumes, 226
    pH, 226–227
    recommended animal species, 226
    tonicity, 226–227
    viscosity, 226–227
  injection site irritation, 218–219
  intramuscular, 224–225
  intraperitoneal, 223–224
  intravenous, 220–223
  subcutaneous, 225

Particle size
optimum, 195
test material, 210
Particle size analyzers, inhalation studies, 195
Partition coefficient, test material, 210
Penile excipients, 262
PET, in radiolabeled excipient deposition studies, 191
Petrolatum, 13
Petrolatum USP, 13
PH
ophthalmic excipient studies, 233
parenteral tests, 226–227
Pharmaceutical excipients
defined, 2, 59
dermal exposure routes, 141–176
literature search, 146–147
pharmacokinetic/toxicokinetic study design, 148–155
safety study design, 156–173
special studies, 173–175
species selection, 147–148
and direct food additives, 30
new
chemical and physical properties, 125–126
exposure assessment, 126–127
oral exposure tests, 123–124
oral administration method, 128–129
oral bioavailability, 287–288
oral exposure routes, 123–137
background information assessment, 127
exposure assessment, 126–127
literature review, 126
study design, 127–134
toxicology tests, 123–124
regulation, 73–97
drug product application requirements, 79–84
European Union for prescription medicinal products, 88–96
general requirements, 73–79
manufacturing and quality requirements, 87–88

[Pharmaceutical excipients]
nomenclature requirements, 84–85
official pharmacopeial standards, 86–87, 95–96
Pharmaceutical excipients, new toxicology tests
oral exposure route, 123–124
Pharmaceutical necessities, 61
Pharmaceutical preparations, excipient ratio to active ingredient, 31t
Pharmacokinetics, defined, 268
Pharmacokinetic studies
dermal excipients, 148–155
ophthalmic excipients, 234
Pharmacology-toxicology inhalation studies, 188–189
Pharmacopeial ethanol monograph, tests, 35t
Pharmacopeial forum, excipient standards, 322
Pharmacopeial standards, pharmaceutical excipient regulation, 86–87, 95–96
Pharmacopoeial Discussion Group (PDG), 322–325
Pharmeuropa, 323
Photocarcinogenicity, dermal excipients, 174–175
Photoirritation, dermal excipients, 174–175
Phototoxicity, dermal excipients, 174
Pigmented rabbits, ophthalmic excipient studies, 235
Pills, history, 63
Polyethylene glycols, 14
Polymers, wet binders, 8–9
Polyols, 7
Polysorbate toxicity, 69
Positron emission tomography (PET)
in radiolabeled excipient deposition studies, 191
Povidone USP (PVP), 9
Pregelatinized starch, 6
Pregelatinized Starch NF, 9
Preservatives, 14
Primary eye irritation studies, 237–238

Primary vaginal irritation studies, 252–254
Primates, intranasal studies, 199–202
Product-specific data, dosage estimation, 288–289
Product usage data, dosage estimation, 289–300
Propellant, 17
Propylene glycol toxicity, 69
PTTIL-based lead limits, 348
Pulmonary hypersensitivity, intranasal studies, 198
Pulmonary irritation, assessment, 190
Purity
  excipients, 21–35
    aspects, 22–25
    standardization, 25–35
PVP, 9

Qualitative fever response test in rabbits, 213
Quartz Crystal Microbalance, 195

Rabbits
  albino
    ophthalmic excipient studies, 234–236, 238
    injection site irritation, 218
    intranasal studies, 201
  pigmented, ophthalmic excipient studies, 235
  qualitative fever response test, 213
  vaginal epithelium, 248–250
  vaginal excipient studies, 248–252, 257
Radioimaging, in radiolabeled excipient deposition studies, 191
Radiolabeled excipient deposition studies, 191–192
Rats
  ophthalmic excipient studies, 236
  vaginal excipient studies, 252
Recipient, risk analysis, 310
Rectal excipients, 258–261
Rectal excipient studies
  administration, 261
  animal models, 260

[Rectal excipient studies]
  dose selection, 260–261
  evaluation criteria, 261–262
  safety evaluation, 260–261
  study design, 260
Rectal preparations, 17–18
Reference listed drug (RLD), 82
Repeat-dose studies
  data interpretation, 278
  dermal excipient, 167–170
  intranasal studies, 196–197, 201
  parenteral testing, 215
Reproduction studies
  data interpretation, 278–279
  dermal excipients, 171–173
  ophthalmic excipients, 236
  vaginal excipients, 252, 255
Reproductive outcome, intranasal studies, 197
Residual monomers, test materials, 213
Residual solvents, 331
Respiratory physiology, parameters, 190
Rhesus monkeys
  ophthalmic excipient studies, 235–236
  vaginal excipient studies, 252
Ringer's solution, 15
Risk, defined, 305–306
Risk analysis, 307–316
  dose, 310
  recipient, 310
  steps, 307–308
  substance, 309–310
  toxicity, 310–311
Risk assessment, 305–319
  risk analysis, 307–316
  risk communication, 317–318
  risk estimation, 312–313
  risk management, 316–317
  risk prediction, 311–312
  safe dose, 314–315
  safe exposure prediction, 315–316
  weight of evidence assessment, 313–314
Risk characterization, 308
Risk communication, 308, 317–318
  defined, 307

Risk estimation, 308, 312–313
humans, 313
NOAEL, 312–313
NOEL, 312–313
toxic effects, 312
Risk identification, 308
Risk management, 308, 316–317
Risk prediction, extrapolation, 311–312
RLD, 82
Rodents (*see also* Rats)
intranasal studies, 200
oral pharmaceutical excipient tests,
131
Roller compaction, 3–4

Safe dose, 314–315
Safe exposure, prediction, 315–316
Safety, defined, 306
Safety evaluation
glyoxal limits, 337–338
inhalation studies, 188
magnesium stearate, 345–352
nitrate limits, 339–400
ophthalmic excipients, 233–246
rectal excipients, 260–261
Safety evaluation guidelines, 101–121
animal models, 105
base set tests, 108–117
data, 102
dose selection, 105–108
exposure route, 105
level I and II tests, 117–120
additional studies rationale, 118–
120
exposure duration, 117–118
literature search, 102–103
parameters evaluated, 107t, 108
principles, 104–105
test material characterization, 103–
104
Salts, inorganic, 7
Septal window, guinea pig, 201
Sheffield, 6
Six-month toxicity dermal excipients,
170
Skin flux, 149, 151t

Skin scoring
acute dermal toxicity assay, 158–159
28-day toxicity, 169
Skin sensitization, 159–163
acute toxicity studies, 112–113
positive assay results, 163
protocol design, 161–162
Slugging, 3–4
Sodium chloride injection, 15
Sodium lauryl sulfate, 12
Sodium starch glycolate, 6–7
Solid excipients, 186–187
Solid particulates, generating atmo-
spheres, 187
Solubility, test material, 210
Solvents
residual, 331
test materials, 213
Sorbitan tioleate, 186
Sorbitol, diarrhea, 67–68
Sorbitol NF, 7
Spinning top generators, 187
Spirits, impurities, 33t
Standards, excipient harmonization (*see*
Harmonization)
Starch, 6–7
Sterile Water for Injection USP (SWFI),
15
Subchronic studies, parenteral testing,
216
Subcutaneous parenteral tests, 225
Sublingual excipients, 262
Sucrose NF, 9
Sugar alcohols, diarrhea, 67–68
Sulfate limits, magnesium stearate, 343
Sulfites, 68–69
Super disintegrant, 10
Super-Tab, 6
Suppositories, 17–18
Suppository bases, list, 17–18
Surfactants, 186
Suspensions, 16
Sweeteners, 12
Systemic toxicity, acute
dermal excipient safety program,
156–167

Tablets, 12, 63–64
Talc, 29
   pharmacopeial requirements, 28t
Tallow derivatives, production routes,
      23f
Tartrazine, hypersensitivity, 69
Technetium 99m, in radiolabeled excipi-
      ent deposition studies, 191–192
Teratology studies, 119
   data interpretation, 278–279
   intranasal studies, 197, 201
   ophthalmic excipients, 236
   parenteral testing, 216
   vaginal excipients, 252, 255–256
Terra alba, 7
Test atmosphere monitoring, inhalation
      studies, 195
Test material characterization, safety
      evaluation guidelines, 103–104
Test materials
   analytical data, 209–210
   biological background, 214
   compatibility with blood, 211
   nutritional effects, 214
   pharmacological effects, 214
   purity, 212–214
      specifications, 212–213
   solution ph, 211
   solution tonicity, 211
   sterility and pyrogenicity, 213–214
Theobroma oil, 18
Tiered-testing strategy, base set tests,
      108–109
Timbrell Dust Generator, 187
Titanium dioxide, 24
   analysis, 25t
   production, 24t
   uses, 24t
Tonicity
   ophthalmic excipient studies, 233
   parenteral tests, 226–227
Topical (*see also* Dermal)
Topical delivery, defined, 144
Topical delivery system, 13–15
Topical drugs, drug product application
      requirements, 81–84

Topical exposure routes, pharmaceutical
      excipients (*see* Pharmaceutical
      excipients, dermal exposure
      routes)
Topical formulations, defined, 144
Toxicity
   defined, 306
   risk analysis, 310–311
Toxicity data, acute
   new excipients, 109–113
Toxicity studies
   acute
      data interpretation, 277–278
      eye irritation, 112
      ophthalmic excipients, 233–234,
         237–238
      parenteral testing, 215
      vaginal excipient studies, 252–254
   chronic
      in two species, 120
      vaginal excipient studies, 254–256
   data interpretation, 277–279
   limit doses, 106t
   90-day, 118–119
   parameters evaluated, 107t
   repeat-dose
      data interpretation, 278
      dermal excipient, 167–170
   reproduction
      data interpretation, 278–279
      dermal excipient safety program,
         171–173
   risk estimation, 312
Toxicity tests
   acute, parenteral testing, 215, 217–218
   new pharmaceutical excipients, oral
      exposure route, 123–124
Toxicokinetics, 267–280
   absorption, 269–271
   age, 269
   base set tests, 114–117
   data interpretation, 275–280
   defined, 114
   distribution, 271
   excretion, 273–275
   metabolism, 271–273

Transdermal (*see also* Dermal)
Transdermal delivery system, 13–15
Transdermal drug delivery
defined, 144
Transdermal exposure routes
pharmaceutical excipients (*see* Pharmaceutical excipients, dermal exposure routes)
Transdermal formulation, defined, 144
Transdermal therapeutic systems (TTS), defined, 144
Tritium, in radiolabeled excipient deposition studies, 191
TSS excipients, safety assessment, 165–167
TTS, defined, 144
Twelve-month toxicity dermal excipients, 170
28-day studies, 119
28-day toxicity dermal excipients
animal models, 167
dose selection, 167–168
skin scoring, 169
test articles application, 168–169

U. S. Code of Federal Register, 74t
U. S. Food and Drug Administration (FDA), 78–79
United States Pharmacopeia/National Formulary (USP/NF), excipient standardization, 25
United States Pharmacopeia (USP), 86
excipient definition, 1
USP, 86
excipient standards, 322–323
USP 23, Biological Reactivity Tests, In Vivo, 166
USP elution assay, TSS excipients, 166

Vaginal epithelium, rabbits, 248–250
Vaginal excipient studies, 248–258
acute toxicity studies, 252–254
administration, 256–257

[Vaginal excipient studies]
animal models, 248–252
carcinogenicity studies, 255
chronic toxicity studies, 254–256
cynomolgus monkeys, 252
dose selection, 256
evaluation criteria, 257–258
interpretation, 257–258
rabbits, 248–252, 257
rats, 252
reproduction studies, 252, 255
rhesus monkeys, 252
safety evaluation studies, 248–258
study design, 252–256
teratology studies, 252, 255–256
Vaginal irritation studies, primary, 252–254
Vaginal preparations, 17–18
Vaginal tissue, microscopic examination, 257–258
Valodon, 64
Valproic acid syrup, diarrhea, 67–68
Vanishing creams, 13–14
Viscosity, parenteral tests, 226–227

Water for Injection USP, 15
Water-soluble bases, 14
Water-soluble suppository bases, 18
Water-washable bases, 13
Weight of evidence assessment, 313–314
Wet binders, 8–9
Wet granulation, 3, 4–7
Wet granulation excipients, 7–9
binders used, 8t
Wetting agents, 12
Whole-body chambers, 194
Worker exposure, new excipients, 109–113
Wright Dust Feeder, 187

Xylitol, 7